Application of Membranes in the Petroleum Industry

This book focuses on the advantageous features of membrane technology in petroleum industries, with an emphasis on membrane materials and the application of membranes in the separation of olefin–paraffin, oil–water, aliphatic–aromatics, heavy metals, etc., along with other applications like waste management, sulphur emission, enhanced oil recovery and so forth. It also discusses the design and development of membranes from novel materials, the challenges of new materials for membrane applications, membrane-based processes and the application of novel membrane-based processes in the petroleum industry.

Features:

- Addresses the fundamental applications of membranes in petroleum industrial separation processes.
- Highlights the role of membrane technology in waste management in petroleum industries.
- Includes novel engineered membrane materials.
- Discusses methods of extracting valuable substances from produced water and membrane fouling control.
- Emphasises solving industrial problems pertinent to membrane usage.

This book is aimed at researchers and graduate students in chemical and petroleum engineering and membrane technology.

Application of Membranes in the Petroleum Industry

Edited by
Swapnali Hazarika, Achyut Konwar and
G. Narahari Sastry

CRC Press
Taylor & Francis Group
Boca Raton London New York

CRC Press is an imprint of the
Taylor & Francis Group, an **informa** business

Designed cover image: © Swapnali Hazarika

First edition published 2024
by CRC Press
2385 NW Executive Center Drive, Suite 320, Boca Raton FL 33431

and by CRC Press
4 Park Square, Milton Park, Abingdon, Oxon, OX14 4RN

CRC Press is an imprint of Taylor & Francis Group, LLC

© 2024 selection and editorial matter, Swapnali Hazarika, Achyut Konwar and G. Narahari Sastry; individual chapters, the contributors

ISBN: 9781032528342 (hbk)
ISBN: 9781032578712 (pbk)
ISBN: 9781003441359 (ebk)

DOI: 10.1201/9781003441359

Typeset in Times
by Newgen Publishing UK

Contents

Preface

The petroleum industry, with a market potential of about 5 trillion US$, is arguably the most important in the world from an economical point of view. Besides, the industry's future has strong implications on the sustainability, environment, energy, chemicals and other allied areas associated with human development. While the petroleum industry is very robust, it has been facing several challenges that impact not only the economical and industrial aspects but also the socio-political perspective globally. Some of the major challenges are increasing productivity, especially by value addition of the products (downstream) and environmental aspects, such as effluent treatment.

The world is now looking for advanced and sophisticated technologies to transform from lab to production plant. To increase the petroleum industry's production efficiency, it is essential to work with updated technologies in various production processes like refining oil and natural gases, removing sour gases for safe transportation, enhanced oil recovery, minimisation of environmental hazards, etc. In recent days, membrane technology has played a critical role in overcoming the technological challenges of the petroleum industry by providing energy-efficient, cost-effective and environmentally friendly solutions.

The petroleum industry encompasses various sectors, including oil production, refining, natural gas separation and purification, waste management and recycling, etc., where membrane separation processes can play an important role. However, membrane design plays the most vital role in achieving high efficiency in the process. Therefore, designing new membranes to enhance the efficiency of a separation process has always been a major concern for researchers. To develop a superior membrane, it is important to have deep knowledge of the principles and chemistry related to a particular process. It is also important to correlate the properties of membrane materials and guest molecules to design an industrially efficient novel membrane. This can be done through computational modelling under the industry 5.0 strategy. Therefore, in this book, the authors provide a detailed discussion of membrane-based research, including principles and mechanisms, design and development, computational modelling and recent advances for the petroleum industry.

This book will help the researchers acquire thorough concepts of novel membrane materials, their working principle, the transport behaviour of membrane-based processes, their applications in the petroleum industry and computational work on the design of materials. Hence, it will benefit the researchers in multidisciplinary research areas, from membrane design to process development and industrial implementation.

The contents of this book are designed to address a large section of the scientific community in the areas of material science, chemical science, environmental science and engineering science, including petroleum technology. Here, the authors have endeavoured to cover the basics and the advanced level of membrane technology and its application in the petroleum industry, which will benefit engineers, scientists, environmentalists and policymakers.

This book comprises 15 chapters. Chapter 1 gives a brief overview and current status of membrane technology. Chapter 2 is a description of the conventional as well as newly used materials for designing membranes. Chapter 3 is on the application of membranes for natural gas processing. Chapter 4 describes the principles and membranes used for separation of olefins and paraffins. Chapter 5 explains the separation of aromatic and aliphatic hydrocarbons by membrane. Chapter 6 is about the characteristics of produced water in petroleum industries. Chapter 7 is a description of produced water management and technologies. Chapter 8 explains the design of membranes for the separation of oil–water mixtures. Chapter 9 explains the principle and membrane design for removal of phenolic compounds from petroleum industry wastewater. Chapter 10 discusses membrane-based treatment for removal of heavy metals from oil industry wastewater. Chapter 11 explains the principle of enhanced oil recovery and the importance of membranes in this technology. Chapter 12 explains the petroleum industry sulphur emission and removal using membrane technology. Chapter 13 describes the application of membranes in oil and gas fields. Chapter 14 describes the scope of computational work on designing membrane materials. Finally, Chapter 15 discusses the future perspective of membrane technology in the petroleum industry.

To our knowledge, this is the first book to date that covers the application of membranes in different processes of petroleum industry. The book describes the principles of membrane technology in a very elaborate way. Also, it illustrates different types, functions, properties of novel materials and computational modelling generally employed to design high-performance membranes. Further, this book describes membrane technologies in the petroleum industry, which may provide the bridge between academic, translational and industrial research.

Contributors

Sudeepta Baruah
Chemical Engineering Group and
 Centre for Petroleum Research,
 CSIR-North East Institute of Science
 and Technology, Jorhat, Assam, India

Diganta Bhusan Das
Department of Chemical Engineering,
 Loughborough University,
 Loughborough, UK

Amrendra Bhushan
National Institute of Hydrology,
 Roorkee, Uttarakhand, India

Sanjay Bhutani
University of Petroleum and Energy
 Studies (UPES), Dehradun,
 Uttarakhand, India

Chinmoy Bhuyan
Chemical Engineering Group and
 Centre for Petroleum Research,
 CSIR-North East Institute of Science
 and Technology, Jorhat, Assam, India
Academy of Scientific and Innovative
 Research (AcSIR), Ghaziabad, India

Prarthana Bora
Chemical Engineering Group and
 Centre for Petroleum Research,
 CSIR-North East Institute of Science
 and Technology, Jorhat, Assam, India
Academy of Scientific and Innovative
 Research (AcSIR), Ghaziabad, India

Akhil Ranjan Borah
Chemical Engineering Group and
 Centre for Petroleum Research,
 CSIR-North East Institute of Science
 and Technology, Jorhat, Assam, India

Alimpia Borah
Chemical Engineering Group and
 Centre for Petroleum Research,
 CSIR-North East Institute of
 Science and Technology, Jorhat,
 Assam, India

Abhijit Gayan
Chemical Engineering Group and
 Centre for Petroleum Research,
 CSIR-North East Institute of
 Science and Technology, Jorhat,
 Assam, India

Monti Gogoi
Chemical Engineering Group and
 Centre for Petroleum Research,
 CSIR-North East Institute of
 Science and Technology, Jorhat,
 Assam, India

Rajiv Goswami
Chemical Engineering Group and
 Centre for Petroleum Research,
 CSIR-North East Institute of
 Science and Technology, Jorhat,
 Assam, India

Krishna Kamal Hazarika
Chemical Engineering Group and
 Centre for Petroleum Research,
 CSIR-North East Institute
 of Science and Technology, Jorhat,
 Assam, India

Swapnali Hazarika
Chemical Engineering Group and
 Centre for Petroleum Research,
 CSIR-North East Institute of Science
 and Technology, Jorhat,
 Assam, India

Pratyashi Kondoli
Chemical Engineering Group and
  Centre for Petroleum Research,
  CSIR-North East Institute of Science
  and Technology, Jorhat, Assam, India

Achyut Konwar
Chemical Engineering Group and
  Centre for Petroleum Research,
  CSIR-North East Institute of Science
  and Technology, Jorhat, Assam, India

Ravi Kumar Lingam
Chemical Engineering Group and
  Centre for Petroleum Research,
  CSIR-North East Institute of Science
  and Technology, Jorhat, Assam, India

Anwesh Pandey
Advance Computation and Data
  Sciences Division
CSIR-North East Institute of Science
  and Technology, Jorhat, Assam, India

Parashmoni Rajguru
Chemical Engineering Group and
  Centre for Petroleum Research,
  CSIR-North East Institute of Science
  and Technology, Jorhat, Assam, India
Academy of Scientific and Innovative
  Research (AcSIR), Ghaziabad, India

Shrisha S. Raj
CSIR-Indian Institute of Chemical
  Technology, Uppal Road,
  Hyderabad, India

Ananya Saikia
Chemical Engineering Group
  and Centre for Petroleum Research,
  CSIR-North East Institute of
  Science and Technology,
  Jorhat, Assam, India
Academy of Scientific and
  Innovative Research (AcSIR),
  Ghaziabad, India

Nazia Shaik
CSIR-Indian Institute of Chemical
  Technology, Uppal Road,
  Hyderabad, India

Sundergopal Sridhar
CSIR-Indian Institute of Chemical
  Technology, Uppal Road,
  Hyderabad, India

Ankush Thakur
Forest Research Institute (Deemed
  to be) University, Dehradun,
  Uttarakhand, India

Joshua Tipple
Department of Chemical Engineering,
  Loughborough University,
  Loughborough, UK

Nikita Yadav
Forest Research Institute (Deemed
  to be) University, Dehradun,
  Uttarakhand, India

About the Editors

Swapnali Hazarika is a dedicated researcher in the field of separation science and technology, designing adsorbents, customised membranes and membrane separation studies, including chiral separations, biomolecule separations, gas separations, wastewater treatment, etc. Her study also includes membrane-based isolation and purifications of natural products, biochemical reaction engineering and mechanism, kinetic analysis of reaction-diffusion, modelling and simulation, QSAR–QSPR study, etc. Under her supervision, three technologies have been developed, whereas two technologies have already been transferred to the industries. She has published 60 research papers and granted 5 patents in her credit. She has presented 81 seminar papers and delivered over 100 talks at national and international conferences/workshops. She also has one book and nine book chapters in her credit. Hazarika has guided 13 PhD students and many project fellows. She has developed three technologies and transferred them to private parties. Hazarika has a collaborative research program with petroleum industries on some specific research problems. She has handled 23 R&D projects to date.

Achyut Konwar is a post-doctoral researcher in the Centre for Petroleum Research, CSIR-North East Institute of Science and Technology (NEIST). He received an MSc in Chemistry and MTech in Polymer Science and Technology. Konwar received his PhD in Chemistry. He is a passionate researcher in the field of polymeric hybrid materials. His research includes the development of polymeric composites for different applications like packaging, membrane design, environmental remediation, biomedical, electronic, etc. Konwar has published more than 15 research articles in different reputed journals and three book chapters with international publications. He has two number of patents granted in his name.

G. Narahari Sastry is currently the Director of CSIR-North East Institute of Science and Technology, Jorhat. Sastry is a chemist working in the interdisciplinary areas spanning chemistry, biology, modelling and informatics. Sastry effectively employed computational and theoretical methods to solve problems in chemistry, biology and allied areas. These efforts are embedded to provide a robust platform for carrying CADD research and to inculcate the culture of developing software packages. G. Narahari Sastry has made fundamental contributions in the areas: (a) computational and theoretical chemistry; (b) theoretical organic chemistry and reaction mechanism; (c) software and database development for drug discovery (molecular property diagnostic suite; (d) non-covalent interactions); (e) cooperativity of non-covalent interactions and (f) computer-aided drug design. Under his guidance, 29 were awarded PhD, 20 post-doctoral fellows and 200 students have done internships or short-term projects. His work was published in more than 330 research papers and reviews, which received about 12,800 citations, with an h-index of 54. Sastry was awarded the Shanti Swarup Bhatnagar Prize in Chemical Sciences (2011) (considered as one of the highest prize for science and engineering in India), the National Bioscience

Award, DBT 2009 (one of the highest for Life sciences in India), the Swarnajayanthi Fellowship 2005 (DST), B.M. Birla award for 2001, BC Deb Memorial award (2009), CRSI silver medal in 2023 (bronze medal in 2010) and Alexander von Humboldt Fellowship. He is also a Sir J. C. Bose National Fellow. He was a visiting professor at IMS, Japan; LMU, Munich, Germany; Jackson State University, United States and Kyushu University, Japan. He is a Fellow of INSA (FNA), National Academy of Sciences (FNASc), Fellow of the Indian Academy of Sciences (FASc), Fellow of Royal Society of Chemistry (FRSC), Fellow of Association of Biotechnology and Pharmacy, Telangana State Academy of Sciences (Founder Fellow) and Andhra Pradesh Academy of Sciences (FAPAS) and also Fellow of Biotech Research Society, India (BRSI).

Acknowledgements

The editors sincerely thank the authors of each chapter for their contributions to this book. Their sincere efforts, dedication, hard work and analytical approach are highly acknowledged. The editors also acknowledge the publishers and associated teams for offering continuous support, guidance and motivation, which constantly pushed them forward to complete the book. Swapnali Hazarika thanks her parents, husband, son and daughter for their everlasting love, enthusiasm for science and encouragement to pursue every task successfully.

G. Narahari Sastry and Achyut Konwar sincerely thank all those who have directly or indirectly rendered valuable input to the book.

Swapnali Hazarika
Achyut Konwar
G. Narahari Sastry

1 An Overview of Membrane Technology

Joshua Tipple and Diganta Bhusan Das

1.1 INTRODUCTION

Membrane technologies for separation processes have been utilised since the mid-1800s [1]. The application of membrane technologies is becoming increasingly attractive due to the versatility of the technique. Currently, membranes are proving to have broad applications in many industries, e.g., within the petroleum industry, where they have seen the use for a variety of purposes, including the separation of compounds, such as olefins and paraffins, aromatic and aliphatic hydrocarbons as well as membrane-based wastewater treatment [2–5].

Membrane technologies face several challenges, such as combating the effects of membrane fouling and the relatively high cost of membrane modules. Despite this, they are expected to play an even increasingly important role in the future, aided by technological advancements in the characterisation and modelling of membrane separation process.

In addressing these points, this chapter provides a brief overview of membrane technology, membrane construction materials and the various applications of the technique. These are followed by a brief analysis of how computational techniques can enable continual improvement in the applications of the technology, e.g., characterisation and modelling of the membrane-based process. This will be demonstrated through computational membrane image processing and a nanofiltration membrane filtration system modelling.

1.2 MEMBRANE TECHNOLOGY

Membrane technologies are classified into several categories based on pore size. The most common categories currently used in industry are reverse osmosis, ultrafiltration, microfiltration and nanofiltration membranes [6]. Their applications rely on the pore size required for the separation process, as shown in Figure 1.1.

The numerous membrane techniques available for separation process all provide different benefits depending on the system requirements in which they are implemented. As each configuration requires various operating conditions, the correct technique must be applied to a system to ensure maximum process efficiency.

DOI: 10.1201/9781003441359-1

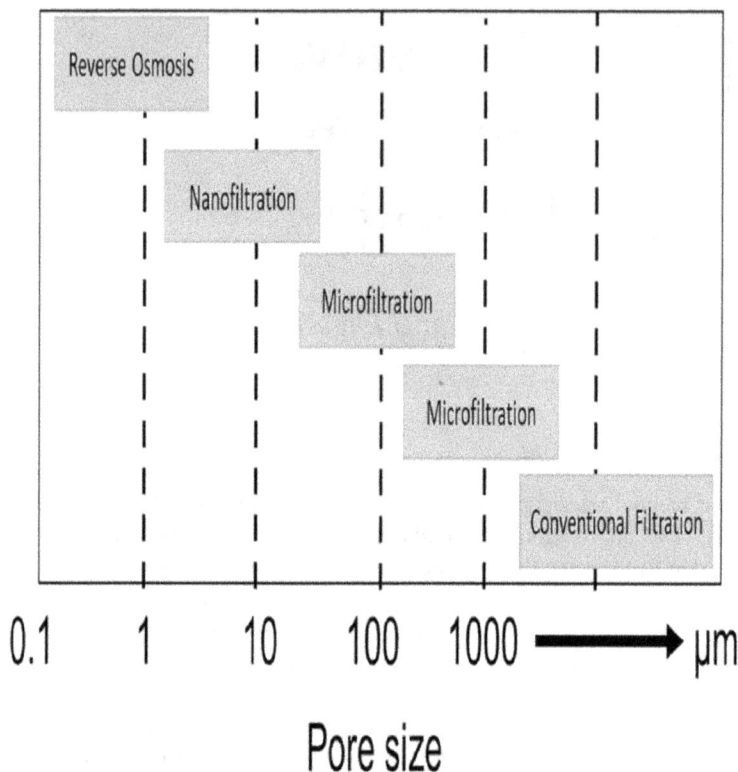

FIGURE 1.1 Membrane technologies and the particle size ranges they operate in.

Source: Adapted from [6].

Characteristics of a given system, such as the scale required, will determine the technique that should be used for the process. It is recommended that for small-scale operations with low volumes of water, dead-end filtration will provide a simple solution to operate and maintain. However, for large-scale operations, such as removing particulates from large bodies of water, cross-flow filtration may provide a better alternative, with reduced operating costs for long-term applications.

1.2.1 DEAD-END FILTRATION

The conventional method typically associated with batch membrane technology applications is dead-end filtration. A typical setup is illustrated in Figure 1.2. This setup operates by supplying the feed perpendicular to the membrane surface so the permeate output flow exits the module in the same direction. The method sees great success in laboratory experimentation and small-scale medical applications due to the high recovery rates of solutes and ease of operation [7].

Due to the lack of a retentate output stream, the retained substances remain in the system and increase the cake layer's thickness over time in solid particulate

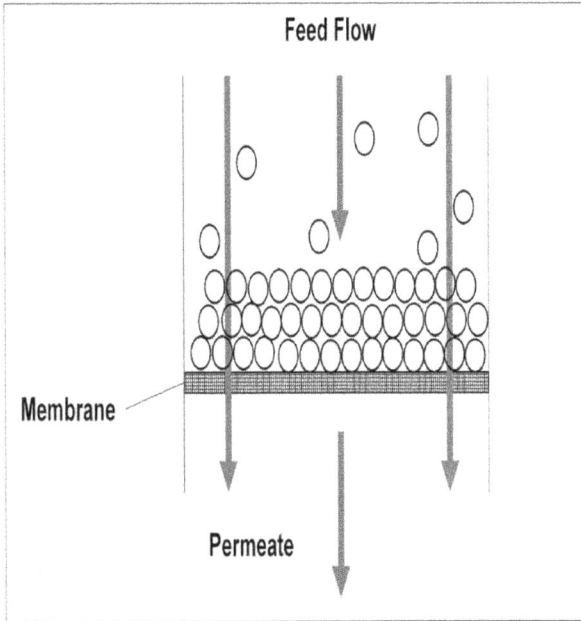

FIGURE 1.2   Typical dead-end filtration setup.

separations. Developing a thick cake reduces the efficiency of the process through reductions in permeate flux. The primary method to negate these effects is to halt operation so that during process downtime, the membrane is returned to its original state. Frequent cleaning and the replacement of membranes result in increased production volume and disposal costs, increasing the negative environmental impacts of the technique and reducing the likelihood of success for the technique in large-scale operations. Alternative methods to reduce the impact of these effects are being introduced that investigate the possibility of recycling flow back through the membrane or passing the flow in parallel to the surface of the membrane in a technique known as crossflow or tangential flow filtration.

1.2.2   CROSSFLOW FILTRATION

Crossflow filtration configurations can also be referred to as tangential filtration, as the method utilises a feed flow that tangentially passes over the membrane surface (Figure 1.3). This results in two output flows: a permeate stream passed through the membrane and a separate retentate stream. The shear forces generated by tangential flow allow the continuous removal of the cake layer (foulant) during the process. While this does not eliminate the possibility of solid particle cake build-up, higher permeate flow rates are maintained longer than dead-end filtration [8]. This enables the configuration to operate continuously with a reduced frequency of membrane replacement. The process can handle significant flow rates to produce a high-concentration retentate stream and, therefore, sees various applications in separation

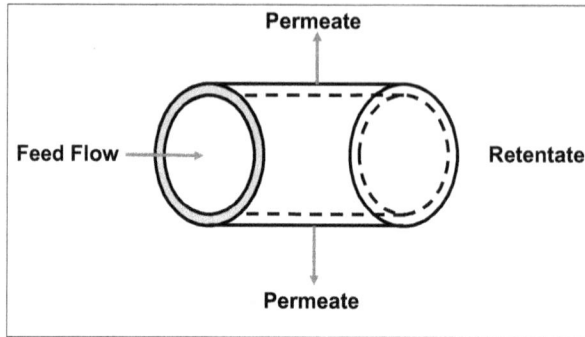

FIGURE 1.3 Typical crossflow filtration setup.

process, including the desalination of seawater and purification stages in biofuel production process [9]. While providing great operational benefits to dead-end filtration techniques, crossflow filtration typically requires a more significant initial capital cost investment, so it may not benefit smaller-scale process.

Hollow fibre membrane modules are a typical crossflow filtration membrane configuration used widely in many industries, e.g., desalination and biomedical industries [10–12]. They comprise many cylindrical fibres arranged in parallel and enclosed within a supporting structure. This gives the system great structural integrity and protects the fibres from external damage. The method is advantageous over other configurations due to its relatively low energy consumption and high surface area to volume ratio. The downsides of using hollow fibre membranes are primarily related to their complicated structure. Complex techniques are required to construct and install hollow fibre membranes, resulting in substantial costs. The membrane itself is highly susceptible to fouling, especially in feeds containing a variety of contaminants [13].

1.2.3 DECENTRALISED MEMBRANE PROCESSES

Decentralised membrane processes are a form of membrane system that is installed at the point of use, as opposed to the industry standard central treatment plant. Decentralised membrane processes are typically used for smaller-scale applications, such as treating water for individual households, small communities or small-scale industrial processes [14]. Decentralised filtration membranes can reduce the overall cost of a separation process, as they typically have reduced energy requirements compared to central treatment techniques. They are incredibly versatile and are used to separate a wide range of contaminants from several solvents. Decentralised filtration membranes offer a more sustainable solution to other techniques by reducing membrane processes' energy and resource requirements.

Like many industrial processes, decentralised membrane processes may require a regular maintenance schedule to ensure they operate as designed. The membrane must be repaired or replaced over time, especially if the process is demanding. This is challenging when operating in remote areas. In combination with this, decentralised membranes may require specialist technical knowledge to perform such maintenance, which may be difficult to access when operating in remote areas. The disadvantages of decentralised systems should be carefully considered when deciding whether to implement them into a process. Sometimes, a combination of centralised and decentralised treatment methods may be the most effective solution.

1.3 MEMBRANE MATERIALS

Membrane separation processes have broad applications within chemical process industries (CPIs), and as a result, many membranes have been designed for different applications. Some of the existing commercially available nanoporous membranes are shown in Table 1.1. As a result of varying feed and membrane characteristics, certain materials may not apply to a specific process. Therefore, several different materials are used to produce these nanoporous membranes.

For example, polyamide membranes are commonly used in wastewater treatment and seawater desalination. The material is formed via an interfacial polymerisation (IP) reaction between aromatic diamines and organic phase acyl chloride monomers. They have relatively low-pressure requirements and flexible operating conditions, but polyamide membranes are highly susceptible to oxidation [27]. Thin-film composite (TFC) membranes are composed of a three-layer structure. A thin barrier layer, a porous polymeric substrate, and mechanical support keep the membrane's structure

---

TABLE 1.1
Examples of Nanoporous Membranes and Their Operating Conditions

Type of Nanofiltration Membrane	Minimum Operating Temperature	pH Range		Composition of Top Layer	Supporting Studies
	C	Min	Max		
NF270	45	2.0	11.0	Polyamide thin film	[15–17]
NF200	45	3.0	10.0	composite	
NF90	45	3.0	10.0		
TS80	45	2.0	11.0	Polyamide	[18–20]
DK	50	3.0	9.0		
XN45	45	2.0	11.0		
UTC20	35	3.0	10.0	Polypiperazineamide	[21,22]
TS40	50	3.0	10.0		
TR60	35	3.0	8.0	Cross-linked polyamide composite	[23]
NFX	50	3.0	10.5	Proprietary polyamide thin-film composite	[24,25]
NFW	50	3.0	10.5		
NFG	50	4.0	10.0		
TFC SR100	50	4.0	10.0	Proprietary thin-film composite polyamide	[26]
SR3D	50	4.0	10.0		
SPIRAPRO	50	3.0	10.0		

Source: [15].

stable under strenuous conditions. They usually have a high permeability to water and compared to other materials have a wide pH range [28]. However, like polyamide membranes, TFC membranes are highly susceptible to oxidation.

Inorganic nanoparticles, referred to as nanofillers, are integrated into membrane polymers to improve the structural integrity of the membrane, offering more protection against thermal, chemical and mechanical damage [29]. Incorporation of nanofillers into membrane systems, especially at an industrial scale, is complex, with large amounts of material going to waste [30].

1.4 MEMBRANE FOULING

Membrane fouling is primarily caused by the accumulation of substances altering the size and characteristics of the membrane pores [31]. The extent to which membrane pore fouling occurs are different depending on three key groups, as shown in Figure 1.4 for fouling of nanofiltration membranes. Fouling can dramatically decrease the membrane's efficiency and produce an increase in operational costs [32]. Solutions to the issue are yet to be fully explored with possible methods for performance recovery being backwashing of the membrane using appropriate cleaning methods [33].

1.5 MODELLING OF MEMBRANE-BASED SEPARATION PROCESSES

To successfully mathematically model a membrane separation process, the chemical and physical properties of the feed's content must be known or estimated accurately. A diffusion coefficient (D) calculated using an empirical expression such as the Wilke and Chang equation (Equation 1.1) are beneficial in analysing the behaviour of the feed. It is an essential equation for understanding and predicting transport properties in porous membranes. It can provide valuable insights into the performance of these systems, where x is the membrane thickness, T is the temperature, μ is the fluid viscosity and V is the solute molar volume.

$$D = 7.4 \times 10^{-8} \frac{(xM)^{1/2} T}{\mu V^{0.6}}. \tag{1.1}$$

Membrane structures are considerably complicated, with various pore sizes and shapes. As a result, the distance between a fluid molecule is likely to be significantly larger than the straight-line distance across the membrane. Therefore, to model the process accurately, this must be considered. One way to do so is by introducing a term called tortuosity (τ), as discussed below.

τ measures the extent of winding in a fluid molecule's path that must travel to permeate a porous material. In the case of modelling a membrane separation process, the tortuosity is an important parameter that determines the permeate flux through a membrane. A membrane with high tortuosity typically indicates that the fluid components must travel more to permeate the membrane, reducing the transport rate. High tortuosity can result in decreased filtration efficiency and an increase in

- Type of foulant present
- Concentration of foulant
- pH of feed
- Ionic strength
- Presence of ions in feed such as Ca^{2+}

Composition of Feed

Factors affecting Nanofiltration Fouling

Properties of nanoporous membrane

Operation Conditions

- Membrane Roughness
- Charge
- Extent of hydrophobicity
- Functional groups on membrane surface

- Flux
- Velocity
- Recovery Rate
- Temperature

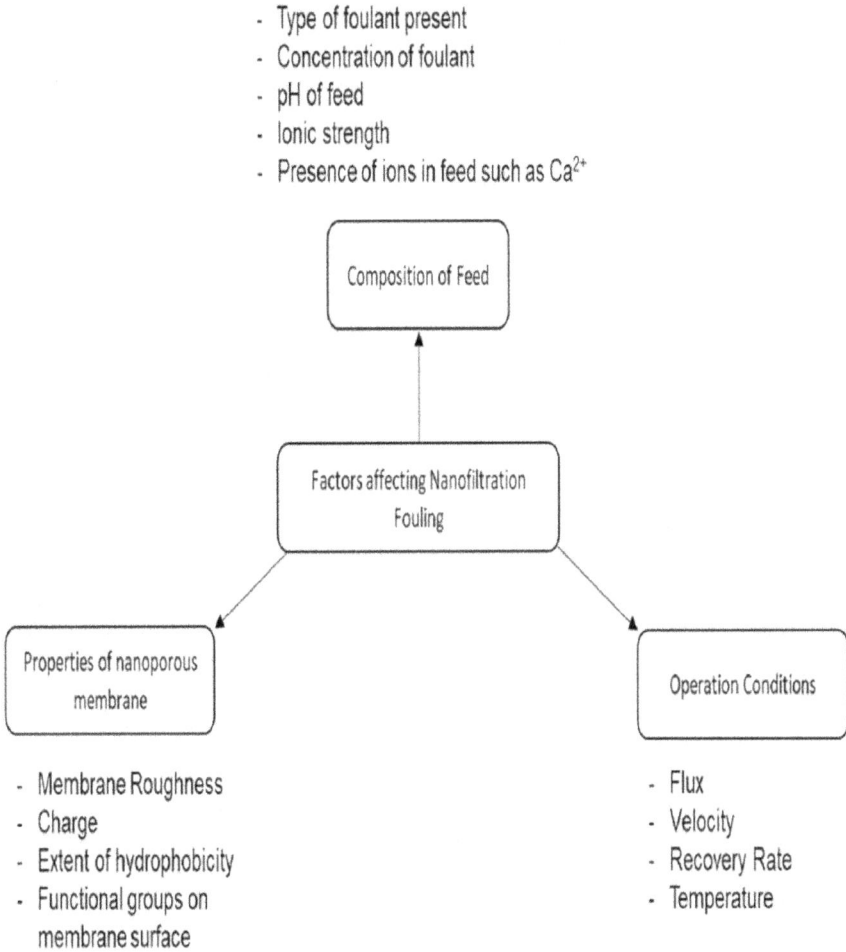

FIGURE 1.4 Factors affecting nanofiltration fouling.

the pressure drop across the membrane. Inversely, a low degree of tortuosity results in improved filtration efficiency as the fluid can permeate the membrane at a greater rate. Understanding the tortuosity of a filtration membrane is important for further optimisation and improving the process efficiency.

The diffusional tortuosity τ of fluid flow through a filtration membrane is approximated from porosity ε using Equation (1.2). This equation operates under the assumption that the solute molecules are perfectly spherical and are equal in size throughout the mixture.

$$\tau = 1 - 0.5\ln(\varepsilon), \tag{1.2}$$

τ is influenced by several physical factors of the membrane. These typically include the pore morphology, such as the pore shape and size, and the mechanical properties of the membrane material.

The effective diffusion coefficient (D_e) of the membrane is a crucial parameter in the design and analysis of membrane processes. D_e can measure how efficiently a membrane allows the transport of solutes from one side to the other through the membrane. Understanding D_e is essential for optimising membrane processes and achieving high-performance regarding solute rejection, flux and energy consumption. The equations used to express the fluid flow through a semi-permeable membrane are given in two parts. The free flow regime is governed by mass continuity (Equation 1.3) and Navier–Stokes equations (Equation 1.4), and the porous flow regime is governed by mass continuity (Equation 1.5) and Darcy's law (Equation 1.6).

The Navier–Stokes equations derived from the laws of conservation of mass, momentum and energy, which are widely used in various engineering fields to study fluid dynamics and the behaviour of fluid systems, see Equations (1.3) and (1.4). In membrane applications, the Navier–Stokes equations are used to model the flow of fluid through the membrane and to predict the behaviour of the system under different operating conditions, where u_f represents the fluid velocity, P_f is the fluid pressure, μ is the dynamic viscosity of the fluid, ρ is the fluid density and g is the gravitational acceleration vector.

$$\nabla \cdot u_f = 0. \tag{1.3}$$

$$\rho\left(u_f \cdot \nabla\right)u_f = \nabla \cdot \left(-P_f I + \mu\left(\nabla \cdot u_f + \left(\nabla \cdot u_f\right)^T\right)\right) + \rho g. \tag{1.4}$$

Darcy's law is a fundamental relationship in fluid mechanics that describes fluid flow velocity through a porous medium u_p. When coupled with mass continuity in membrane applications, Darcy's law represents the non-inertial incompressible flow occurring within the porous media. Expressed as (1.5) and (1.6), respectively, where ∇P_p is the pressure gradient across the porous medium and ρ is the fluid density. The permeability $\left(k\right)$ is a measure of the difficulty with which a fluid can flow through the porous medium and is highly dependent on the size and shape of the pores, the fluid viscosity $\left(\mu\right)$ and the orientation of the pores relative to the direction of fluid flow.

$$\nabla \cdot u_p = 0. \tag{1.5}$$

$$u_p = -\frac{k}{\mu}\left(\nabla P_p + \rho g\right). \tag{1.6}$$

One approach to assess the performance of a modelled or experimental membrane process is observing the rejection of solute in the membrane. Rejection of solute in membrane filtration refers to separating a solute from the solvent. Determining the rejection is vital to assess the likelihood of successfully taking a lab-scale membrane process to an industrial scale. Rejection (R) is calculated by measuring the initial and final concentrations of the solute, $C_{i,p}$ and $C_{i,w}$, respectively (Equation 1.7). Rejection can also be approximated using the solute–membrane mass transfer coefficient $K_{i,c}$, where Φ is the solute volume fraction in the feed solution, and Pe_m is the Péclet number for the solute in the membrane. The equation assumes that a combination of diffusion and convection governs solute rejection, and the solute concentration in the membrane is negligible compared to the solute concentration in the feed.

$$R = 1 - \frac{C_{i,p}}{C_{i,w}} = 1 - \frac{K_{i,c}\Phi}{1 - exp\left(-Pe_m\right)\left[1 - K_{i,c}\Phi\right]}. \qquad (1.7)$$

1.6 COMPUTATIONAL TECHNIQUES IN THE CHARACTERISATION OF MEMBRANES

The membrane pore structure must be known to specialise a membrane for different feeds and rejection requirements. Therefore, membrane characterisation is an important part of the research and development of membrane technologies. Understanding the membrane characteristics allows industry specialists to apply a membrane to the most appropriate process, as membrane performance highly depends on feed characteristics, hydrodynamics and surface chemistry [34]. The key membrane parameters required for characterisation are pore size, thickness and charge density [35].

When analysing the functionality of a membrane, several key characteristics need to be measured to decide on its applications. Various spectrometry techniques are applied to determine the chemical structure of the membrane, including X-ray photoelectron spectroscopy and nuclear magnetic resonance [36]. One factor that can affect the susceptibility of a membrane to fouling is the functional groups that comprise the surface of the membrane. Spectrometry allows these functional groups to be analysed, determining possible membrane–solvent–solute interactions [37]. Measuring the water contact angle can estimate the surface hydrophilicity of a membrane.

Computers and their applications have dramatically changed the methods used to characterise and model membranes. Advancements in artificial intelligence and the general limits of computers are enabling new techniques to be developed that revolutionise the field. As the tendency to work from home increases in the field of engineering, computational analysis provides versatility that is otherwise unavailable in laboratory experimentation.

1.6.1 IMAGE PROCESSING

Image processing techniques are used to estimate the pore size distribution of a selected membrane, which can determine a value for membrane porosity. A code developed by Rabbani and Salehi [38] can be used to analyse membrane scanning electron microscope (SEM) images. The tool enables quantitative analysis of the pore structure, operating in three-key stages. The code processes images under the assumption that darker spaces represent porous spaces capable of allowing fluid

TABLE 1.2
Image Processing of Selected Membranes

	PVDF	PVDF/MOF
Scale	≈82 × 50 μm	≈82 × 50 μm
Original SEM image		
Binary segmented image		
Depth map		
Segmented pore space image		
Calculated porosity	61.9%	69.1%

Source: [39–42].

PES

≈132 × 70 μm

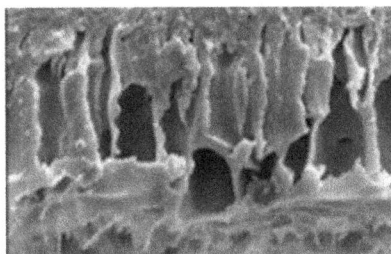

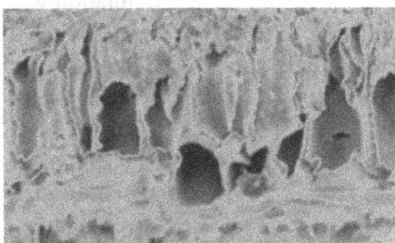

64.1%

PTT

≈15 × 7 μm

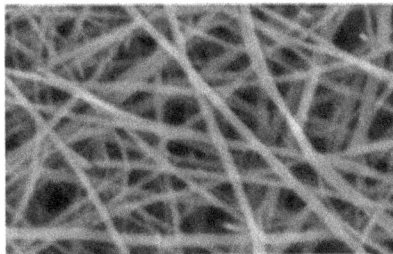

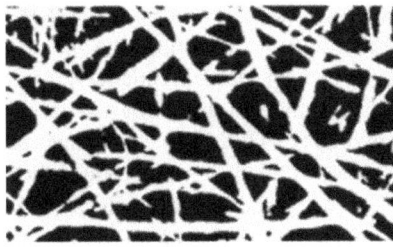

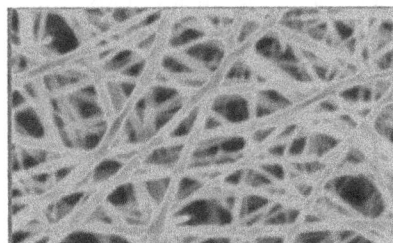

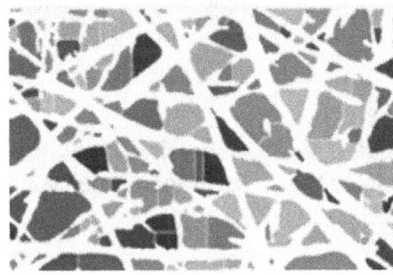

53.2%

permeation. Multi-level thresholding of greyscale SEM images enables the detection of dark spaces within the image, simulating the membrane structure by producing a depth map. The approach can differentiate pixels that have similar colour values into distinct sections. Binary segmentation allows the areas with the most depth to be shown as black, and the areas with least depth are highlighted in white. A watershed algorithm is then introduced to analyse the binary segmented image and is capable of effectively detect the possible overlap of the complex porous geometry. From this information, the algorithm counts the number of pores and estimates the size of each pore space. The porosity of the membrane ε is calculated using Equation (1.8), where A_p is the area of pore space, and A_t is a total area of the SEM image:

$$\varepsilon = \frac{\int_0^h \left(A_p\right) dz}{\left(A_t\right)}. \tag{1.8}$$

This technique is applied to membranes of a wide range of pore sizes and structures as the system operates on a ratio basis. One potential application is in the characterisation of nanofiltration membranes. We demonstrate this approach using SEM images of polyvinylidene fluoride (PVDF), polytrimethylene terephthalate (PTT), polyethersulfone (PES) and polyvinylidene fluoride mixed with metal-organic frameworks (PVDF/MOF) membranes were collected from various studies [39–42], and processed by the tool to produce the results shown in Table 1.2.

In our work, the porosity values estimated from the approach underestimated the experimental values by up to 25.1%. The method has potential for improvement, however. As image-capturing techniques improve in quality, a wider range of higher-definition images taken at various membrane depths can provide the simulation with a more accurate representation of the actual membrane structure. While this method is effective, black spots produced in the image can often be an image anomaly and not an indication of the presence of a pore. In these instances, a subjective decision is required on which black spots to include in the calculation.

Using calculated values for tortuosity, the diffusion coefficient D and the membrane porosity obtained from image processing (ε), Equation (1.9) is utilised to estimate the effective diffusivity D_e of a solute in water through various membrane structures. An example of these calculations is shown in Table 1.3 for the four types

---

TABLE 1.3
Estimated Average Porosity (ε), Solute Diffusivity in Water (D) and Solute Effective Diffusivity in Individual Membranes (D_e) at 20 °C

Membrane (–)	ϵ (–) (Table 1.2)	D (m²/s) (Equation 1)	D_e (m²/s) (Equation 9)
PVDF	0.619	1.46×10^{-9}	7.29×10^{-10}
PVDF/MOF	0.691		8.51×10^{-10}
PES	0.641		7.65×10^{-10}
PTT	0.532		5.90×10^{-10}

of membrane: PVDF, PVDF/MOF, PES and PTT. The calculations are performed assuming diffusivity is constant across all regions of the membranes.

$$D_e = \frac{\varepsilon}{\tau} D. \tag{1.9}$$

1.6.2 Basic Steps of Computational Modelling of Filtration Systems

Filtration membranes are complex systems with many interacting components. This makes modelling them accurately to ensure all relevant interactions are incorporated difficult. Computational modelling of membrane separation process allows for increased detail and a more accurate representation of the complex interactions within filtration membranes. These are developed at a wide range of scales, enabling macroscopic and molecular analyses. Models can predict the performance of a membrane under various operating conditions, allowing researchers and engineers to make informed decisions when considering which membrane to use. As the simulations are performed in a virtual environment, expensive and time-consuming physical experimental processes are avoided.

Boundary conditions in the computational modelling of filtration systems are divided into two categories: inlet boundary conditions and outlet boundary conditions. The governing equations (Equations 1.3–1.6) are solved to account for these boundary conditions. At the inlet, the fluid velocity and pressure are specified; at the outlet, the fluid velocity and pressure are calculated based on the fluid properties and the flow conditions within the system. The Navier–Stokes equations represent the laminar, incompressible free flow that exists outside the porous media when coupled with the mass continuity equation. As discussed earlier, they are expressed as Equations (1.3) and (1.4).

In many computational models, parameters are easily adjusted to represent various types of membranes and operating conditions. This enables researchers to predict and compare membrane performance of membranes, configurations and operating conditions with minimal experimental costs. This reduces the financial risk of membrane development, as new prototype designs and configurations are modelled with relative ease. Once a base case model is developed, design aspects are manipulated and the model are used to optimise the process for improved filtration efficiency and other performance indicators. Models can perform sensitivity analysis on a given membrane module to better understand, which parameters significantly affect performance and costs.

While the benefits of computational modelling are impressive, the technique has some drawbacks that should be considered when implementing the model into a design process. Typically, computational models require a significant amount of computational resources and time. Poor computational performance can result in lengthy modelling times, making the method impractical for some applications. Computational models typically require assumptions regarding the behaviour of components within the system. These implement an error element into the model-making predictions not completely accurate. As the results of a

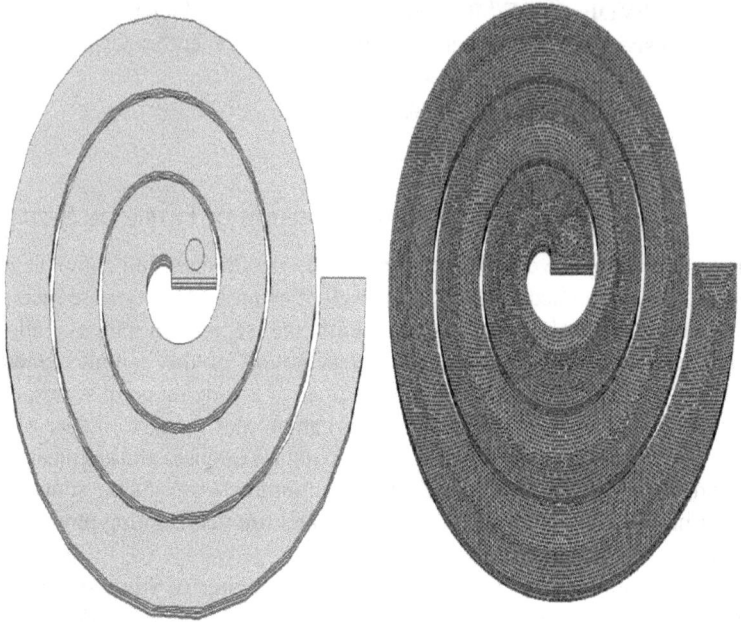

FIGURE 1.5 (a) 3D system geometry of a circular cross flow system from Ref. [8]; (b) 3D system geometry of a circular cross flow system from Ref. [8] with the mesh applied.

model may not always replicate potential real experimental results, validation and refinement are required. This is difficult for niche or new experimental techniques as the accurate and comprehensive data needed for validation may not always be available. Models are limited by the algorithms and software used in the development process. The developed model must have an element of robustness so that unexpected behaviour or a parameter change does not lead to potentially incorrect predictions. Developing a model of this quality is considered a significant barrier to their use, as a high level of specialist expertise is required to create and improve these models effectively.

To illustrate how computational modelling is used in the membrane process, we have developed a three-dimensional system geometry of a circular crossflow module in COMSOL Multiphysics 5.5 based on a physical experimental rig [8]. Due to computational restrictions, the external structure and sample reservoir were removed to simplify and reduce the computational time requirements of the modelling process, as they are not required for this simulation. Figure 1.5(a) shows the basic system geometry constructed to represent the crossflow membrane module. The membrane is placed under the flow channels, as shown in Figure 1.5(a). Figure 1.5(b) shows the system with the mesh applied, enabling a nodal representation of the geometry and a functional representation of the domain.

Using the system geometry shown in Figure 1.5, a computational fluid dynamics model was developed in COMSOL Multiphysics 5.5 to study a nanofiltration water treatment process to remove an antimicrobial agent, triclosan, from wastewater. The

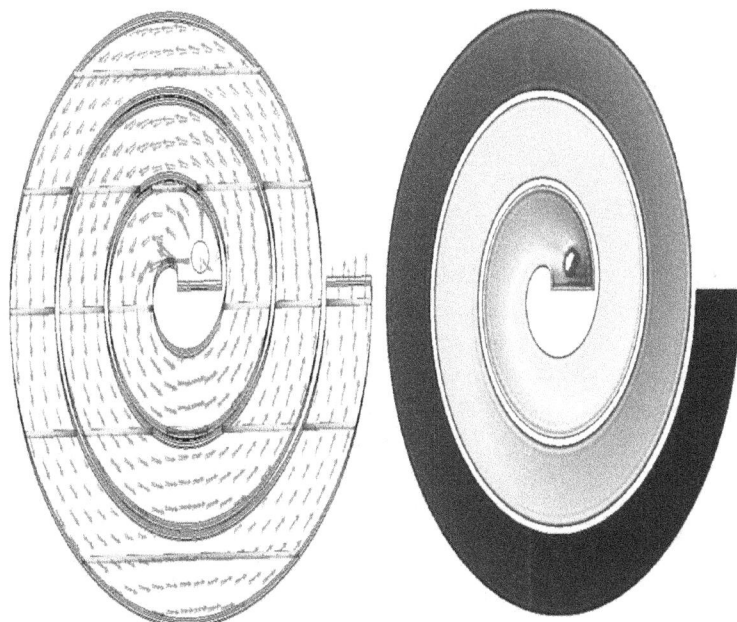

FIGURE 1.6 (a) Arrow fluid velocity profile of a circular cross flow system; (b) fluid pressure contour plot for the circular cross flow nanofiltration system.

base-case simulation was developed using common operating parameters seen in previous experimentation with an inlet fluid pressure of 7 Bar. The membrane selected for modelling comprised PVDF/MOF. The computational model was designed to simulate a velocity profile and pressure contour plot for the circular cross-flow nanofiltration system, and the results are shown in Figure 1.6(a). The fluid flow is tangential to the membrane surface. The initial fluid pressure at the inlet (centre) dissipated gradually throughout, becoming minimal at the exit of the spiral channel, as seen in Figure 1.6(b).

1.7 THE ROLE OF MEMBRANE TECHNOLOGY IN PROCESS INTENSIFICATION

Process intensification is at the forefront of innovation in chemical engineering significantly improving equipment and techniques through rigorous research and development. Process intensification aims to substantially change industrial processes making them safer, more economical and sustainable [43]. Commonly, the techniques used to achieve these goals involve reducing the number of apparatuses involved in a process, usually by integrating several functionalities into a single device. Thus, it reduces overall investment requirements for a new system and increases the sustainability of an otherwise demanding process.

The membrane separation process provides opportunities for process intensification in many industries, providing low-energy alternatives to costly separation techniques. For example, nanofiltration has already been implemented in various process to reduce energy consumption, notably offering alternatives to complex and expensive evaporation techniques in coffee production [44].

1.8 CONCLUSION

Membrane technologies are proving to be a valuable tool for various applications. When applied to remove contaminants selectively, membranes offer a cost-effective and energy-efficient solution that can address many industrial and environmental challenges. This introductory chapter has provided an overview of the basics of membrane technology, including the existing technology and configurations, materials, operating principles and potential applications. The potential for computational modelling and characterisation techniques has been explored and shows great promise for the future of membrane technologies. As the need for sustainable and versatile separation process continues to grow, membranes are predicted to play an increasingly important role in meeting this demand. The continued development and refinement of these technologies are essential to ensure that they remain a vital tool for meeting future challenges.

REFERENCES

1. Mulder, M., & Mulder, J. (1996). *Basic Principles of Membrane Technology*. Springer Science & Business Media.
2. Goh, P. S., Wong, K. C., & Ismail, A. F. (2022). Membrane technology: A versatile tool for saline wastewater treatment and resource recovery. *Desalination, 521*, 115377. https://doi.org/10.1016/j.desal.2021.115377
3. Hamlil, A., & Aouinti, L. (2023). Preparation and characterization of dense membrane-based metal organic networks (MOF-5) for separation: Aromatic–aliphatic mixtures. *Polymer-Plastics Technology and Materials, 62*(7), 909–920. https://doi.org/10.1080/25740881.2023.2172683
4. Nunnes, S. P., & Peinemann, K. V. (2006). *Membrane Technology in the Chemical Chemistry Industry*. Wiley.
5. Roy, A., Venna, S. R., Rogers, G., Tang, L., Fitzgibbons, T. C., Liu, J., ... & Fish, B. (2021). Membranes for olefin–paraffin separation: An industrial perspective. *Proceedings of the National Academy of Sciences, 118*(37), e2022194118. https://doi.org/10.1073/pnas.202219411
6. Shon, H. K., Phuntsho, S., Chaudhary, D. S., Vigneswaran, S., & Cho, J. (2013). Nanofiltration for water and wastewater treatment—A mini review. *Drinking Water Engineering and Science, 6*(1), 47–53. https://doi.org/10.5194/dwes-6-47-2013
7. Hu, R., Zhao, G., He, Y., & Zhu, H. (2020). The application feasibility of graphene oxide membranes for pressure-driven desalination in a dead-end flow system. *Desalination, 477*, 114271. https://doi.org/10.1016/j.desal.2019.114271
8. Shamsuddin, N., Das, D. B., & Starov, V. M. (2015). Filtration of natural organic matter using ultrafiltration membranes for drinking water purposes: Circular cross-flow compared with stirred dead-end flow. *Chemical Engineering Journal, 276*, 331–339. https://doi.org/10.1016/j.cej.2015.04.075
9. Gao, Y., Zhang, Y., Dudek, M., Qin, J., Øye, G., & Østerhus, S. W. (2021). A multivariate study of backpulsing for membrane fouling mitigation in produced water

treatment. *Journal of Environmental Chemical Engineering*, *9*(2), 104839. https://doi.org/10.1016/j.jece.2020.104839

10. Verma, S. K., Modi, A., & Bellare, J. (2019). Polyethersulfone-carbon nanotubes composite hollow fiber membranes with improved biocompatibility for bioartificial liver. *Colloids and Surfaces B: Biointerfaces*, *181*, 890–895. https://doi.org/10.1016/j.colsurfb.2019.06.051

11. Xiao, L., Yang, M., Yuan, W. Z., & Huang, S. M. (2020). Coupled heat and mass transfer of cross-flow random hollow fiber membrane tube bundle used for seawater desalination. *International Journal of Heat and Mass Transfer*, *152*, 119499. https://doi.org/10.1016/j.ijheatmasstransfer.2020.119499

12. Ye, H., Cui, Z. F., Ellis, M. J., Macedo, H., & Mantalaris, A. (2010). Hollow fiber membrane bioreactor technology for tissue engineering and stem cell therapy. In *Chemical and Biochemical Transformations in Membrane Systems* (Vol. 3, pp. 213–227). Elsevier.

13. McKeen, L. W., 2012. *Permeability Properties of Plastics and Elastomers*. 3rd ed. s.l.:s.n. Elsevier.

14. Sharma, M., Yadav, A., Dubey, K. K., Tipple, J., & Das, D. B. (2022). Decentralized systems for the treatment of antimicrobial compounds released from hospital aquatic wastes. *Science of the Total Environment*, 156569. https://doi.org/10.1016/j.scitotenv.2022.156569

15. Dang, 2020. Mathematical modelling for predicting rejection of trace organic contaminants by the nanofiltration membrane NF270. *Journal of Environmental Treatment Techniques*, 8(3), 2309–1185.

16. Kumar, R., Ahmed, M., Ok, S., Garudachari, B., & Thomas, J. P. (2019). Boron selective thin film composite nanofiltration membrane fabricated via a self-assembled trimesic acid layer at a liquid–liquid interface on an ultrafiltration support. *New Journal of Chemistry*, *43*(9), 3874–3883. DOI: 10.1039/C8NJ05670F

17. Zhao, Y. Y., Kong, F. X., Wang, Z., Yang, H. W., Wang, X. M., Xie, Y. F., & Waite, T. D. (2017). Role of membrane and compound properties in affecting the rejection of pharmaceuticals by different RO/NF membranes. *Frontiers of Environmental Science & Engineering*, *11*, 1–13. https://doi.org/10.1007/s11783-017-0975-x

18. Chen, P., Ma, X., Zhong, Z., Zhang, F., Xing, W., & Fan, Y. (2017). Performance of ceramic nanofiltration membrane for desalination of dye solutions containing NaCl and Na_2SO_4. *Desalination*, *404*, 102–111. https://doi.org/10.1016/j.desal.2016.11.014

19. Košutić, K., Kaštelan-Kunst, L., & Kunst, B. (2000). Porosity of some commercial reverse osmosis and nanofiltration polyamide thin-film composite membranes. *Journal of Membrane Science*, *168*(1–2), 101–108. https://doi.org/10.1016/S0376-7388(99)00309-9

20. Weinman, S. T., Fierce, E. M., & Husson, S. M. (2019). Nanopatterning commercial nanofiltration and reverse osmosis membranes. *Separation and Purification Technology*, *209*, 646–657. https://doi.org/10.1016/j.seppur.2018.09.012

21. Sidik, D. A. B., Hairom, N. H. H., & Mohammad, A. W. (2019). Performance and fouling assessment of different membrane types in a hybrid photocatalytic membrane reactor (PMR) for palm oil mill secondary effluent (POMSE) treatment. *Process Safety and Environmental Protection*, *130*, 265–274. https://doi.org/10.1016/j.psep.2019.08.018

22. Damtie, M. M., Woo, Y. C., Kim, B., Hailemariam, R. H., Park, K. D., Shon, H. K., … & Choi, J. S. (2019). Removal of fluoride in membrane-based water and wastewater treatment technologies: Performance review. *Journal of Environmental Management*, *251*, 109524. https://doi.org/10.1016/j.jenvman.2019.109524

23. Gündoğdu, M., Jarma, Y. A., Kabay, N., Pek, T. Ö., & Yüksel, M. (2019). Integration of MBR with NF/RO processes for industrial wastewater reclamation and water reuse-effect of membrane type on product water quality. *Journal of Water Process Engineering, 29*, 100574. https://doi.org/10.1016/j.jwpe.2018.02.009

24. Premachandra, A., O'Brien, S., Perna, N., McGivern, J., LaRue, R., & Latulippe, D. R. (2021). Treatment of complex multi-sourced industrial wastewater—New opportunities for nanofiltration membranes. *Chemical Engineering Research and Design, 168*, 499–509. https://doi.org/10.1016/j.cherd.2021.01.005

25. Schmidt, C. M., Sprunk, M., Löffler, R., & Hinrichs, J. (2020). Relating nanofiltration membrane morphology to observed rejection of saccharides. *Separation and Purification Technology, 239*, 116550. https://doi.org/10.1016/j.seppur.2020.116550

26. Madhura, L., Kanchi, S., Sabela, M. I., Singh, S., & Bisetty, K. (2018). Membrane technology for water purification. *Environmental Chemistry Letters, 16*, 343–365. https://doi.org/10.1007/s10311-017-0699-y

27. Nitto, 2020. *RO/NF Polyamide Membrane Feedwater Requirements*. Hydranautics Nitto Group Company.

28. Goh, P. S., & Ismail, A. F. (2018). Flat-sheet membrane for power generation and desalination based on salinity gradient. In *Membrane-Based Salinity Gradient Processes for Water Treatment and Power Generation* (pp. 155–174). Elsevier. https://doi.org/10.1016/B978-0-444-63961-5.00005-5

29. Althues, H., Henle, J., & Kaskel, S. (2007). Functional inorganic nanofillers for transparent polymers. *Chemical Society Reviews, 36*(9), 1454–1465. https://doi.org/10.1039/B608177K

30. Lee, T. H., Park, I., Oh, J. Y., Jang, J. K., & Park, H. B. (2019). Facile preparation of polyamide thin-film nanocomposite membranes using spray-assisted nanofiller predeposition. *Industrial & Engineering Chemistry Research, 58*(10), 4248–4256. https://doi.org/10.1021/acs.iecr.9b00029

31. Yu, W., Liu, T., Crawshaw, J., Liu, T., & Graham, N. (2018). Ultrafiltration and nanofiltration membrane fouling by natural organic matter: Mechanisms and mitigation by pre-ozonation and pH. *Water Research, 139*, 353–362. https://doi.org/10.1016/j.watres.2018.04.025

32. Guo, W., Ngo, H. H., & Li, J. (2012). A mini-review on membrane fouling. *Bioresource Technology, 122*, 27–34. https://doi.org/10.1016/j.biortech.2012.04.089

33. Liu, D., Cabrera, J., Zhong, L., Wang, W., Duan, D., Wang, X., … & Xie, Y. F. (2021). Using loose nanofiltration membrane for lake water treatment: A pilot study. *Frontiers of Environmental Science & Engineering, 15*, 1–11. https://doi.org/10.1007/s11783-020-1362-6

34. Rosa, M. J., & De Pinho, M. N. (1997). Membrane surface characterisation by contact angle measurements using the immersed method. *Journal of Membrane Science, 131*(1–2), 167–180. https://doi.org/10.1016/S0376-7388(97)00043-4

35. Bowen, W. R., & Mohammad, A. W. (1998). Characterization and prediction of nanofiltration membrane performance—A general assessment. *Chemical Engineering Research and Design, 76*(8), 885–893. https://doi.org/10.1205/026387698525685

36. Khulbe, K. C., Matsuura, T., Khulbe, K. C., & Matsuura, T. (2021). Membrane characterization. *Nanotechnology in Membrane Processes, 89*–133. DOI: 10.1007/978-3-030-64183-2_3

37. Misdan, N., Lau, W. J., Ismail, A. F., Matsuura, T., & Rana, D. (2014). Study on the thin film composite poly (piperazine-amide) nanofiltration membrane: Impacts of physicochemical properties of substrate on interfacial polymerization formation. *Desalination, 344*, 198–205. https://doi.org/10.1016/j.desal.2014.03.036

38. Rabbani, A., & Salehi, S. (2017). Dynamic modeling of the formation damage and mud cake deposition using filtration theories coupled with SEM image processing. *Journal of Natural Gas Science and Engineering, 42*, 157–168. https://doi.org/10.1016/j.jngse.2017.02.047

39. Cao, X., Ma, J., Shi, X., & Ren, Z. (2006). Effect of TiO_2 nanoparticle size on the performance of PVDF membrane. *Applied Surface Science, 253*(4), 2003–2010. https://doi.org/10.1016/j.apsusc.2006.03.090

40. Gnanasekaran, G., Balaguru, S., Arthanareeswaran, G., & Das, D. B. (2019). Removal of hazardous material from wastewater by using metal organic framework (MOF) embedded polymeric membranes. *Separation Science and Technology, 54*(3), 434–446. https://doi.org/10.1080/01496395.2018.1508232

41. Li, M., Wang, D., Xiao, R., Sun, G., Zhao, Q., & Li, H. (2013). A novel high flux poly (trimethylene terephthalate) nanofiber membrane for microfiltration media. *Separation and Purification Technology, 116*, 199–205. https://doi.org/10.1016/j.seppur.2013.05.046

42. Shen, L., Bian, X., Lu, X., Shi, L., Liu, Z., Chen, L., … & Fan, K. (2012). Preparation and characterization of ZnO/polyethersulfone (PES) hybrid membranes. *Desalination, 293*, 21–29. https://doi.org/10.1016/j.desal.2012.02.019

43. Keil, F. J. (2018). Process intensification. *Reviews in Chemical Engineering, 34*(2), 135–200. https://doi.org/10.1515/revce-2017-0085

44. Laurio, M. V. O., & Slater, C. S. (2020). Process scale-up, economic, environmental assessment of vibratory nanofiltration of coffee extracts for soluble coffee production process intensification. *Clean Technologies and Environmental Policy, 22*, 1891–1908.https://doi.org/10.1007/s10098-020-01931-x

2 Novel Membrane Materials

*Achyut Konwar, Akhil Ranjan Borah and
Swapnali Hazarika*

2.1 INTRODUCTION

Based on their principal constituting materials, membranes are categorised as organic and inorganic membranes. Organic membranes are generally formed from organic polymeric materials. Due to their flexibility, cost-effectiveness, ease of fabrication and tunability, polymeric membranes are commonly used for various applications. Pore size and its distribution are the key factors in the use of membranes for different purposes. Depending upon the pore size, membranes are divided into three types: nanofiltration membranes (solute with a size ≤2 nm can be retained), ultrafiltration membranes (solute with a size of 2–100 nm can be retained) and microfiltration membranes (solute with a size ≥100 nm can be retained). Using polymeric membranes, it becomes easy to tune the pore size of the membranes [1]. Polymeric materials also contain functional groups, which sometimes help to retain the solutes through physical interactions [2]. Such membranes can also be fabricated through varied processes such as non-solvent-induced phase inversion, thermally induced phase inversion, electrospinning, stretching, etc. [3–5]. They can also be designed in various shapes and forms, such as flat sheets, hollow fibre, spiral, etc. [6].

However, polymeric membranes have some issues, like low heat and organic solvent resistance [7]. Because most polymeric materials are soluble in common organic solvents, they are prone to fouling when applied to organic solvent-mixed feed [8]. Inorganic membranes possess high-chemical stability and high-temperature resistance are durable and autoclavable [9]. However, these membranes are expensive and difficult to manufacture. It is also challenging to achieve variable shapes with inorganic membranes [1]. Based on the design and function, inorganic membranes are again divided into two categories: (i) porous inorganic membranes and (ii) dense or non-porous inorganic membranes. Porous inorganic membranes are composed of a porous active layer on top of another porous support. This porous support maintains the mechanical stability of the whole membrane. The porous support generally has a larger pore size than the active layer and has a minimal role in membrane permeability. Dense or non-porous membranes also have a porous support that contains a metal or alloy (generally Pb and Ag alloys). Such membranes find application in separating and purifying hydrogen or oxygen from a mixture of gases [9,10].

DOI: 10.1201/9781003441359-2

2.2 SYNTHETIC POLYMERIC MATERIALS USED FOR MEMBRANE DESIGN

Conventional synthetic polymers such as polysulfone (PSf), polyamide (PA), cellulose acetate (CA), polyvinylidene fluoride (PVDF) and polyethersulfone (PES), etc., commonly used in different membrane-based separation processes like microfiltration (MF), ultrafiltration (UF), nanofiltration (NF), reverse osmosis (RO) and other new membrane-based processes. PES, PSf and PVDF are generally used to design MF and UF membranes; PA, having high chemical and thermal resistance, is used to develop membranes for a wide range of membrane-based processes [11]. CA is one of the oldest, most widely used and extensively researched polymeric materials for designing UF, MF and RO membranes. CA membranes are also used in dialysis [12]. Cellulose acetate is derived from cellulose, and being of natural origin, it is proven to be biodegradable. Although cellulose acetate is resistant to water, it has some other issues, such as low thermal resistance, poor chemical resistance and unsatisfactory mechanical properties [13,14]. Therefore, additional modifications, such as the incorporation of reinforcing agents and additives, modification of the surface, etc., are required to enhance the performance of cellulose acetate membranes [15–17]. PSf is the most widely used polymeric material in membrane design. Commercial availability, as well as the ease of processing, make PSf the most popular one. PSf possesses relatively high mechanical strength, thermal resistance and chemical resistance [18]. Polyether sulfone (PES) is a member of the PSf group and has many superior properties compared to other PSf materials [19]. PVDF exhibits high mechanical strength, chemical and thermal resistance while being significantly hydrophobic in nature [20]. The hydrophobic nature of PVDF membranes makes them suitable for membrane distillation [21,22]. Polytetrafluoroethylene (PTFE) is the most suitable polymer for membrane distillation because of its highly hydrophobic in nature. Polypropylene and polyethylene membranes have also been tried for membrane distillation [23] (Figure 2.1).

2.3 BIOPOLYMERS FOR MEMBRANE

The trend towards using polymers of natural origin for membrane fabrication is not new. During the initial developmental stage of membrane-based separation, the experiments were performed on the membranes produced from cellulose acetate, cellulose nitrate, etc. In the 1960s, Loeb and Sourirajan used cellulose acetate to develop a reverse osmosis membrane with high salt rejection as well as high permeability at moderate operating pressure. This new inversion by these two scientists, Loeb and Sourirajan, paved the way for desalination technology to get potable water from both underground and seawater [24]. Apart from separation and purification, bio-based membranes, due to their unique physical, chemical and biocompatible properties, find wide applications as sensors, wound dressing materials, etc. This section will discuss various types of membranes produced from materials of biological origin and their applications [25,26].

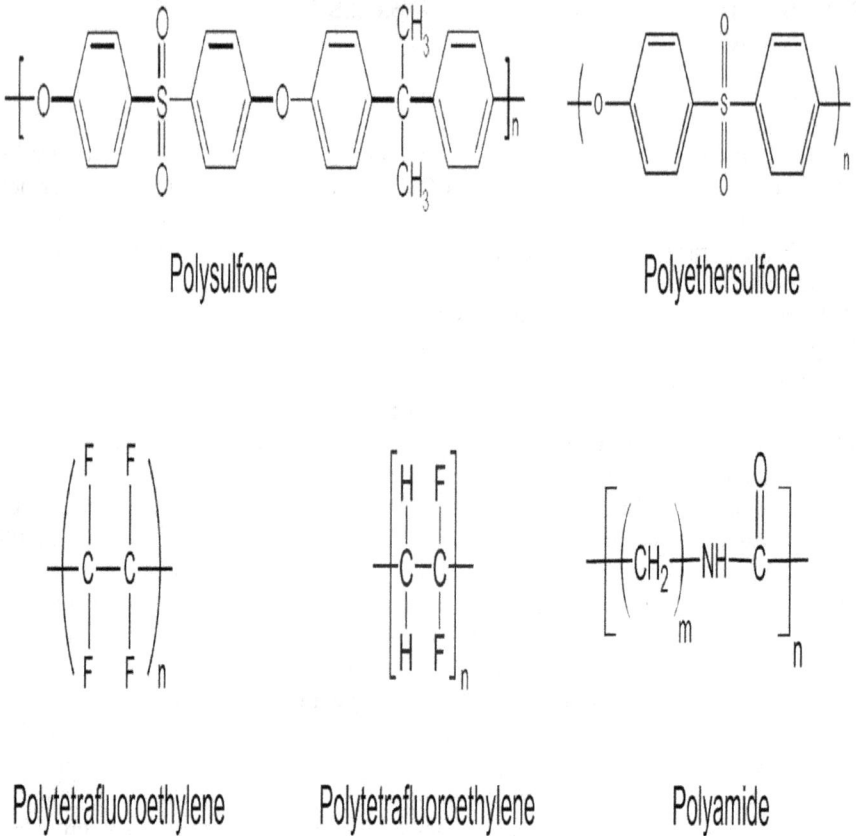

Polysulfone

Polyethersulfone

Polytetrafluoroethylene

Polytetrafluoroethylene

Polyamide

FIGURE 2.1 Chemical structure of synthetic plastic materials used for membrane fabrication.

2.3.1 CHITOSAN-BASED MEMBRANES

Chitosan, the cationic (1–4)-2-amino-2-deoxy-β-D-glucan, is produced by partial deacetylation of chitin. In mildly acidic conditions, it dissolves and becomes positively charged. In an acidic medium, the NH_2 group of chitosan acquires a positive charge by protonation and gets converted to NH_3^+. The presence of this amine group on the polymeric chain of chitosan is responsible for its unique chemical and physical properties. The solubility of chitosan in an aqueous medium is also pH-dependent. As a result, dissolution of chitosan-made products in aqueous medium at pH <6.5 becomes a major concern during application. Embedding compatible nanomaterials and using suitable cross-linking agents can improve the physical and mechanical properties of chitosan membranes [27].

Various innovative techniques have been applied to fabricate chitosan-based self-standing membranes. The most common technique is the use of pore generators. Pore generators (porogens) are initially used as an additive in the polymeric matrix, followed by dissolution using a suitable solvent medium. Poly(ethylene oxide), poly(vinyl pyrrolidone), poly(ethylene glycol), zirconia, alumina, silica and inorganic salts, such as NaCl and $CaCO_3$, are some of the examples of pore generators used in the fabrication of membranes from chitosan. In the fabrication process, porogens are first blended with chitosan in a mild aqueous acid medium. Then, solvent casting followed by drying forms thin films. Once the solid films are formed, the porogens inside the chitosan matrix are made soluble using proper solvent systems such as aqueous NaOH solutions, water, etc.

Polymer blending is an effective technique for mixing two or more polymeric materials to obtain a hybrid material with a new set of improved properties. Miscibility between the component polymeric materials is the major requirement to get the best properties from blending. Miscible polymers show effective interaction between the polymer chains and thus ensure homogeneity and reinforcement in the blend structure. Chitosan and its derivatives, such as carboxymethyl chitosan, can introduce amine and hydroxyl functional groups into the polymeric matrix of the common synthetic polymers used for membrane fabrication, like PES, CA, PVDF, PS, etc., through blending. In addition of hydrophilic groups into a hydrophobic polymeric matrix can enhance the water permeability of the hybrid membrane. Chitosan can also affect the membrane surface charge, as it introduces –OH and $–NH_2$ groups in the hybrid polymeric membranes. Chitosan-based nanoparticles prepared using various techniques have also been used for the fabrication of composite membranes with various other polymeric materials [28–30].

2.3.2 ALGINATE-BASED MEMBRANES

They are an unbranched copolymer family composed of (1,4) linked b-D-mannuronic acid (M) and -L-guluronic acid (G) groups. Alginate is extracted from seaweeds and possesses $–COO^-$ functionality on the polymeric chain. As a result, alginate can bind electrostatically with different multivalent metal ions to form a three-dimensional cross-linked structure. This polymer's unique ability causes it to convert into membrane form; however, few reports on alginate-based membranes have been published [31]. Aburabie et al. reported one of the most significant works on alginate-based membrane fabrication. By spin coating at 1000 rpm and cross-linking with Ca^{2+} ions, they were able to create coated thin film composites (TFC) as well as self-standing alginate-based membranes. Alginate-based membranes could also be formed by the electrospinning method [32]. Recently, Dodero et al. reported the fabrication of ZnO nanoparticle-incorporated fibrous electrospun alginate membranes using a 3.5% polymeric solution with alginate and polyethylene oxide in the ratio 70:30. In the electrospinning technique, a polymeric solution is ejected from a syringe by applying high voltage. The application of high voltage creates a high charge inside the solution that splits the solution, leading to the formation of fibrous structures constituted by nanofibers on the collector surface. The electrospinning operation was carried out at 12.5 kV, the distance between the spinneret and collector was maintained at 15 cm, whereas the flow rate of the polymeric solution was fixed at 0.75 mL/h. The inner diameter of the needle used was 0.4 mm. After the formation of the fibrous sheet structure on a flat collector, it was again cross-linked with different divalent metal ions [33].

2.3.3 POLYLACTIC ACID (PLA)-BASED MEMBRANES

Polylactic acid (PLA), with the molecular formula $(C_3H_4O_2)_n$, is a biodegradable thermoplastic. PLA is synthesised from L-lactic acid, which, on the other hand, is obtained from glucose. Glucose can be primarily obtained from corn, sugarcane, cassava, etc. PLA behaves as a thermoplastic polymer, which facilitates the melt processing of this polymer into various products. Techniques used for membrane fabrication from PLA include electrospinning, solution blow spinning, non-solvent-induced phase separation, temperature-induced phase separation, etc. [34].

Electrospinning is a popularly used method for the fabrication of fibrous polymeric membranes from PLA. Electrospun PLA membrane shows high hydrophobicity because the nanofibers present in the membrane structure led to an increase in the superficial area of the surface. Solution-blown spinning is another method used for the fabrication of fibrous membranes from PLA. In this technique, the polymer solution mixture is blown through an inner nozzle and pressurised air is blown through an outer nozzle. It is collected on a rotating collector, over which the nonwoven fibre mat is formed. Unlike the electrospinning process, the solution-blow spinning process does not use high voltage.

The non-solvent-induced phase inversion technique can also be used to create PLA membranes. This technique involves the casting of a polymeric solution on a substrate, followed by immersion in a non-solvent. In this method, the polymer solution prepared in a suitable organic solvent is allowed to solidify in a non-solvent bath containing water. The most common solvents used for PLA are chloroform, dioxane, methylene chloride, acetonitrile, dichloroacetic acid, 1,1,2-trichloroethane, etc. Although this is the most common method for the fabrication of polymeric membranes, it generates a large amount of organic waste, making it an environmental concern [35–38].

3D printing is another recent and advanced technique used for the fabrication of PLA membranes. The most common principle involves heating a polymeric material to a temperature above its melting point and then extruding it according to a computer-designed program. This method is called 'fused deposition modelling' (FDM). PLA is often blended with other polymeric materials to improve its rheological properties. However, the pore size of the membranes fabricated by the 3D printing method is generally in the macroporous range [39].

2.3.4 POLYHYDROXYALKANOATE-BASED MEMBRANES

Polyhydroxyalkanoates (PHAs) belong to the biodegradable polyester family and are produced by a group of microorganisms. The general chemical structure of PHAs is provided in Figure 2.2. The diverse production conditions lead to wide variation in the side chain length and composition of PHA, which ultimately influence the final thermomechanical properties of the biopolymer. The homopolymer polyhydroxybutyrate (PHB) is the most studied and applied biopolymer of the

PHA family. Relatively poor mechanical properties are one of the main drawbacks of PHB when compared to other common polymers. However, current solutions to this issue include copolymerisation of PHB to obtain poly(hydroxybutyrate-co-hydroxyvalerate) (PHBHV) (or even longer-chain length PHA) or blending with (bio-based) plasticisers [40].

Reported methods for the fabrication of PHA-based membranes are mainly electrospinning and non-solvent-induced phase inversion. Electrospinning membranes of PHA are mainly formed by blending with other polymers like PLA, sodium alginate, etc. Since PHA also behaves as a thermoplastic, there is a possibility of using other techniques, such as melt spinning, 3D printing, etc., for the fabrication of PHA-based membranes [41–42].

2.3.5   MEMBRANES OF CELLULOSE AND CELLULOSE-BASED NANOMATERIALS

Cellulose is the most abundant and widely used natural polymer on the Earth. It can be extracted from cotton, grass, bamboo, wood, plants, algae, tunicates and bacteria. Cellulose has a very good chemical resistance property. It does not dissolve in common solvents due to the large number of intermolecular and intramolecular hydrogen bonds. So far, only a limited number of solvents have been found to be effective in dissolving this biopolymer. For example, N-methylmorpholine-N-oxide (NMMO), LiCl/N, N-dimethylacetamide (DMAc), ammonia/ethylenediamine/thiocyanate salts, NaOH/urea solution, ionic liquids (ILs), etc. Thus, regenerated cellulosic membranes can be formed with good hydrophilic surface properties and mechanical stability. Cellulosic membranes easily get contaminated with microbes. Hence, modification with different antimicrobial additives is generally done to mitigate this issue.

Cellulosic materials converted to their nanoform are now commonly used in membranes. Cellulose nanocrystals (CNC) can be fabricated from the cellulosic source material by acid hydrolysis. Cellulose molecules in nature do not exist individually. They exist in an assembled state, forming fibres with crystalline and amorphous zones. As a result, the fibres break down in the amorphous regions, forming crystalline one-dimensional nanostructures. CNC have even been found to exhibit antimicrobial activity. The addition of CNC to the polymer matrix improves the hydrophilic properties as well as the mechanical properties of the membranes. Functionalisation of cellulose nanocrystals is also done to add new functionality to the membranes [43–49].

2.3.6   MEMBRANES OF CELLULOSE DERIVATIVES

Cellulose derivatives, such as cellulose acetate and nitrocellulose are widely used for membrane fabrication because of their easy availability and good film-forming abilities. Cellulose derivatives are the oldest raw materials used for the fabrication of membranes. As mentioned above, the first reverse osmosis membrane fabricated was cellulose acetate. The main disadvantages of cellulose derivative membranes are their low mechanical strength in comparison to other synthetic polymeric membranes, their limited operational range (<30 °C), and their susceptibility to microbial attack.

In recent days, a lot of research has been conducted on the modification of cellulose acetate membranes by using various types of additives. Cellulose acetate has very good solubility in common organic solvents like N-methyl pyrrolidone (NMP), N,N-dimethyl acetamide (DMAc), mixed solvents composed of N,N-dimethylformamide (DMF), acetone or 2-propanol, etc. making them suitable for both phase inversion and the electrospinning method of membrane fabrication. On the other hand, nitrocellulose is soluble in an ether:alcohol (2:1) solvent mixture. Cellulose-derived membranes are largely produced and commercially available [50–52].

2.3.7 OTHER BIO-BASED MEMBRANES

Apart from the biopolymeric materials discussed above, other biopolymeric materials, such as guar gum, xanthan, protein, etc., are also used for membrane fabrication. These materials are mainly used as additives or functional materials to improve selectivity or hydrophilicity and hence the permeability of the membranes [53–58].

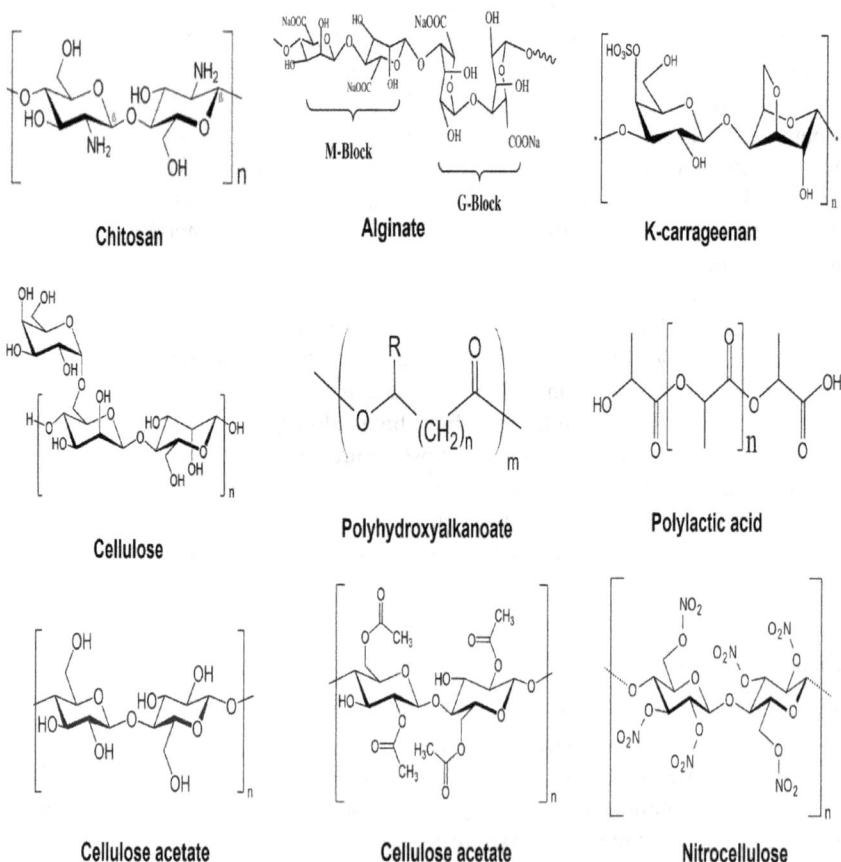

FIGURE 2.2 Biopolymeric materials used for membrane fabrication.

Lignin is another bio-based material used as a functional material and an additive with the polymeric matrix to create superior membranes [59]. In recent years, advanced biobased nanomaterials like carbon dots have also been proven to be an attractive choice to tune various properties of membranes. Carbon dots can be synthesised from any natural source. They have a size <10 nm and a higher carbon percentage. Carbon dots have their characteristic fluorescent behaviour. Carbon dots are cheap, easy to synthesise, non-toxic and highly effective due to their small size [60].

2.3.8 APPLICATIONS AND CHALLENGES OF BIO-BASED MEMBRANES

Depending on their physical properties and functionality, biobased membranes have the potential to be used in almost every application. For example, water purification includes the removal of metal ions and bacteria, oil–water separation, dyes and other organic substances, gas separation and purification, etc. Because of their biodegradability, biocompatibility and non-cytotoxicity, bio-based membranes are widely applied in various biological and biomedical research and applications. Some of them include biosensors, wound dressing, tissue engineering, cell culture, haemodialysis, etc. Bio-based membranes can also be used in fuel cells and as photocatalytic membranes [61]. Functional groups present in different biopolymers and biobased materials help the membranes become especially efficient for some applications. For example, chitosan with the $-NH_2$ group can interact with different metal ions, which allows the chitosan-based membranes to be used as an absorption membrane for the separation of metal ions [62]. Sodium alginate can exchange its Na^+ ions with other divalent or trivalent metal ions. This property of alginate prevents them from adhering to the wounded skin. Therefore, alginate-based membranes are proven to be efficient wound dressing materials [63]. Natural polymers have high organic solvent resistance properties, so polymeric membranes with such properties have a high potential for use in organic solvent purification. Because most synthetic polymers are soluble in common organic solvents, it is challenging to use the synthetic polymer membranes to separate or purify organic substances [1].

Although natural polymers are abundantly available, environmental friendly, and have many superior physical and chemical properties, the solubility issue becomes a significant barrier to fabricating membranes entirely of natural polymers. As mentioned above, the most commonly used phase inversion method cannot be applied to natural polymers because of their insolubility in organic solvents. Thus, it also becomes difficult to convert them into the hollow fibre form. In an attempt to make organic solvents soluble, modification of biopolymers can be done, but the process again uses different chemicals or organic solvents. Moreover, after modification, although the biodegradability remains intact, the original functionality of the natural polymers generally changes. Electrospinning can be a good option for the fabrication of flat sheets of biopolymeric membranes. Another issue with bio-polymeric films or membranes is the loss of mechanical stability when they come in contact with water. This issue can be addressed by finding the proper cross-linking agent, blending or composite formation with suitable materials.

With global environmental concerns growing, it has become critical to find environmental friendly and sustainable solutions to our day-to-day demands. Due to its

high efficiency, membrane-based separation is one of the most promising technologies in different industrial sectors. Therefore, we must think of sustainable growth for
this technology. Membrane materials are a crucial part of this technology. Therefore,
it is a very good opportunity as well as a responsibility for membrane technologists,
biotechnologists, material scientists and engineering scientists to actively participate
in developing new materials and methods for developing bio-based membranes for a
sustainable future.

2.4 INORGANIC MATERIALS FOR MEMBRANE

Inorganic membranes are mainly composed of metals, their oxides, elementary
carbon, etc. They can be operated in harsh conditions and have high permeability
and selectivity. Inorganic membranes are made in the form of self-standing sheets
or tubes depending on their permeability and mechanical strength. Otherwise, a thin
layer or coating, is formed on porous multi-layer supporting structures. Fabrication
methods like slip casting, chemical vapour deposition (CVD), pyrolysis and sol–gel
are commonly used for the synthesis of such membranes. Inorganic membranes are
widely used for the separation of H_2 and O_2 from air, the separation of CO_2 from natural gas and the purification of water and organic solvents. Based on the pore size (θ_p)
membranes are categorised as dense (0.5 nm > θ_p), microporous (2 nm > θ_p > 0.5),
meso-porous (50 > θ_p > 2 nm) and macro-porous (50 nm < θ_p). Dense and microporous membranes are generally used for the separation of gaseous molecules, while
they are impermeable to liquids. They are constructed using either a solid layer of
metals (Pd, Ag, alloys, etc.) or solid electrolytes. Such membranes are generally
composed of a semipermeable support layer with an immobilised liquid, such as
molten salts immobilised in porous ceramic or steel supports.

The most commonly used ceramic membranes include alumina (Al_2O_3), glass
(SiO_2), titania (TiO_2), silicon carbide (SiC), zirconia (ZrO_2) or a combination of multiple metal oxides, etc. Other ceramic materials used in inorganic membranes include
non-oxides (borides, carbides, silicides and nitrides) and composites of oxides and
non-oxides. Zeolites are another widely used inorganic material in membrane design
[64,65].

2.5 COMPOSITE MEMBRANES

Composite membranes are designed to enjoy the advantages of inorganic materials,
like high thermal stability, mechanical strength, chemical resistance, etc., while at
the same time enjoying the properties of organic polymeric materials, like flexibility,
ease of processing and low cost. Composite membranes can be divided into: mixed
matrix membranes (MMM) and TFC membranes. MMM are formed by mixing the
additives with the polymeric matrix. Then this hybrid material is used for the fabrication of membranes using different membrane fabrication techniques, such as phase
inversion, electrospinning, stretching, etc. In MMMs, both the additive and polymeric
phases determine the efficiency of the membrane. Additives in the polymeric matrix
modify the surface properties as well as the porosity of the membranes.

On the other hand, thin-film composite membranes (TFC) are formed by the deposition of inorganic or organic membrane materials on the surface of a porous support membrane. The top layer is called the active layer because it controls the properties of the membrane. It is a thin layer of organic or inorganic materials, either deposited by crosslinking or by physical interaction. The support layer has a comparatively larger pore size and therefore does not affect the permeability of the membrane [1].

2.6 NANOMATERIALS FOR MEMBRANE

2.6.1 2D NANOMATERIALS

2.6.1.1 Graphene Oxide

Graphene oxides have rapidly emerged as a promising 2D material for designing superior gas and water purification membranes. They can be easily synthesised by the oxidation and exfoliation of graphite flakes. Unlike graphene, graphene oxides are dispersible in aqueous media because of the polar groups attached to their surfaces. The interlayer spacing between the layers also increases in the case of graphene oxide, which is estimated to be around 1 nm. Both mixed matrix and TFC membranes have been fabricated using graphene oxides. Nair et al. (2012) reported that graphene oxide membranes can allow unimpeded water permeation at least 10^{10} times faster than liquid He. Graphene's near-frictionless 2D surface and high chemical resistance make it a promising material for producing high-quality, efficient membranes.

The advantages of graphene-based membranes are: high solvent flux; enhanced antifouling and antimicrobial properties; thermal and chemical stability; and very high mechanical strength. In a graphene-based membrane, solvent permeation occurs mainly through the interlayer spacing between the graphitic layers (Figure 2.3). So, the solvent flux can be increased by tuning the interlayer spacing or the d-spacing of graphene layers. But this approach has a limit. Therefore, the researchers are now trying to create nanopores on the surface of the graphene to increase the permeability. The performance of these porous graphene-based membranes is very encouraging. However, the techniques used to create pores on graphene-based membranes are physical methods like plasma treatment, bombardment of ionic and electron beams, etc. These techniques have some disadvantages, such as being time-consuming, expensive and not suitable for pore formation on a large area. Recently, some reports have demonstrated the synthesis of porous graphene and GOs by some chemical or solution-based methods. These methods can obtain graphene sheets with in-plane pores of different sizes [66–68].

2.6.1.2 Metal–Organic Framework (MOF) Nanosheets

MOFs are porous crystalline hybrid materials where organic linkers interconnect the metal ions or clusters to form bulk crystals. In recent years, MOF-based membranes have gained importance for separating gases and liquids due to their advantageous properties, such as ultrahigh surface area, superior polymer affinity and highly regular and tuneable pore structures. Their design has also attracted increased attention for

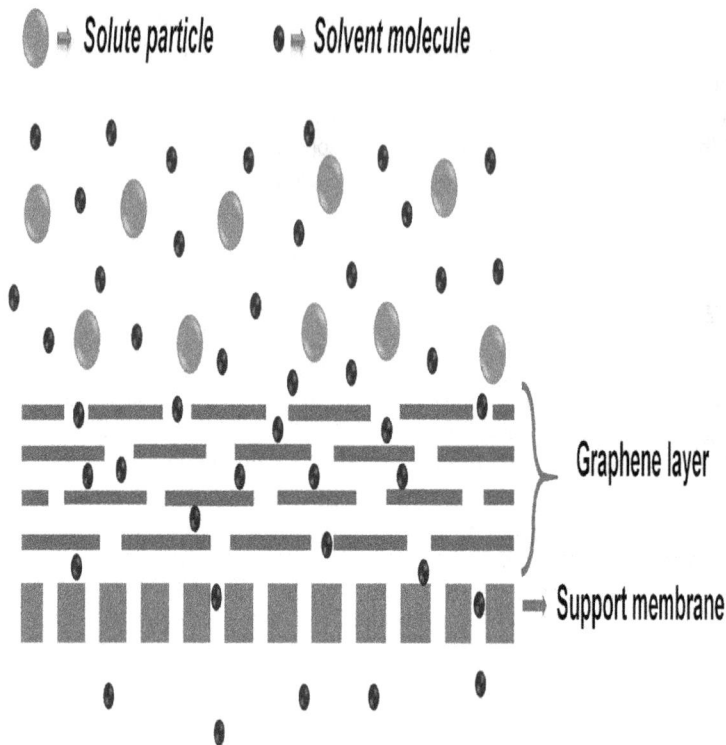

FIGURE 2.3 Pictorial representation showing filtration of solvent through the interlayer pores in a graphene-based filtration membrane.

liquid separation. MOFs can be synthesised with different morphologies, structures and functionalities depending on the mode of coordination between ligands and metal centres. Both top-down and bottom-up processes can obtain MOF nanosheets. The top-down process involves the formation of 2D MOF nanosheets from a bulk crystal using mechanical exfoliation. On the other hand, in a bottom-up process, 2D MOF nanosheets are directly formed from their raw materials [69].

2.6.1.3　Covalent Organic Framework (COF) Nanosheets

COF nanosheets are two-dimensional materials formed by reactions between organic precursors that result in strong, covalent bonds with a porous, stable and crystalline structure. COFs are structurally predesignable and functionally manageable. The ultimate design of the COF depends on the geometry of the monomer. The building blocks, or monomers, usually have backbones and a rigid conformation to maintain the 2D planarity of the extended polygons. Because of their adjustable pore size and uniform pore distribution, COF membranes can become highly selective for the separation of ions, gases and organic molecules. The high compatibility of COF with other polymeric materials makes easy fabrication of TFC membranes using porous polymeric supports possible [70,71].

2.6.1.4 Transition Metal Dichalcogenides (TMDs)

TMDs consist of multiple monolayers stacked together with the help of van der Waals forces. These 2D nanomaterials are represented by the chemical formula MX_2, where M indicates a transition metal element and X is a chalcogen (e.g., S, Se or Te). Monolayers of TMD consist of three atomic layers, where a transition metal layer is sandwiched between two chalcogen layers. TMDs have some advantages over GOs, for example: (a) TMD sheets can resist disintegration in aqueous media with a wide pH range due to the stability provided by the strong van der Waals force that exists between the atomic layers; (b) a single-layer TMD sheet is more rigid compared to that of GO because of the presence of three atomic layers; (c) the low water affinity as well as low surface roughness of TMD nanosheets compared to GO forms rapid molecular nanochannels for the passage of water molecules. Besides, the antibacterial activity of some TMD nanosheets like MoS_2 single layers as well as their non-toxic nature are beneficial to water treatment [72].

2.6.1.5 MXenes

The popularity of MXenes is growing fast. They are 2D-layered transition metal carbides and/or nitrides with the general formula $Mn+1XnTz$, where M represents an early transition metal (e.g., V, Ti, Cr, Nb, etc.), X is C and/or N, $n = 1, 2$ or 3 and Tz represents the surface terminal functional groups (e.g., –OH, =O, –F). To date, the most studied membrane of the MXene family is $Ti_3C_2T_x$, where T is O, OH and/or F. High mechanical strength, 2D structure, negative surface charge and hydrophilicity are the intrinsic advantages that make $Ti_3C_2T_x$ a strong candidate for the construction of membranes for various applications [72,73].

2.6.1.6 Graphitic Carbon Nitride

Like graphite, graphitic carbon nitride ($g-C_3N_4$) is a laminar structure consisting of carbon, nitrogen and some impurities. 2D $g-C_3N_4$ can be obtained by exfoliating its bulk structure, where the $g-C_3N_4$ monolayers are connected via covalent linkage. Additionally, the high aspect ratio and regular nanopores (3.11 A) of exfoliated tris-triazine-based $g-C_3N_4$, along with possible surface defects, provide the possibility of molecular (e.g., H_2 and water) transport across these nanopores. Graphitic carbon nitrides are proven to be rigid and stable enough to resist a wide range of pH and mechanical stress, confirming their excellent overall properties and making them suitable for constructing highly efficient membranes for water treatment under harsh conditions [74].

2.6.1.7 2D Zeolite

Zeolites are aluminium silicates in the form of microporous crystalline solids. Pore sizes of zeolite typically range from 0.25 to more than 1 nm. Such 2D zeolite layers can be obtained by either a top-down or bottom-up process. In a bottom-up process, 2D zeolite layers are obtained as an intermediate during the formation of 3D zeolites. On the other hand, in the top-down process, the 2D layered sheets are obtained by disassembling a 3D zeolite framework. Due to its thickness being in the range of unit-cell dimensions of their corresponding structures, 2D zeolite is a promising material for designing ultrathin molecular separation membranes [75,76].

2.6.1.8 Layered Double Hydroxide (LDH)

They are one of the most popular two-dimensional inorganic nanomaterials, having a typical formula of $[M^{2+}_{1-x}M^{3+}_{x}(OH)_{2}]^{x+}(A^{n-})_{x/n}\cdot mH_{2}O$, where M^{2+} represent divalent metal cations and M^{3+} represent trivalent metal cations. A^{n-} indicates an interlayer anion with n-valence. Where the values of x lie between 0.22 and 0.33. The positively charged metal-hydroxide layer is the basic constituent of LDH, whereas the negatively charged anions are positioned inside the interlayer galleries. The electrostatic charge gets balanced by the water molecules. Properties that make LDHs excellent materials for gas and liquid purification are: (i) high surface area; (ii) high anion exchange capacity; (iii) interlayer galleries; (iv) tunable composition; (v) functionalisation as per requirement, etc. LDH is an intriguing 2D material due to its simple synthetic procedures and tunability of physicochemical properties via metal or intercalated anion changes. Moreover, further tuning of properties can be achieved by chemical modification using different materials. Different approaches, including hydrothermal, mechanochemical, co-precipitation, etc., have been used for the synthesis of LDH.

2.6.1.9 Nanoclay

Nanoclay are layered silicates consisting of sheets of SiO_4^{4-} tetrahedra or $[AlO_3(OH)_3]^6$ octahedra. Individual sheets are stacked together by van der Waals force of attraction. Ionic substitutions in nanosheet interlayers generate a net negative charge, which is generally balanced by cations like Na^+, Mg^{2+}, K^+ and Ca^{2+}. Organic cations, commonly $R\text{-}NH^{3+}$ groups (R = long aliphatic chains), can replace these interlayer metallic cations. Such organically modified clays become compatible with organic polymers. Thus, nano clay has become the most widely used nanoparticle for the formation of polymer nanocomposites. Exfoliated nanoclays, having a thickness in the nanometer range as well as good compatibility with polymeric materials, serve as a high-potential material for membranes [77,78] (Figure 2.4).

2.6.2 1D NANOMATERIALS

2.6.2.1 Carbon Nanotube (CNT)

CNTs are allotropes of carbon. They are cylinder-shaped structures of graphitic layers containing sp2 hybridised carbon with a diameter of less than 100 nm. CNTs can be of two types: single-wall carbon nanotubes (SWCNT) and multi-wall carbon nanotubes (MWCNT), depending upon the number of graphitic layers in the cylinder. CNTs are widely used for membrane fabrication because of their high strength and chemical resistance. CNTs can be functionalised with polar functional groups by oxidation, followed by reactions with other functional groups. MMM and TFC membranes are designed using CNT and various polymeric materials [79].

2.6.2.2 Cellulose Nanocrystals (CNC)

CNC are one of the sustainable nanomaterials widely used for membrane design these days. CNCs can be prepared from different types of cellulosic biomass.

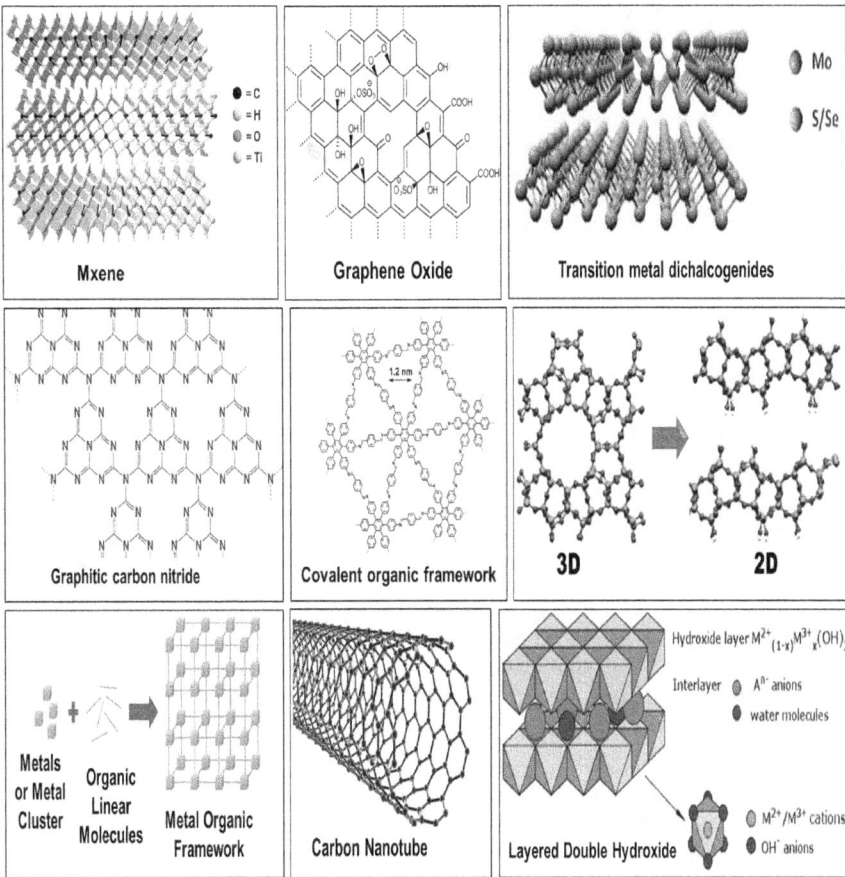

FIGURE 2.4 Pictures showing the structures of different types of 2D nanomaterials.

Cellulose occurs in nature in the form of fibres that contain both crystalline and amorphous regions. The nanostructured crystalline regions of the cellulose fibres can be separated by hydrolysis. CNCs are one-dimensional, non-toxic nanomaterials with a high concentration of polar functional groups. Being organic in nature, CNC can have excellent compatibility with other polymeric materials. CNCs can increase the hydrophilicity of the membrane surface by having a large number of polar functional groups on their surfaces [80].

2.6.3 0D NANOMATERIALS

2.6.3.1 Carbon Dots

Carbon dots are fluorescent nanoprobes with a size <10 nm. They can be synthesised from any natural source with a high carbon ratio. CDs are now gaining much attention from researchers because of their easy preparation method, low-cost raw materials, non-toxic nature and high surface area. The fluorescent property of the CDs arises because of their quantum confinement effect and structural defect. Depending on their source, carbon dots can have different functional groups. These functional groups control the ultimate properties of the carbon dots [81].

2.6.3.2 Metal and Metal Oxide Nanoparticles

Various metal and metal oxide nanoparticles have been used as fillers inside the polymeric matrix. Metal and metal oxide nanoparticles have excellent thermal stability. They can also improve the mechanical strength of the membrane. Some of them, for example, Ag nanoparticles, also possess excellent antimicrobial activity, which can be crucial for mitigating membrane fouling by microbial growth. They may also contribute to water purification. Metal and metal oxide nanoparticles can also resist the fouling caused by organic molecules. Metal oxide nanoparticles, in general, show excellent absorption of metal ions, and therefore, as a membrane component, they can contribute to metal ion absorption [82,83].

2.7   CONCLUSION

With the growing expansion of membrane technology for wider applications, the demand for high-performance membranes as well as membranes with specific properties is also increasing. Therefore, it has become of the utmost importance to find novel materials to improve the quality of the membranes and tune their properties for specific applications. From the above discussion, we can conclude that nanomaterials can play a very important role in the future for developing high-quality membranes. On the other hand, with the increase in production, disposal of used membranes will also be an issue in the near future; therefore, it is also important to address the biodegradability of the membrane. Biopolymeric membranes can play a significant role in developing biodegradable and biocompatible membranes. The biocompatibility of the membranes is also important in different biomedical and pharmaceutical applications.

REFERENCES

1. Marchetti, P., Jimenez Solomon, M. F., Szekely, G., & Livingston, A. G. (2014). Molecular separation with organic solvent nanofiltration: A critical review. *Chemical Reviews*, *114*(21), 10735–10806. https://doi.org/10.1021/cr500006j
2. Mansourpanah, Y., & Emamian, F. (2020). Membrane and bioseparation. In *Advances in Membrane Technologies*. IntechOpen. http://dx.doi.org/10.5772/intechopen.82587
3. Uragami, T. (Ed.). (2017). *Science and Technology of Separation Membranes*. John Wiley & Sons.| https://doi/book/10.1002/9781118932551
4. Tan, X., & Rodrigue, D. (2019). A review on porous polymeric membrane preparation. Part I: Production techniques with polysulfone and poly (vinylidene fluoride). *Polymers*, *11*(7), 1160. https://doi.org/10.3390/polym11071160
5. Tan, X., & Rodrigue, D. (2019). A review on porous polymeric membrane preparation. Part I: Production techniques with polysulfone and poly (vinylidene fluoride). *Polymers*, *11*(7), 1160. https://doi.org/10.3390/polym11071160

6. Cui, Z. F., Jiang, Y., & Field, R. W. (2010). Fundamentals of pressure-driven membrane separation processes. In *Membrane Technology* (pp. 1–18). Butterworth-Heinemann. https://doi.org/10.1016/B978-1-85617-632-3.00001-X

7. Weigelt, F., Escorihuela, S., Descalzo, A., Tena, A., Escolástico, S., Shishatskiy, S., ... & Brinkmann, T. (2019). Novel polymeric thin-film composite membranes for high-temperature gas separations. *Membranes, 9*(4), 51. https://doi.org/10.3390/membranes 9040051

8. Ren, D., Ren, S., Lin, Y., Xu, J., & Wang, X. (2021). Recent developments of organic solvent resistant materials for membrane separations. *Chemosphere, 271*, 129425. https://doi.org/10.1016/j.chemosphere.2020.129425

9. Kayvani Fard, A., McKay, G., Buekenhoudt, A., Al Sulaiti, H., Motmans, F., Khraisheh, M., & Atieh, M. (2018). Inorganic membranes: Preparation and application for water treatment and desalination. *Materials, 11*(1), 74. https://doi.org/10.3390/ma11010074

10. Verweij, H. (2012). Inorganic membranes. *Current Opinion in Chemical Engineering, 1*(2), 156–162. https://doi.org/10.1016/j.coche.2012.03.006

11. Benedetti, F. M., De Angelis, M. G., Degli Esposti, M., Fabbri, P., Masili, A., Orsini, A., & Pettinau, A. (2020). Enhancing the separation performance of glassy PPO with the addition of a molecular sieve (ZIF-8): Gas transport at various temperatures. *Membranes, 10*(4), 56. https://doi.org/10.3390/membranes10040056

12. Idris, A., & Yet, L. K. (2006). The effect of different molecular weight PEG additives on cellulose acetate asymmetric dialysis membrane performance. *Journal of Membrane Science, 280*(1–2), 920–927. https://doi.org/10.1016/j.memsci.2006.03.010

13. M'barki, O., Hanafia, A., Bouyer, D., Faur, C., Sescousse, R., Delabre, U., ... & Pochat-Bohatier, C. (2014). Greener method to prepare porous polymer membranes by combining thermally induced phase separation and crosslinking of poly (vinyl alcohol) in water. *Journal of Membrane Science, 458*, 225–235. https://doi.org/10.1016/j.mem sci.2013.12.013

14. Guillen, G. R., Pan, Y., Li, M., & Hoek, E. M. (2011). Preparation and characterization of membranes formed by nonsolvent induced phase separation: A review. *Industrial & Engineering Chemistry Research, 50*(7), 3798–3817. https://doi.org/10.1021/ ie101928r

15. Kutowy, O., & Sourirajan, S. (1975). Cellulose acetate ultrafiltration membranes. *Journal of applied polymer science, 19*(5), 1449–1460. https://doi.org/10.1002/ app.1975.070190525

16. Zhao, C., Xue, J., Ran, F., & Sun, S. (2013). Modification of polyethersulfone membranes–A review of methods. *Progress in Materials Science, 58*(1), 76–150. https://doi.org/10.1016/j.pmatsci.2012.07.002

17. Otitoju, T. A., Ahmad, A. L., & Ooi, B. S. (2018). Recent advances in hydrophilic modification and performance of polyethersulfone (PES) membrane via additive blending. *RSC Advances, 8*(40), 22710–22728. https://doi.org/10.1039/C8RA03296C

18. Guillen, G. R., Pan, Y., Li, M., & Hoek, E. M. (2011). Preparation and characterization of membranes formed by nonsolvent induced phase separation: A review. *Industrial & Engineering Chemistry Research, 50*(7), 3798–3817. https://doi.org/10.1021/ ie101928r

19. Van der Bruggen, B. (2009). Chemical modification of polyethersulfone nanofiltration membranes: A review. *Journal of Applied Polymer Science, 114*(1), 630–642. https:// doi.org/10.1002/app.30578

20. Liu, F., Hashim, N. A., Liu, Y., Abed, M. M., & Li, K. (2011). Progress in the production and modification of PVDF membranes. *Journal of Membrane Science, 375*(1–2), 1–27. https://doi.org/10.1016/j.memsci.2011.03.014

21. Eykens, L., De Sitter, K., Dotremont, C., Pinoy, L., & Van der Bruggen, B. (2017). Membrane synthesis for membrane distillation: A review. *Separation and Purification Technology, 182*, 36–51. https://doi.org/10.1016/j.seppur.2017.03.035

22. Alkhudhiri, A., Darwish, N., & Hilal, N. (2012). Membrane distillation: A comprehensive review. *Desalination, 287*, 2–18. https://doi.org/10.1016/j.desal.2011.08.027

23. Kang, G. D., & Cao, Y. M. (2014). Application and modification of poly (vinylidene fluoride)(PVDF) membranes: A review. *Journal of Membrane Science, 463*, 145–165. https://doi.org/10.1016/j.memsci.2014.03.055

24. Uragami, T. (Ed.). (2017). *Science and Technology of Separation Membranes*. John Wiley & Sons. https://doi.org/10.1002/9781118932551

25. Fan, J., Zhang, S., Li, F., Yang, Y., & Du, M. (2020). Recent advances in cellulose-based membranes for their sensing applications. *Cellulose, 27*, 9157–9179. https://doi.org/10.1007/s10570-020-03445-7

26. de Souza, R. F. B., de Souza, F. C. B., Bierhalz, A. C. K., Pires, A. L. R., & Moraes, Â. M. (2020). Biopolymer-based films and membranes as wound dressings. In *Biopolymer Membranes and Films* (pp. 165–194). Elsevier. https://doi.org/10.1016/B978-0-12-818134-8.00007-9

27. Konwar, A., Kandimalla, R., Kalita, S., & Chowdhury, D. (2018). Approach to fabricate a compact cotton patch without weaving: A smart bandage material. *ACS Sustainable Chemistry & Engineering, 6*(5), 5806–5817. https://doi.org/10.1021/acssuschemeng

28. Salehi, E., Daraei, P., & Shamsabadi, A. A. (2016). A review on chitosan-based adsorptive membranes. *Carbohydrate Polymers, 152*, 419–432. https://doi.org/10.1016/j.carbpol.2016.07.033

29. Vedula, S. S., & Yadav, G. D. (2021). Chitosan-based membranes preparation and applications: Challenges and opportunities. *Journal of the Indian Chemical Society, 98*(2), 100017. https://doi.org/10.1016/j.jics.2021.100017

30. Khoerunnisa, F., Kulsum, C., Dara, F., Nurhayati, M., Nashrah, N., Fatimah, S., ... & Opaprakasit, P. (2021). Toughened chitosan-based composite membranes with antibiofouling and antibacterial properties via incorporation of benzalkonium chloride. *RSC Advances, 11*(27), 16814–16822. https://doi.org/10.1039/d1ra01830b

31. Konwar, A., & Chowdhury, D. (2015). Property relationship of alginate and alginate–carbon dot nanocomposites with bivalent and trivalent cross-linker ions. *RSC Advances, 5*(77), 62864–62870. https://doi.org/10.1039/c5ra09887d

32. Aburabie, J. H., Puspasari, T., & Peinemann, K. V. (2020). Alginate-based membranes: Paving the way for green organic solvent nanofiltration. *Journal of Membrane Science, 596*, 117615. https://doi.org/10.1016/j.memsci.2019.117615

33. Dodero, A., Scarfi, S., Pozzolini, M., Vicini, S., Alloisio, M., & Castellano, M. (2019). Alginate-based electrospun membranes containing ZnO nanoparticles as potential wound healing patches: Biological, mechanical, and physicochemical characterization. *ACS Applied Materials & Interfaces, 12*(3), 3371–3381. https://doi.org/10.1021/acsami.9b17597

34. Casalini, T., Rossi, F., Castrovinci, A., & Perale, G. (2019). A perspective on polylactic acid based polymers use for nanoparticles synthesis and applications. *Frontiers in Bioengineering and Biotechnology, 7*, 259. https://doi.org/10.3389/fbioe.2019.00259

35. Oliveira, J. E., Medeiros, E. S., Cardozo, L., Voll, F., Madureira, E. H., Mattoso, L. H. C., & Assis, O. B. G. (2013). Development of poly (lactic acid) nanostructured membranes for the controlled delivery of progesterone to livestock animals. *Materials Science and Engineering: C, 33*(2), 844–849. https://doi.org/10.1016/j.msec.2012.10.032

36. Bonan, R. F., Bonan, P. R., Batista, A. U., Perez, D. E., Castellano, L. R., Oliveira, J. E., & Medeiros, E. S. (2017). Poly (lactic acid)/poly (vinyl pyrrolidone) membranes

produced by solution blow spinning: Structure, thermal, spectroscopic, and microbial barrier properties. *Journal of Applied Polymer Science*, *134*(19),44802. https://doi.org/ 10.1002/app.44802

37. Ye, B., Jia, C., Li, Z., Li, L., Zhao, Q., Wang, J., & Wu, H. (2020). Solution-blow spun PLA/SiO2 nanofiber membranes toward high efficiency oil/water separation. *Journal of Applied Polymer Science*, *137*(37), 49103. https://doi.org/10.1002/app.49103

38. Yue, M., Zhou, B., Jiao, K., Qian, X., Xu, Z., Teng, K., … & Jiao, Y. (2014). Switchable hydrophobic/hydrophilic surface of electrospun poly (l-lactide) membranes obtained by CF_4microwave plasma treatment. *Applied Surface Science*, *327*(BNL-108030-2015-JA). https://doi.org/10.1016/j.apsusc.2014.11.149

39. Zhang, H. Y., Jiang, H. B., Ryu, J. H., Kang, H., Kim, K. M., & Kwon, J. S. (2019). Comparing properties of variable pore-sized 3D-printed PLA membrane with conventional PLA membrane for guided bone/tissue regeneration. *Materials*, *12*(10), 1718. https://doi.org/10.3390/ma12101718

40. Li, Z., Yang, J., & Loh, X. J. (2016). Polyhydroxyalkanoates: Opening doors for a sustainable future. *NPG Asia Materials*, *8*(4), e265. https://doi.org/10.1038/am.2016.48

41. Tomietto, P., Loulergue, P., Paugam, L., & Audic, J. L. (2020). Biobased polyhydroxyalkanoate (PHA) membranes: Structure/performances relationship. *Separation and Purification Technology*, *252*, 117419. https://doi.org/10.1016/j.sep pur.2020.117419

42. Tomietto, P., Russo, F., Galiano, F., Loulergue, P., Salerno, S., Paugam, L., Audic, L. J., Bartolo, D. L., & Figoli, A. (2022). Sustainable fabrication and pervaporation application of bio-based membranes: Combining a polyhydroxyalkanoate (PHA) as biopolymer and Cyrene™ as green solvent. *Journal of Membrane Science*, *643*, 120061. https://doi.org/ 10.1016/j.memsci.2021.120061

43. Weng, R., Huang, X., Liao, D., Xu, S., Peng, L., & Liu, X. (2020). A novel cellulose/ chitosan composite nanofiltration membrane prepared with piperazine and trimesoyl chloride by interfacial polymerization. *RSC Advances*, *10*(3), 1309–1318. https://doi. org/10.1039/c9ra09023a

44. Karim, Z., & Monti, S. (2021). Microscopic hybrid membranes made of cellulose-based materials tuned for removing metal ions from industrial effluents. *ACS Applied Polymer Materials*, *3*(8), 3733–3746. https://doi.org/10.1021/acsapm.1c00105

45. Guccini, V., Carlson, A., Yu, S., Lindbergh, G., Lindström, R. W., & Salazar-Alvarez, G. (2019). Highly proton conductive membranes based on carboxylated cellulose nanofibres and their performance in proton exchange membrane fuel cells. *Journal of Materials Chemistry A*, *7*(43), 25032–25039. https://doi.org/10.1039/c9ta04898g

46. Li, X., Li, H. C., You, T. T., Wu, Y. Y., Ramaswamy, S., & Xu, F. (2019). Fabrication of regenerated cellulose membranes with high tensile strength and antibacterial property via surface amination. *Industrial Crops and Products*, *140*, 111603. https://doi.org/ 10.1016/j.indcrop.2019.111603

47. Halim, A., Ernawati, L., Ismayati, M., Martak, F., & Enomae, T. (2022). Bioinspired cellulose-based membranes in oily wastewater treatment. *Frontiers of Environmental Science & Engineering*, *16*(7), 94. https://doi.org/10.1007/s11783-021-1515-2

48. Woffindin, C., Hoenich, N. A., & Matthews, J. N. S. (1992). Cellulose-based haemodialysis membranes: Biocompatibility and functional performance compared. *Nephrology Dialysis Transplantation*, *7*(4), 340–345. https://doi.org/10.1093/oxfordj ournals.ndt.a092139

49. Zhou, Q., Bao, Y., Zhang, H., Luan, Q., Tang, H., & Li, X. (2020). Regenerated cellulose-based composite membranes as adsorbent for protein adsorption. *Cellulose*, *27*, 335–345. https://doi.org/10.1007/s10570-019-02761-x

50. Vatanpour, V., Pasaoglu, M. E., Barzegar, H., Teber, O. O., Kaya, R., Bastug, M., Khataee, M., & Koyuncu, I. (2022). Cellulose acetate in fabrication of polymeric membranes: A review. *Chemosphere*, 133914. https://doi.org/10.1016/j.chemosph ere.2022.133914

51. Kaiser, A., Stark, W. J., & Grass, R. N. (2017). Rapid production of a porous cellulose acetate membrane for water filtration using readily available chemicals. *Journal of Chemical Education*, *94*(4), 483–487. https://doi.org/10.1021/acs.jchemed.6b00776

52. Asiri, A. M., Petrosino, F., Pugliese, V., Khan, S. B., Alamry, K. A., Alfifi, S. Y., Marwani, H. M., Alotabi, M. M., Algieri, C., & Chakraborty, S. (2021). Synthesis and characterization of blended cellulose acetate membranes. *Polymers*, *14*(1), 4. https://doi.org/10.3390/polym14010004

53. Mofradi, M., Karimi, H., Dashtian, K., & Ghaedi, M. (2021). Processing Guar Gum into polyester fabric based promising mixed matrix membrane for water treatment. *Carbohydrate Polymers*, *254*, 116806. https://doi.org/10.1016/j.carbpol.2020.116806

54. Wu, P., & Imai, M. (2011). Food polymer pullulan-κ-carrageenan composite membrane performed smart function both on mass transfer and molecular size recognition. *Desalination and Water Treatment*, *34*(1–3), 239–245. https://doi.org/10/5004/dwt.2011.2872

55. Coelhoso, I. M., Ferreira, A. R. V., & Alves, V. D. (2014). Biodegradable barrier membranes based on nanoclays and carrageenan/pectin blends. *International Journal of Membrane Science and Technology*, *1*(1), 23–30. https://doi.org/10.15379/2410-1869.2014.01.01.3

56. Zakaria, Z., Kamarudin, S. K., Kudus, M. H. A., & Wahid, K. A. A. (2022). κ-carrageenan/polyvinyl alcohol–graphene oxide biopolymer composite membrane for application of air-breathing passive direct ethanol fuel cells. *Journal of Applied Polymer Science*, *139*(22), 52256. https://doi.org/10.1002/app.52256

57. Kononova, S. V., Volod'ko, A. V., Petrova, V. A., Kruchinina, E. V., Baklagina, Y. G., Chusovitin, E. A., & Skorik, Y. A. (2018). Pervaporation multilayer membranes based on a polyelectrolyte complex of λ-carrageenan and chitosan. *Carbohydrate Polymers*, *181*, 86–92. https://doi.org/ 10.1016/j.carbpol.2017.10.050

58. Kumar, R. S., Arthanareeswaran, G., Paul, D., & Kweon, J. H. (2015). Effective removal of humic acid using xanthan gum incorporated polyethersulfone membranes. *Ecotoxicology and Environmental Safety*, *121*, 223–228. https://doi.org/10.1016/j.ecoenv.2015.03.036

59. de Haro, J. C., Tatsi, E., Fagiolari, L., Bonomo, M., Barolo, C., Turri, S., Bella, F., & Griffini, G. (2021). Lignin-based polymer electrolyte membranes for sustainable aqueous dye-sensitized solar cells. *ACS Sustainable Chemistry & Engineering*, *9*(25), 8550–8560. https://doi.org/10.1021/acssuschemeng.1c01882

60. Yan, F., Xu, M., Xu, J., Zang, Y., Sun, J., Yi, C., & Wang, Y. (2021). Advances in integrating carbon dots with membranes and their applications. *ChemistrySelect*, *6*(29), 7443–7462. https://doi.org/ 10.1002/slct.202101957

61. Galiano, F., Briceño, K., Marino, T., Molino, A., Christensen, K. V., & Figoli, A. (2018). Advances in biopolymer-based membrane preparation and applications. *Journal of Membrane Science*, *564*, 562–586. https://doi.org/10.1016/j.memsci.2018.07.059

62. Omer, A. M., Dey, R., Eltaweil, A. S., Abd El-Monaem, E. M., & Ziora, Z. M. (2022). Insights into recent advances of chitosan-based adsorbents for sustainable removal of heavy metals and anions. *Arabian Journal of Chemistry*, *15*(2), 103543. https://doi.org/ 10.1016/j.arabjc.2021.103543

63. Dodero, A., Scarfi, S., Pozzolini, M., Vicini, S., Alloisio, M., & Castellano, M. (2019). Alginate-based electrospun membranes containing ZnO nanoparticles as potential

wound healing patches: Biological, mechanical, and physicochemical characterization. *ACS Applied Materials & Interfaces*, *12*(3), 3371–3381. https://doi.org/10.1021/acsami.9b17597

64. Verweij, H. (2012). Inorganic membranes. *Current Opinion in Chemical Engineering*, *1*(2), 156–162. https://doi.org/10.1016/j.coche.2012.03.006

65. Kayvani Fard, A., McKay, G., Buekenhoudt, A., Al Sulaiti, H., Motmans, F., Khraisheh, M., & Atieh, M. (2018). Inorganic membranes: Preparation and application for water treatment and desalination. *Materials*, *11*(1), 74. https://doi.org/10.3390/ma11010074

66. Marchetti, P., Jimenez Solomon, M. F., Szekely, G., & Livingston, A. G. (2014). Molecular separation with organic solvent nanofiltration: A critical review. *Chemical Reviews*, *114*(21), 10735–10806. https://doi.org/10.1021/cr500006j

67. Homaeigohar, S., & Elbahri, M. (2017). Graphene membranes for water desalination. *NPG Asia Mater*, *9*, 427. https://doi.org/10.1038/am.2017.135

68. Nair, R. R., Wu, H. A., Jayaram, P. N., Grigorieva, I. V., & Geim, A. K. (2012). Unimpeded permeation of water through helium-leak–tight graphene-based membranes. *Science*, *335*(6067), 442–444. https://doi.org/10.1126/science.1211694

69. Cheng, Y., Datta, S. J., Zhou, S., Jia, J., Shekhah, O., & Eddaoudi, M. (2022). Advances in metal–organic framework-based membranes. *Chemical Society Reviews*. https://doi.org/10.1039/d2cs00031h

70. Wang, H., Zhai, Y., Li, Y., Cao, Y., Shi, B., Li, R., ... & Jiang, Z. (2022). Covalent organic framework membranes for efficient separation of monovalent cations. *Nature Communications*, *13*(1), 7123. https://doi.org/10.1038/s41467-022-34849-7

71. Geng, K., He, T., Liu, R., Dalapati, S., Tan, K. T., Li, Z., ... & Jiang, D. (2020). Covalent organic frameworks: Design, synthesis, and functions. *Chemical Reviews*, *120*(16), 8814–8933. https://doi.org/10.1021/acs.chemrev.9b00550

72. Zhu, J., Hou, J., Uliana, A., Zhang, Y., Tian, M., & Van der Bruggen, B. (2018). The rapid emergence of two-dimensional nanomaterials for high-performance separation membranes. *Journal of Materials Chemistry A*, *6*(9), 3773–3792. https://doi.org/10.1039/c7ta10814a

73. Gogotsi, Y., & Anasori, B. (2019). The rise of MXenes. *ACS Nano*, *13*(8), 8491–8494. https://doi.org/10.1021/acsnano.9b06394

74. Inagaki, M., Tsumura, T., Kinumoto, T., & Toyoda, M. (2019). Graphitic carbon nitrides (g-C3N4) with comparative discussion to carbon materials. *Carbon*, *141*, 580–607. https://doi.org/ 10.1016/j.carbon.2018.09.082

75. Xu, L., & Sun, J. (2016). Recent advances in the synthesis and application of two-dimensional zeolites. *Advanced Energy Materials*, *6*(17), 1600441. https://doi.org/10.1002/aenm.201600441

76. Roth, W. J., Nachtigall, P., Morris, R. E., & Cejka, J. (2014). Two-dimensional zeolites: Current status and perspectives. *Chemical Reviews*, *114*(9), 4807–4837. https://doi.org/10.1021/cr400600f

77. Peng, L., Tuantuan, Z., Qiang, W., & Yanshuo, L. (2018). Recent advances in layered double hydroxides (LDHs) as two-dimensional membrane materials for gas and liquid separations. *Journal of Membrane Science*, *567*, 89–103. https://doi.org/10.1016/j.memsci.2018.09.041

78. Sajid, M., Jillani, S. M. S., Baig, N., & Alhooshani, K. (2022). Layered double hydroxide-modified membranes for water treatment: Recent advances and prospects. *Chemosphere*, *287*, 132140. https://doi.org/10.1016/j.chemosphere.2021.132140

79. Ma, L., Dong, X., Chen, M., Zhu, L., Wang, C., Yang, F., & Dong, Y. (2017). Fabrication and water treatment application of carbon nanotubes (CNTs)-based composite

membranes: A review. *Membranes*, *7*(1), 16. https://doi.org/10.3390/membranes 7010016

80. Dai, Z., Ottesen, V., Deng, J., Helberg, R. M. L., & Deng, L. (2019). A brief review of nanocellulose based hybrid membranes for CO_2 separation. *Fibers*, *7*(5), 40. https://doi.org/10.3390/fib7050040

81. Yan, F., Xu, M., Xu, J., Zang, Y., Sun, J., Yi, C., & Wang, Y. (2021). Advances in integrating carbon dots with membranes and their applications. *ChemistrySelect*, *6*(29), 7443–7462. https://doi.org/10.1002/slct.202101957

82. Khan, A. A., Maitlo, H. A., Khan, I. A., Lim, D., Zhang, M., Kim, K. H., ... & Kim, J. O. (2021). Metal oxide and carbon nanomaterial based membranes for reverse osmosis and membrane distillation: A comparative review. *Environmental Research*, *202*, 111716. https://doi.org/10.1016/j.envres.2021.111716

83. Lu, P., Liu, Y., Zhou, T., Wang, Q., & Li, Y. (2018). Recent advances in layered double hydroxides (LDHs) as two-dimensional membrane materials for gas and liquid separations. *Journal of Membrane Science*, *567*, 89–103.https://doi.org/10.1016/j.memsci.2018.09.041

3 Membranes for Natural Gas Processing

Achyut Konwar, Prarthana Bora and Abhijit Gayan

3.1 INTRODUCTION

Natural gases are gaseous hydrocarbon mixtures formed under the earth's surface. World natural gas consumption is estimated to be roughly around 3.8 trillion cubic meters. The primary constituent of natural gas is methane (80–90%). It also contains C_{2+} hydrocarbons, N_2, CO_2, He, H_2S and noble gases with varying concentrations depending upon their origin. The treatment of natural gas before it enters the pipeline is a necessary step to take. Conventionally available techniques for separation of gases are membrane, absorption, adsorption and cryogenic distillation [1]. Among these techniques, membrane technology has been proven to be one of the best-established technologies for separation-based treatment of molecule mixtures, either in liquid or gaseous form [2]. A membrane's main advantages are that it can be operated at a low cost, with little energy consumption and with ease and high efficiency [2,3]. Membrane technology plays a crucial role in the processing of natural gas. Industrial application of membrane-based technology for natural gas processing started in the 1980s. Almost all types of gases can be separated or purified using suitable membranes [4,5]. In spite of having these advantageous properties, membranes still face some disadvantages, including low permeability and fouling [6]. To overcome such hurdles, continuous attention has been paid to the improvement of membrane material to acquire antifouling properties with high performance values. Membrane materials commonly used for gas separation are polymeric membranes, molecular sieve membranes, hybrid membranes, etc. Nowadays, membranes used for separation of natural gases are compiled into a spiral-wound module by packaging hollow fibres or flat sheet membranes. The main advantage of preparing hollow fibre modules is that they allow a large membrane area to compact into smaller modules. In practical use, a membrane should be resistant when exposed to harsh separation environments [1]. To meet such criteria, many improvements have been made in the membrane development area to make a durable membrane.

DOI: 10.1201/9781003441359-3

3.1.1 PERMEATION THEORY

Transport of a gases via a dense polymer membrane can be expressed by Equation (3.1):

$$J_x = D_x K_x \frac{\left(p_{xf} - p_{ip}\right)}{l} \tag{3.1}$$

where J_x is the volumetric flux (cm³/(cm² s)), l represents membrane thickness, p_{xf} and p_{xp} are the partial pressures of component x on feed side and permeate side, respectively. D_x represents diffusion coefficient that indicates individual molecule's mobility through membrane material and K_x stands for gas sorption coefficient (cm³ (STP) of x per cm³ of polymer per unit pressure) that indicates the numbers of gas molecules got dissolved in membrane material. The product $(D_x K_x)$ is called the membrane permeability and can be written as P_x; this term refers to the ability of the membrane to permeate gas. Thus, Equation (3.1) can be rewritten as:

$$\frac{P_x}{l} = \frac{J_x}{\Delta P_x}, \tag{3.2}$$

where, permeance (P_x/l) is defined as the gas volume that penetrate through a unit area while an unit pressure difference is there across the membrane. GPU is the unit of permeance and 1 GPU = 10^{-6} cm³ (STP)/cm² s cm Hg.

The ideal gas selectivity of binary mixture of two gas molecule is the ratio of permeance of fast one (P_x/l) to that of the slow one (P_y/l), the representation is shown in Equation (3.3).

$$\alpha = \frac{P_x/l}{P_y/l} = \frac{D_x/D_y}{K_x/K_y} \tag{3.3}$$

where D_x and D_y are the diffusion coefficients and K_x and K_y represent sorption coefficients of gases x and y.

In the case of transport through polymeric materials, the diffusion coefficient value is inversely related to the gas molecule size. The larger the molecules, the better the interaction with more polymer segments, which is not observed in the case of smaller molecules. Hence, the mobility selectivity D_x/D_y is always favourable for the smaller molecules like H_2O and CO_2 (kinetic diameters of 2.65 and 3.30 Å, respectively) over the comparatively larger molecules such as methane CH_4 (kinetic diameter 3.80 Å). The mobility selectivity or diffusion selectivity D_x/D_y also depends on the nature of the polymer, especially on its glassy or rubbery nature, which, on the other hand, depends on the polymer's glass transition temperature. Below this temperature, polymers become rigid and glassy hard because of the higher chain mobility, which then increases the effect of size differences on relative gas permeability. The effect of size differences on the relative gas permeability reduces above the glass transition temperature, which is attributed to the higher chain mobility leading to a rubbery state.

On the other hand, the sorption coefficient indicates the energy needed by the polymer for gas molecule sorption. The value of the sorption coefficient increases with the increase in condensability of gas molecules. The condensability of methane (boiling point 113 K) is very poor compared to the other natural gas components. Thus, the sorption selectivity K_x/K_y favours the permeation of the non-methane components. H_2O, having a small size and good condensability, can be separated from a mixture with methane by glassy and rubbery polymeric membranes, whereas CO_2 and H_2S are best separable by glassy and rubbery polymeric membranes, respectively. Nitrogen is generally separated from methane by glassy, polymeric-based membranes. The rubbery membranes can effectively separate the gaseous hydrocarbons due to their higher condensability [4–5,7–8].

3.2   MEMBRANES FOR SEPARATION OF COMPONENTS FROM NATURAL GAS

3.2.1   CARBON DIOXIDE SEPARATION

Carbon dioxide present in natural gas needs to be separated to a concentration of <2% to avoid the corrosion of pipelines during transportation. Moreover, natural gas should contain a minimal amount of CO_2 before being released into the environment. Before the development of membrane-based technology, the amine absorption technique was used as an alternative for the removal of CO_2. The amine absorption technique is complex and requires a well-monitored operating procedure. This process is suitable for the separation of natural gas with a lower concentration of CO_2. On the other hand, being an easy-to-operate technology, membrane-based separation is suitable for natural gas samples having a higher concentration of CO_2. The gradual development of membranes of superior quality also minimises the loss of natural gas compared to the amine absorption technique. In practise, a membrane-based separation for bulk removal of CO_2 and an amine absorption technique as a polishing system can offer a low-cost and efficient method compared to that of only a membrane or only an amine absorption method [4,9–10].

The most commonly used polymeric materials for the fabrication of CO_2 separation membranes are cellulose acetate, polyamide, polyimide and block polymers of polyamides like Pebax. Despite their low permeability, cellulose acetate membranes are the most widely used membranes for CO_2 separation. The selectivity of membranes is significantly reduced by the hydrocarbons present in the feed solution. Membrane plasticisation is another phenomenon that influences membrane performance. Plasticisation of the polymeric membrane material occurs because of the absorption of CO_2, leading to swelling and dilation of the polymer chains. As a result, the glass transition temperature (T_g) is observed to be dropping in a sharp manner. Consequently, there is a reduction in membrane selectivity towards the separation of the CO_2/CH_4 gas mixture [4]. For a bare cellulose acetate membrane, the CO_2 permeability is found to be around 6 bars, while the selectivity reaches up to more than 20. Polyimides are another group of polymers that perform better than cellulose acetate for CO_2/CH_4 and are widely investigated for gas separation applications. Matrimid 5218, a copolymer of polyimide that gives carbon dioxide permeability ~8.5 barrer [10,11] with the selectivity of around 28 and 36.7 for CO_2/CH_4 and CO_2/N_2, respectively [12,13]. Fluorination of polyimides could improve the CO_2 permeability up to 456 barrer [14,15]. Polymers like polytetra-fluoroethylene (PTFE) and polydimethylsiloxane (PDMS) also show very high CO_2 permeability (~3900 and 4000 barrer, respectively) but low selectivity (~6.5 and 2.6, respectively) [14,16]. The most common polymer, i.e. polysulfone, used for membrane fabrication generally shows similar performances with cellulose acetate [10,17–18]. However, Almuhtaseb et al. could achieve CO_2 permeability around 30 barrer with CO_2/CH_4 selectivity ~50 using THF as the solvent for polysulfone membrane fabrication [19]. Recently, many different types of blend and composite membranes have been fabricated that showed improved performance compared to pristine polymeric membranes. 2D nanomaterials like GO, clay, LDH, etc., can form highly efficient CO_2/CH_4 separation membranes [20–26]. Recently, Xu et al. reported a thin film composite membrane system fabricated by layer-by-layer (LBL) deposition of layered double hydroxide (LDH) nanosheets and formamidine sulfinic acid (FAS) on the surface of PTFE. Finally, the LBL assembly was coated with a poly(dimethylsiloxane) (PDMS) layer. They claimed that this type of membrane demonstrated all-time high selectivity and permeability (CO_2/H_2), (CO_2/N_2) and (CO_2/CH_4) mixtures with selectivity factors of 43, 86 and 62, respectively [27] (Figure 3.1).

Currently, CO_2 separation work has been widely done commercially using polymeric membranes made of polymers like polysulfone, cellulose acetate, etc. Polyamides, polyimides and polyether sulfone are other examples of polymers that are used extensively. This type of polymer-based membrane provides low main-tenance and a simple operation procedure. However, due to the disadvantages of low selectivity and permeability associated with polymeric membranes, they can still unable to compete with other technologies, such as amine-based CO_2 separ-ation. Another disadvantage of polymeric membranes is their low resistance to high temperatures and tendency to undergo degradation, which limits their extensive use. Plasticisation is a never-ending issue faced by polymeric membranes at high pressure, due to which membranes often lose their separation efficiency. Many modifications have been made to get an improved membrane. One example is a mixed matrix membrane, where porous materials such as molecular sieves are incorporated into a polymer matrix to increase membrane performance and structural properties. The combined properties of both structures can open up many possibilities for designing high-performance membranes. The combination of inorganic and organic materials can improve the membrane's performance and overcome the Robeson upper bound limit. But still, they suffer sometimes in terms of performance due to poor contact between the molecular sieve and polymer interface. A supported liquid membrane is another type of membrane where solvent is filled via capillary forces into the pores of the membrane. This type of membrane has been proven to be one of the best designed for CO_2 separation, showing high permeability and selectivity. The disadvantage associated with supported liquid membranes is their limiting operating conditions. Such membranes face problems when applied at high temperatures, which leads to solvent evaporation from the membrane pores. Moderate pressure can result in the liquid's displacement from pores.

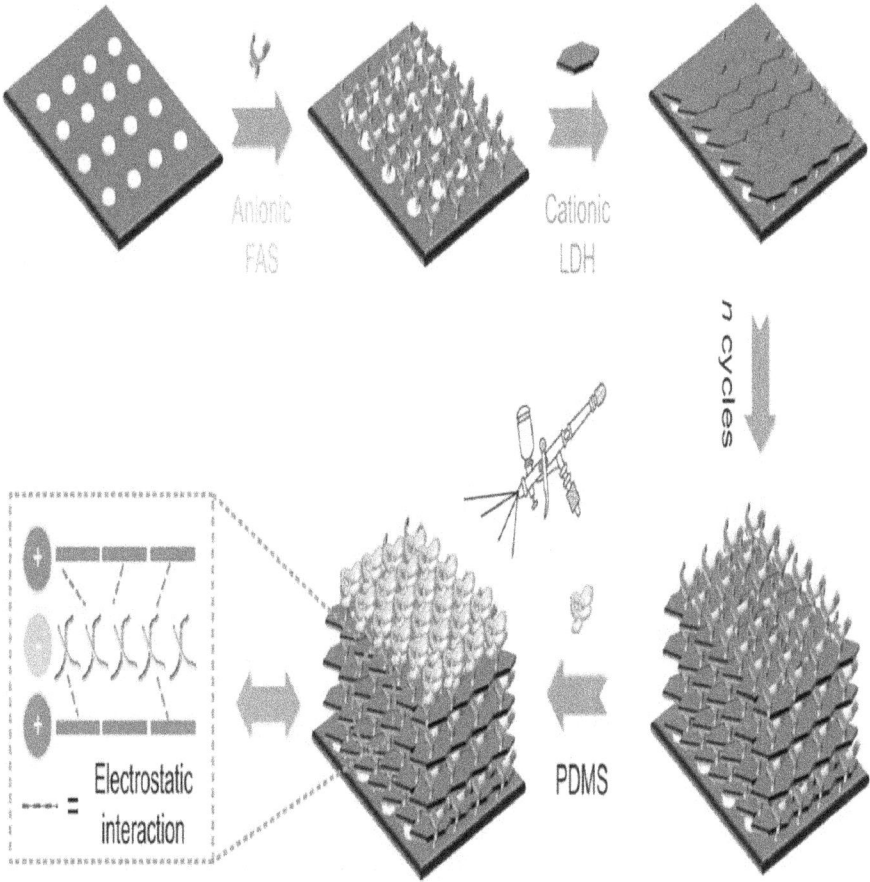

FIGURE 3.1   Schematic presentation of (LDH/FAS)-PDMS membrane fabrication.

Source: [27].

This limitation of polymeric membranes in gas separation triggers the trend of research into searching for other alternative materials. Porous materials can be incorporated into the membrane matrix to prepare inorganic membranes with better selectivity and thermochemical stability than polymeric ones. Thus, porous inorganic membranes are considered a suitable alternative to the polymeric membranes for gas separation purposes. Silica, zeolite and carbon are some examples of porous materials that show molecular sieving properties and are used in inorganic membranes. The most commonly used inorganic membranes are based on alumina, silica, glass, zirconia, ceramics, etc. Besides gas separation, other high-temperature-based separations can be obtained by using inorganic membranes. Some disadvantages are still associated with inorganic membrane fabrication. The cost-effective fabrication of a thin membrane layer on a module while maintaining a crack-free structure is still challenging. Crack formation in inorganic membranes may result in a decrease in separation efficiency.

3.2.2 HYDROGEN SULPHIDE REMOVAL

Similar to CO_2, hydrogen sulphide (H_2S) is also considered a sour gas. Natural gas with more than 4 ppm of H_2S is considered sour gas [28]. So, being corrosive, these gases must be removed from the natural gas before entering the pipeline. Three main techniques have been applied for H_2S separation: (i) absorption; (ii) adsorption and (iii) membranes. Both chemical and physical absorption or adsorption principles are used for such separation. Amine and methanol absorption techniques are two main examples of chemical and physical absorption techniques, respectively [10]. Materials studied for adsorption-based separation are activated carbon [28], zeolite [29], carbon nanotube [30], graphene [31], metal oxides [28,32], etc.

Like CO_2, plasticisation is inevitable for the separation of H_2S using glassy polymeric membranes. Increase in the permeability while selectivity decreases as a result of plasticisation. In general, in the case of H_2S separation, initially the permeability decreases with the increase in partial pressure of H_2S because of the saturation of Langmuir adsorption sites. Then, a critical pressure value is reached, above which the H_2S permeability increases with an increase in feed pressure. A transition in H_2S permeability at the critical pressure indicates the occurrence of plasticisation. Another major issue regarding sweetening natural gas is competitive adsorption. H_2S can replace the CO_2 at the adsorption sites of the polymer, leading to a decrease in both CO_2 permeability as well as CO_2/H_2S selectivity [33] (Figure 3.2).

Cellulose acetate and polyimides (PIs) are the most widely used materials for H_2S separation membranes. However, cellulose acetate membranes seemed to have low permeability to H_2S. The H_2S permeability for a mixed gas permeation experiment on a bare cellulose acetate membrane was found to be ~3.5 and ~37.2 barrer at pressures of 13.8 and 48.3 bar, respectively. In their experiment, Achoundong et al. observed improvement in cellulose acetate membrane performance after silane modification [33,34]. Polyimides are another group of widely used polymers for H_2S separation membranes, thanks to their high chemical stability and free volume. Generally, polyimide membranes have similar H_2S separation performance to that of cellulose acetate. However, in recent developments, 40-(hexafluoroisopropylidene)-diphthalianhydride (6FDA)-based polyimide membranes have outperformed the other types of membranes. Such membranes also possess higher thermal and chemical stability along with easy processibility. A lot of studies have been carried out on 4, 40-hexafluoroisopropylidene diphthalic anhydride (6FDA), 2,4-diamino mesitylene (DAM) and 3,5-diaminobenzoic acid (DABA)-based membranes [34–36]. Kraftschik et al. [37] reported an increase in H_2S permeability as well as the selectivity of a 6FAD-DAM-DABA membrane upon annealing. Annealing at 230 °C, the permeability increased from ~28.1 to ~58.5 barrer at 6.6 bar pressure. The selectivity factor of a mixture of H_2S and CH_4 showed an improvement from 10.4 to 12.7 at 6.9 bar pressure. The presence of CO_2 in the gas mixture also influences the separation performance. The increase in CO_2 or H_2S concentration in the feed reduces both permeability and H_2S/CH_4 selectivity. This is because of the increased competitive absorption of H_2S and CO_2 on polymeric membranes. It was further observed

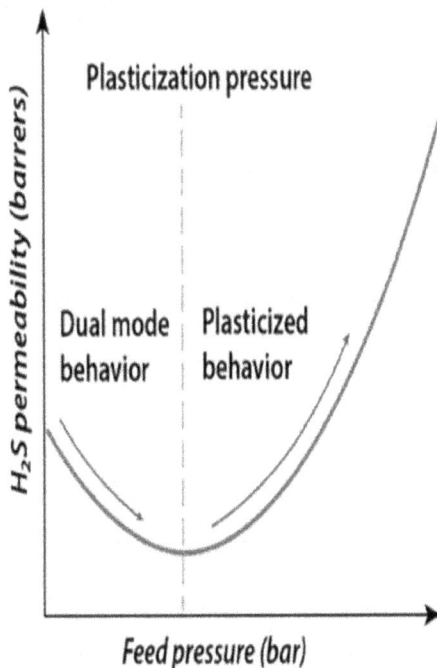

FIGURE 3.2   Typical H$_2$S permeability behaviour of glassy polymers.

Source: Reproduced from [33].

that although H$_2$S-induced plasticisation occurs in 6FAD-DAM-based membranes, it does not significantly reduce the selectivity factor for H$_2$S/CH$_4$, although a significant reduction in selectivity is observed for CO$_2$/CH$_4$ [38]. With an increase in the percentage of DAM in 6FAD-DAM-DABA type membranes, the permeability of H$_2$S increases, maintaining a high selectivity value, which has great significance for industrial applications [39]. Moreover, it was demonstrated that monoesterified crosslinking in 6FAD-DAM-DABA membranes using glycols could provide an anti-plasticisation effect to such membranes [40,41].

Polyurethane [42] and pebax [43] show very good H$_2$S separation performances among rubbery polymeric membranes. Both of these polymers exhibit very good H$_2$S/CH$_4$ selectivity. Pebax shows higher H$_2$S permeability compared to polyurethane polymeric membranes. In a single gas experiment test by Vaughn et al., the Pebax SA01 MV 3000 membrane showed H$_2$S permeability of 744 barrer whereas H$_2$S/CH$_4$ ideal selectivity was 71.9 at 4.2 bar pressure. Many different types of hybrid membranes were also developed using various types of additives, like metal–organic frameworks (MOFs) [44], ILs [45,46], etc., for further improvement in the separation performances of gases via polymeric membranes. One such example is mixed matric membranes (MMMs), which can be prepared by incorporating inorganic additives into a polymeric matrix. Graphene oxide, zeolite, carbon molecular sieve, MOFs and COFs are used as filler additives in the fabrication of mixed matrix membranes where the continuous phase is a polymer. Functional groups with strong H_2S affinity can be utilised in H_2S adsorption. UiO-66 is a zirconium-based MOF, which stands out as one of the best filler materials for MMM to apply in gas separation. The most attractive qualities of this MOF filler are its high-pressure resistance and high chemical resistance. UiO-66 and its derivatives, viz., UiO-66-NH_2 and UiO-66-NH-COCH3, were used in membranes of 6FDA-DAM [33]. The prepared MMMs are subjected to H_2S separation at 20 bar pressure and with 5:30:65 (%) of $H_2S:CO_2:CH_4$ gas feed mixture. UiO-66 membrane shows H_2S permeability of 352 barrer and H_2S/CH_4 selectivity of 13.6. Recent trends of ionic liquid-based membranes have also attracted the attention of researchers towards hybrid membrane-based gas separation. The unique properties of a high solubility range, as well as the thermochemical and electrochemical inertness of ionic liquid made this type of membrane very attractive and suitable for gas separation applications. H_2S shows high solubility in ionic liquids as compared to other gases, such as CH_4, CO_2, etc., and the extent of solubility can be varied by tailoring the anionic and cationic combinations in the ionic liquid structure. Supported ionic liquid membranes (SILMs) are membranes made up of polymers and ionic liquids that suffer problems with ionic liquid loss over time and have limited operational pressure. These disadvantages, along with the high viscosity and cost of ionic liquid, do not allow commercial use of this type of membrane. Ionic liquids of the (BMIM) cationic group with different anionic moieties, viz. [Ac], [PF_6], [CF_3SO_3], [BF_4] and [Tf_2N] are used to prepare different SILMs [33]. Where the membrane with anions [PF_6], [CF_3SO_3], [BF_4] and [Tf_2N] is subjected to H_2S separation, the H2S permeability increases with the alkalinity of the ionic liquid. In the case of [Bmim][Ac]-based SILMs, the gas absorption takes place via complex formation between ionic liquid and H2S, and a permeability of 7304 barrer is obtained. Another membrane-based technology that has been gaining great attention in acid gas treatment is the membrane contactor. This technology has the advantages of high selectivity, chemical tolerance, flexible operation, scalability and a small footprint. Also, this type of contactor does not undergo a reduction in mass transfer efficiency. While typical porous material-based membranes are used in membrane contactors, they show no selectivity. Thus, absorbent materials need to be incorporated, which can absorb the gases and result in gas selectivity. The main areas that need to be focused on are membrane materials, operating conditions and mathematical simulation for the mass transfer of gas. Due to their hydrophobic nature, polymers usually used in membrane reactor preparation are polyvinylidene fluoride, PTFE, etc. However, when they undergo operations at high pressure, these polymers often show wetting issues. Different modifications on the membrane surface, such as $CaCO_3$ nanoparticle incorporation, SiO_2 nanoparticle incorporation, etc., are implemented to decrease membrane wetting and increase H_2S permeability. Some absorbent materials are water, ionic liquids, K_2CO_3, MEA, DEA, MIDEA, TEA, etc., which can be used for absorption-based separation of H_2S. Alkanolamine and NaOH can be great absorbents for H_2S. However, the corrosive nature and low surface tension of alkanolamine and the salt-foaming ability of NaOH with H_2S limit their use in H_2S separation (Figure 3.3).

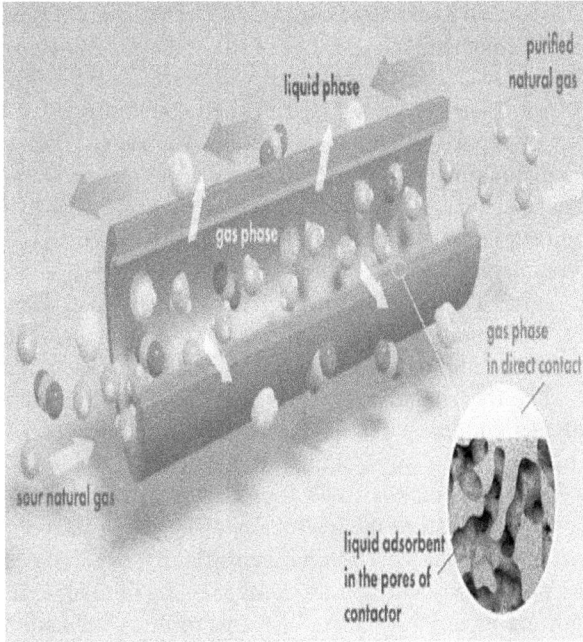

FIGURE 3.3   Schematic diagram of the working principle of membrane contactors.

Source: [33].

3.2.3   REMOVAL OF HYDROCARBONS

Natural gas generally contains gaseous hydrocarbons, i.e. C_2–C_7. The presence of heavier hydrocarbons makes the direct use of natural gas unsuitable for use as fuel in some specific engines. The separated hydrocarbons can also be used for other applications. For example, propane and butane, the most abundantly available hydro-carbon gas present in natural gas, are used as liquefied petroleum gas (LPG) [4]. The traditional method of hydrocarbon separation from natural gas is repeated cooling and condensation ($-40\ °C$) and lean oil absorption. Exxon Mobil introduced the con-cept of membrane technology for separation of LPG gases from refinery off-gases in an industrial setting for the first time in 2006. Using rubbery sorption-selective membranes, C_{3+} hydrocarbons can be separated from methane based on their condensability [4,47]. Silicone-based rubbery polymers are proven to be suitable for hydrocarbon separation from natural gas. The higher permeability for silicone-based membranes is because of the high chain mobility of these rubbery polymeric materials, which is attributed to the Si–O bond [48]. However, the major draw-back of rubbery polymeric membranes for the separation of hydrocarbons is their low methane/hydrocarbon selectivity. One of the major causes of the poor select-ivity of such membranes is the swelling of the polymeric chains after the sorption of hydrocarbons, which leads to an increase in free volume. A comparative selectivity study of membrane-based hydrocarbon separation from natural gas was reported by J. Schultz and K.V. Peinemann in 1995. In this study, the authors used many different types of membranes, including poly(dimethoxy siloxane) (PDMS), which are already used for other types of gas separation applications. Interestingly, in this study, the poly(trimethyl-silyl-propyne) (PTMSP) membrane was found to perform better than the other silicone-based rubbery polymers (more than 40 numbers) used in the study. The butane/methane selectivity (which is taken as a standard for hydrocarbon separation from natural gas) for the PTMSP membrane was found to be near about 30. PTMSP is an example of a glassy polymer with unique properties that show a glass transition temperature at or above 250 °C. The polymer has a stiff backbone with a bulky $Si(CH_3)_3$ side-group and is reported to have a high fractional free volume >25% [49,50]. The mobility of propane through the PTMSP membrane was said to be 33,800 bars, while for butane, it is 53,500 barrer [10]. PIM-1 is a spirobisindane-based glassy polymer with the unique property of intrinsic microporosity. Polymeric membranes formed by PIM-1 show similar performance to that of PTMS. n-C_4H_{10}/ CH_4 selectivity for such membranes was found to be around 24 [48]. The material choice is the most crucial part of the hydrocarbon-separating membrane. The application of polymeric membranes are still not industrialised due to the physical ageing of the membrane when exposed to a multi-component mixture of gases. Glassy polymeric membranes are mostly appreciated for hydrocarbon separations due to their highly scalable property and ability to show challenging performances. In glassy polymeric membranes, physical ageing has become the main problem, which causes selectivity and permeability loss of the membrane. This urges for an economically and energetically efficient ultrathin selective membrane with high flux and selectivity. Besides physical ageing, other issues such as operation at high pressure (8000+ psi), multi-component feed and presence of small amounts of CO_2, N_2 and BTEX as impurities are needed to be faced by a membrane during hydrocarbon separation. Different polymeric membranes have been tested for hydrocarbon separation purposes, and the results obtained showed that the operating condition and composition of feed gas have the highest impact on membrane performances. Thus, in addition to the cost, energy effectiveness and high performance, operating conditions and gas compositions are also need to be considered while preparing a membrane for hydrocarbon separation. Although polymeric membranes have been extensively used in hydrocarbon separation, they have always been suffering from trade-offs between permeability and selectivity. This reason limits the use of polymeric membranes. Thus, attention is paid towards other selective materials, such as inorganic fillers or porous materials. These include zeolite, carbon nanotube, alumina, silica, hybrid silica, etc., with high thermal as well as chemical stability and the ability to overcome the limit faced by polymeric membranes. Thus, much research has been done to establish inorganic membranes as an alternative to polymeric membranes.

3.2.4 Natural Gas Dehydration

One of the major issues with the transportation of natural gas is the formation of solid hydrates by the hydrocarbons in the presence of water, which is facilitated by low temperatures and high pressure. For example, CH_4nH_2O is a hydrate of methane, where n is the hydration number [10]. Thus, to avoid such hydrate formation as well as to prevent corrosion during transportation, water removal from natural gas is of utmost importance. The water present in a high-pressure, saturated natural gas system can be as high as 1000 ppm, while it is required to remove the water vapour below 100 ppm before its transportation [4]. The most widely used technique for the dehydration of natural gas is the absorption of triethylene glycol. The absorption-based method can also be applied using other absorbates like zeolite, silica, alumina, etc. [51]. However, regeneration of such absorbates requires a high amount of energy and frequent replacement. Membrane-based separation is a promising technique for the dehydration of natural gas. Water has higher condensability and hence better permeability in a polymeric matrix, as indicated by its higher critical temperature (647 K) compared to methane's (190 K). The diffusibility of water molecules through a membrane is also high due to their smaller size, with a kinetic diameter of 2.5 Å compared to methanes' kinetic diameter of 3.8 Å. Thus, it is possible to separate water molecules selectively from methane using suitable membranes [52].

Membranes with high water-methane selectivity (>500) can be applied to the separation of water from natural gas. Flow of water vapor to occur through the membrane, the partial pressure of water on the feed side of the membrane ($nH_2O(f)P_{water}(f)$) must be greater than the partial pressure of water on the permeate side ($nH_2O(p)P_{water}(p)$) of the membrane. That is,

$$nH_2O(f)P_{water}(f) \geq nH_2O(p)P_{water}(p).$$

Maximum separation achieved by the membrane can be expressed as:

$$P_{water}(f)/P_{water}(p) = nH_2O(p)/nH_2O(f).$$

Thus, operating at low permeate pressure gives effective separation and also minimises the loss of methane with permeate water [4].

It has been observed that rubbery polymers show higher water vapour diffusivity compared to glassy polymers because of their flexible polymer chains. While the hydrophilic polymers have better permeability as well as solubility selectivity of water vapour compared to a hydrophobic one. On the other hand, glassy polymers show better H_2O/CH_4 selectivity because of their more rigid polymer chain structure. Thus, by considering all these points, it can be concluded that a hydrophilic rubbery polymer membrane can have the best water vapour separation property compared to the other types of polymeric materials [52]. Among various membranes tested till date, PDMS, ethyl cellulose, Pebax® 1074 (55%PEO/45%Nylon-12), PEO-PBT and Nafion® 117 are seemed to show superior water vapour permeability (45×10^3, 20×10^3, 50×10^3, 855×10^2 and 41×10^4 barrer, respectively) while poly(phenylene oxide), polycarbonate, polyimide (Kapton), ethyl cellulose, cellulose acetate, sulphonated poly(amide-imide) (BDSA-40%), Pebax® 1074 (55%PEO/45%Nylon-12), PEO-PBT and Nafion® are reported to show high H_2O/CH_4 selectivity (780, 31×10^2, 14×10^3, 25×10^2, 19×10^4, 16×10^5, 6060, 71×10^2 and 41×10^5, respectively) [53–60].

In membrane-based natural gas dehydration the feed gas is passed through the membrane with high selectivity and permeability under high pressure. The permeate-side pressure is controlled at a lower value than that on the feed side. Among the gas components of natural gas, water vapour has the highest permeability, which results in the selective permeation of water vapour through membranes compared to other feed components. The design of a membrane unit is done in such a way that it allows operation at high pressures up to 1200 psig and water concentrations of 500–2000 vppm in the feed gas. Membrane systems have been proven to be one of the most competitive technologies for natural gas dehydration due to their module forming ability, high turnover ratio, low maintenance needs and lightweight. From an economic point of view, membrane-based natural gas dehydration is the most attractive at low flow rates. Conventional membranes used for natural gas dehydration usually undergo damage when they come into contact with solvents and heavy hydrocarbons. To protect the membrane from these damages, extra care is needed, which is complex as well as costly. A membrane generally needs few characteristics to meet commercial requirements, viz.: (i) robustness, durability and chemical inertness are needed; (ii) a low-cost membrane with no requirement for extra pretreatment; and (iii) the requirement of counter-current flow, which can be achieved by hollow fibre membrane devices. Although membrane processes offer various advantages, some limitations are always present when considering their application in natural gas dehydration. In gas separation, membranes have been widely used for more than 30 years. However, in the case of natural gas dehydration, membrane utilisation was initiated only 10 years ago, and commercialisation is still at an early stage. Membranes based on cellulose acetate are used extensively in natural gas treatment for acid gas removal. A cellulose acetate membrane can permeate water through it, which allows the use of this membrane for natural gas dehydration as well. The only problem associated with this is the membrane damage that occurs when it comes in contact with condensed water. Polyimide membranes can be used to withstand the water contact but they undergo performance loss after drying. PEEK-Sep is a membrane technology based on membranes made up of chemically durable and mechanically stable polyether ether ketone (PEEK) polymer. This technology uses PEEK to make porous hollow fibre supports via the melt extrusion method. On this support, a selective layer of polymer is coated to prepare the target gas selective membrane. The material used in selective layers is changeable based on the target gas molecules.

3.3 CONCLUSION

Membrane technology is one of the fastest-growing areas of research in the field of natural gas processing. According to reports, more than 20 membrane plants for CO_2 separation have been built (including the large one), as well as numerous smaller units. The onshore amine-based CO_2 removal plant can be challenged by membrane technology, which provides CO_2 removal on an offshore platform. With time, newer materials have been innovated and successfully used in the CO_2 separation from natural gas mixtures. A growing area of membrane application is the removal of heavy hydrocarbons, which requires the development of a highly selective and better-performing membrane to compete with existing technologies. Different membranes have also been developed for nitrogen removal in recent years. In the separation of nitrogen from methane, membrane technology is the most preferred method in terms of low cost and easy scalability compared to other technologies. The natural gas dehydration is tried to be done with the help of membrane technology, but still it could not compete with the conventionally used glycol absorption process [2]. In some cases, only membranes are preferred where consideration of size and weight comes into play. The problem is not the membrane's quality but the required pressure ratio, which allows the successful use of the membrane. One can summarise that membrane-based technology is one of the fastest-growing technologies in the natural gas processing area, which includes applications such as CO_2 removal, N_2 removal and heavy hydrocarbon removal. Due to the huge demand for membrane technology in natural gas processing, improvements have been implemented for the modification of membrane properties such as structure, performance, process design, etc. Thus, the application range of membranes has been widening day by day.

REFERENCES

1. Li, Y., & Chung, T. S. (2008). Exploratory development of dual-layer carbon–zeolite nanocomposite hollow fiber membranes with high performance for oxygen enrichment and natural gas separation. *Microporous and Mesoporous Materials*, *113*(1–3), 315–324. https://doi.org/10.1016/j.micromeso.2007.11.038
2. Li, Y., & Chung, T. S. (2008). Exploratory development of dual-layer carbon–zeolite nanocomposite hollow fiber membranes with high performance for oxygen enrichment and natural gas separation. *Microporous and Mesoporous Materials*, *113*(1–3), 315–324. https://doi.org/10.1016/j.micromeso.2007.11.038
3. Ding, L., Wei, Y., Li, L., Zhang, T., Wang, H., Xue, J., ... & Gogotsi, Y. (2018). MXene molecular sieving membranes for highly efficient gas separation. *Nature Communications*, *9*(1), 155. https://doi.org/10.1038/s41467-017-02529-6
4. Baker, R. W., & Lokhandwala, K. (2008). Natural gas processing with membranes: An overview. *Industrial & Engineering Chemistry Research*, *47*(7), 2109–2121. https://doi.org/10.1021/ie071083w
5. Faramawy, S., Zaki, T., & Sakr, A. E. (2016). Natural gas origin, composition, and processing: A review. *Journal of Natural Gas Science and Engineering*, *34*, 34–54. https://doi.org/10.1016/j.jngse.2016.06.030
6. Zhang, Y., Ma, J., Bai, Y., Wen, Y., Zhao, N., Zhang, X., ... & Wei, L. (2018). The preparation and properties of nanocomposite from bio-based polyurethane and graphene oxide for gas separation. *Nanomaterials*, *9*(1), 15. https://doi.org/10.3390/nano9010015
7. Freeman, B. D., & Pinnau, I. (1999). Polymeric materials for gas separations. https://doi.org/10.1021/bk-1999-0733.ch001
8. Koros, W. J., & Fleming, G. K. (1993). Membrane-based gas separation. *Journal of Membrane Science*, *83*(1), 1–80. https://doi.org/10.1016/0376-7388(93)80013-N
9. Yeo, Z. Y., Chew, T. L., Zhu, P. W., Mohamed, A. R., & Chai, S. P. (2012). Conventional processes and membrane technology for carbon dioxide removal from natural gas: A review. *Journal of Natural Gas Chemistry*, *21*(3), 282–298. https://doi.org/10.1016/S1003-9953(11)60366-6
10. Alqaheem, Y., Alomair, A., Vinoba, M., & Pérez, A. (2017). Polymeric gas-separation membranes for petroleum refining. *International Journal of Polymer Science*. https://doi.org/10.1155/2017/4250927

11. Ekiner, O. M., & Hayes, R. A. (1991). *U.S. Patent No. 5,015,270*. Washington, DC: U.S. Patent and Trademark Office.
12. David, O. C., Gorri, D., Nijmeijer, K., Ortiz, I., & Urtiaga, A. (2012). Hydrogen separation from multicomponent gas mixtures containing CO, N_2 and CO_2 using Matrimid® asymmetric hollow fiber membranes. *Journal of Membrane Science, 419*, 49–56. https://doi.org/10.1016/j.memsci.2012.06.038
13. Huang, Y., & Paul, D. R. (2007). Effect of film thickness on the gas-permeation characteristics of glassy polymer membranes. *Industrial & Engineering Chemistry Research, 46*(8), 2342–2347. https://doi.org/10.1021/ie0610804
14. Bernardo, P., Drioli, E., & Golemme, G. (2009). Membrane gas separation: A review/state of the art. *Industrial & Engineering Chemistry Research, 48*(10), 4638–4663. https://doi.org/10.1021/ie8019032
15. Sanders, D. F., Smith, Z. P., Guo, R., Robeson, L. M., McGrath, J. E., Paul, D. R., & Freeman, B. D. (2013). Energy-efficient polymeric gas separation membranes for a sustainable future: A review. *Polymer, 54*(18), 4729–4761. https://doi.org/10.1016/j.polymer.2013.05.075
16. Sadrzadeh, M., Shahidi, K., & Mohammadi, T. (2010). Synthesis and gas permeation properties of a single layer PDMS membrane. *Journal of Applied Polymer Science, 117*(1), 33–48. https://doi.org/10.1002/app.31180
17. Mohamad, M. B., Fong, Y. Y., & Shariff, A. (2016). Gas separation of carbon dioxide from methane using polysulfone membrane incorporated with Zeolite-T. *Procedia Engineering, 148*, 621–629. https://doi.org/10.1016/j.proeng.2016.06.526
18. Abdul Mannan, H., Mukhtar, H., Shima Shaharun, M., Roslee Othman, M., & Murugesan, T. (2016). Polysulfone/poly (ether sulfone) blended membranes for CO_2 separation. *Journal of Applied Polymer Science, 133*(5). https://doi.org/10.1002/app.42946
19. Almuhtaseb, R. M., Awadallah-F, A., Al-Muhtaseb, S. A., & Khraisheh, M. (2021). Influence of casting solvents on CO_2/CH_4 separation using polysulfone membranes. *Membranes, 11*(4), 286. https://doi.org/10.3390/membranes11040286
20. Jo, E. S., An, X., Ingole, P. G., Choi, W. K., Park, Y. S., & Lee, H. K. (2017). CO_2/CH_4 separation using inside coated thin film composite hollow fiber membranes prepared by interfacial polymerization. *Chinese Journal of Chemical Engineering, 25*(3), 278–287. https://doi.org/10.1016/j.cjche.2016.07.010
21. Shafie, S. N. A., Viriya, V., Nordin, N. A. H. M., Bilad, M. R., & Wirzal, M. D. H. (2021, April). Ionic liquid blend thin film composite membrane for carbon dioxide separation. In *IOP Conference Series: Materials Science and Engineering* (Vol. 1142, No. 1, p. 012013). IOP Publishing. https://doi.org/10.1088/1757-899X/1142/1/012013
22. Goh, K., Karahan, H. E., Yang, E., & Bae, T. H. (2019). Graphene-based membranes for CO_2/CH_4 separation: Key challenges and perspectives. *Applied Sciences, 9*(14), 2784. https://doi.org/10.3390/app9142784
23. Sainath, K., Modi, A., & Bellare, J. (2021). CO_2/CH_4 mixed gas separation using graphene oxide nanosheets embedded hollow fiber membranes: Evaluating effect of filler concentration on performance. *Chemical Engineering Journal Advances, 5*, 100074. https://doi.org/10.1016/j.ceja.2020.100074
24. Karunakaran, M., Shevate, R., Kumar, M., & Peinemann, K. V. (2015). CO_2-selective PEO–PBT (PolyActive™)/graphene oxide composite membranes. *Chemical Communications, 51*(75), 14187–14190. https://doi.org/10.1039/C5CC04999G
25. Norahim, N., Faungnawakij, K., Quitain, A. T., & Klaysom, C. (2019). Composite membranes of graphene oxide for CO_2/CH_4 separation. *Journal of Chemical Technology & Biotechnology, 94*(9), 2783–2791. https://doi.org/10.1002/jctb.5999

26. Zhao, Y., Zhou, C., Kong, C., & Chen, L. (2021). Ultrathin reduced graphene oxide/ organosilica hybrid membrane for gas separation. *JACS Au, 1*(3), 328–335. https:// dx.doi.org/10.1021/jacsau.0c00073

27. Xu, X., Wang, J., Zhou, A., Dong, S., Shi, K., Li, B., … & O'Hare, D. (2021). High-efficiency CO2 separation using hybrid LDH-polymer membranes. *Nature Communications, 12*(1), 3069. https://doi.org/10.1038/s41467-021-23121-z

28. Yang, C., Florent, M., de Falco, G., Fan, H., & Bandosz, T. J. (2020). ZnFe2O4/ activated carbon as a regenerable adsorbent for catalytic removal of H2S from air at room temperature. *Chemical Engineering Journal, 394*, 124906. https://doi.org/ 10.1016/j.cej.2020.124906

29. Liu, X., & Wang, R. (2017). Effective removal of hydrogen sulfide using 4A molecular sieve zeolite synthesized from attapulgite. *Journal of Hazardous Materials, 326*, 157– 164. https://doi.org/10.1016/j.jhazmat.2016.12.030

30. Chizari, K., Deneuve, A., Ersen, O., Florea, I., Liu, Y., Edouard, D., … & Pham-Huu, C. (2012). Nitrogen-doped carbon nanotubes as a highly active metal-free catalyst for selective oxidation. *ChemSusChem, 5*(1), 102–108. https://doi.org/10.1002/ cssc.201100276

31. Cortés-Arriagada, D., Villegas-Escobar, N., & Ortega, D. E. (2018). Fe-doped graphene nanosheet as an adsorption platform of harmful gas molecules (CO, CO_2, SO_2 and H_2S), and the co-adsorption in O_2 environments. *Applied Surface Science, 427*, 227–236. https://doi.org/10.1016/j.apsusc.2017.08.216

32. Lonkar, S. P., Pillai, V., Abdala, A., & Mittal, V. (2016). In situ formed graphene/ZnO nanostructured composites for low temperature hydrogen sulfide removal from natural gas. *RSC Advances, 6*(84), 81142–81150. https://doi.org/10.1039/C6RA08763A

33. Ma, Y., Guo, H., Selyanchyn, R., Wang, B., Deng, L., Dai, Z., & Jiang, X. (2021). Hydrogen sulfide removal from natural gas using membrane technology: A review. *Journal of Materials Chemistry A, 9*(36), 20211–20240. https://doi.org/10.1039/D1T A04693D

34. Achoundong, C. S., Bhuwania, N., Burgess, S. K., Karvan, O., Johnson, J. R., & Koros, W. J. (2013). Silane modification of cellulose acetate dense films as materials for acid gas removal. *Macromolecules, 46*(14), 5584–5594. https://doi.org/10.1021/ ma4010583

35. Yi, S., Ma, X., Pinnau, I., & Koros, W. J. (2015). A high-performance hydroxyl-functionalized polymer of intrinsic microporosity for an environmentally attractive membrane-based approach to decontamination of sour natural gas. *Journal of Materials Chemistry A, 3*(45), 22794–22806. https://doi.org/10.1039/C5TA05928C

36. Du, N., Park, H. B., Dal-Cin, M. M., & Guiver, M. D. (2012). Advances in high permeability polymeric membrane materials for CO2 separations. *Energy & Environmental Science, 5*(6), 7306–7322. https://doi.org/10.1039/C1EE02668B

37. Kraftschik, B., Koros, W. J., Johnson, J. R., & Karvan, O. (2013). Dense film polyimide membranes for aggressive sour gas feed separations. *Journal of Membrane Science, 428*, 608–619. https://doi.org/10.1016/j.memsci.2012.10.025

38. Liu, Y., Liu, Z., Liu, G., Qiu, W., Bhuwania, N., Chinn, D., & Koros, W. J. (2020). Surprising plasticization benefits in natural gas upgrading using polyimide membranes. *Journal of Membrane Science, 593*, 117430. https://doi.org/10.1016/ j.memsci.2019.117430

39. Liu, Z., Liu, Y., Qiu, W., & Koros, W. J. (2020). Molecularly engineered 6FDA-based polyimide membranes for sour natural gas separation. *Angewandte Chemie International Edition, 59*(35), 14877–14883. https://doi.org/10.1002/ anie.202003910

40. Babu, V. P., Kraftschik, B. E., & Koros, W. J. (2018). Crosslinkable TEGMC asymmetric hollow fiber membranes for aggressive sour gas separations. *Journal of Membrane Science, 558*, 94–105. https://doi.org/10.1016/j.memsci.2018.04.028

41. Kraftschik, B., & Koros, W. J. (2013). Cross-linkable polyimide membranes for improved plasticization resistance and permselectivity in sour gas separations. *Macromolecules, 46*(17), 6908–6921. https://doi.org/10.1021/ma401542j

42. Chatterjee, G., Houde, A. A., & Stern, S. A. (1997). Poly (ether urethane) and poly (ether urethane urea) membranes with high H_2S/CH_4 selectivity. *Journal of Membrane Science, 135*(1), 99–106. https://doi.org/10.1016/S0376-7388(97)00134-8

43. Vaughn, J. T., & Koros, W. J. (2014). Analysis of feed stream acid gas concentration effects on the transport properties and separation performance of polymeric membranes for natural gas sweetening: A comparison between a glassy and rubbery polymer. *Journal of Membrane Science, 465*, 107–116. https://doi.org/10.1016/j.memsci.2014.03.029

44. Liu, G., Chernikova, V., Liu, Y., Zhang, K., Belmabkhout, Y., Shekhah, O., ... & Koros, W. J. (2018). Mixed matrix formulations with MOF molecular sieving for key energy-intensive separations. *Nature Materials, 17*(3), 283–289. https://doi.org/10.1038/s41563-017-0013-1

45. Zhang, X., Tu, Z., Li, H., Huang, K., Hu, X., Wu, Y., & MacFarlane, D. R. (2017). Selective separation of H_2S and CO_2 from CH_2 by supported ionic liquid membranes. *Journal of Membrane Science, 543*, 282–287. https://doi.org/10.1016/j.memsci.2017.08.033

46. Akhmetshina, A. I., Yanbikov, N. R., Atlaskin, A. A., Trubyanov, M. M., Mechergui, A., Otvagina, K. V., ... & Vorotyntsev, I. V. (2019). Acidic gases separation from gas mixtures on the supported ionic liquid membranes providing the facilitated and solution-diffusion transport mechanisms. *Membranes, 9*(1), 9. https://doi.org/10.3390/membranes9010009

47. Scholes, C. A., Stevens, G. W., & Kentish, S. E. (2012). Membrane gas separation applications in natural gas processing. *Fuel, 96*, 15–28. https://doi.org/10.1016/j.fuel.2011.12.074

48. Yampolskii, Y. (2012). Polymeric gas separation membranes. *Macromolecules, 45*(8), 3298–3311. https://doi.org/10.1021/ma300213b

49. Schultz, J., & Peinemann, K. V. (1996). Membranes for separation of higher hydrocarbons from methane. *Journal of Membrane Science, 110*(1), 37–45. https://doi.org/10.1016/0376-7388(95)00214-6

50. Srinivasan, R., Auvil, S. R., & Burban, P. M. (1994). Elucidating the mechanism (s) of gas transport in poly [1-(trimethylsilyl)-1-propyne](PTMSP) membranes. *Journal of Membrane Science, 86*(1–2), 67–86. https://doi.org/10.1016/0376-7388(93)E0128-7

51. Thomas, S., Pinnau, I., Du, N., & Guiver, M. D. (2009). Pure-and mixed-gas permeation properties of a microporous spirobisindane-based ladder polymer (PIM-1). *Journal of Membrane Science, 333*(1–2), 125–131. https://doi.org/10.1016/j.memsci.2009.02.003

52. Lin, H., Thompson, S. M., Serbanescu-Martin, A., Wijmans, J. G., Amo, K. D., Lokhandwala, K. A., & Merkel, T. C. (2012). Dehydration of natural gas using membranes. Part I: Composite membranes. *Journal of Membrane Science, 413*, 70–81. https://doi.org/10.1016/j.memsci.2012.04.009

53. Potreck, J., Nijmeijer, K., Kosinski, T., & Wessling, M. (2009). Mixed water vapor/gas transport through the rubbery polymer PEBAX® 1074. *Journal of Membrane Science, 338*(1–2), 11–16. https://doi.org/10.1016/j.memsci.2009.03.051

54. Merkel, T. C., Bondar, V. I., Nagai, K., Freeman, B. D., & Pinnau, I. (2000). Gas sorption, diffusion, and permeation in poly (dimethylsiloxane). *Journal of Polymer Science Part B: Polymer Physics, 38*(3), 415–434. https://doi.org/10.1002/(SICI)1099-0488(20000201)38:3%3C415::AID-POLB8%3E3.0.CO;2-Z

55. Barbi, V., Funari, S. S., Gehrke, R., Scharnagl, N., & Stribeck, N. (2003). SAXS and the Gas Transport in Polyether-b lock-polyamide Copolymer Membranes. *Macromolecules, 36*(3), 749–758. https://doi.org/10.1021/ma0213403

56. Metz, S. J., Mulder, M. H. V., & Wessling, M. (2004). Gas-permeation properties of poly (ethylene oxide) poly(butylene terephthalate) block copolymers. *Macromolecules, 37*(12), 4590–4597. https://doi.org/10.1021/ma049847w

57. Watari, T., Wang, H., Kuwahara, K., Tanaka, K., Kita, H., & Okamoto, K. I. (2003). Water vapor sorption and diffusion properties of sulfonated polyimide membranes. *Journal of Membrane Science, 219*(1–2), 137–147. https://doi.org/10.1016/S0376-7388(03)00195-9

58. Chiou, J. S., & Paul, D. R. (1988). Gas permeation in a dry Nafion membrane. *Industrial & Engineering Chemistry Research, 27*(11), 2161–2164. https://doi.org/10.1021/ie00083a034

59. Chen, G., Zhang, X., Wang, J., & Zhang, S. (2007). Synthesis and characterization of soluble poly (amide-imide) s bearing triethylamine sulfonate groups as gas dehumidification membrane material. *Journal of Applied Polymer Science, 106*(5), 3179–3184. https://doi.org/10.1002/app.26819

60. He, X., Kumakiri, I., & Hillestad, M. (2020). Conceptual process design and simulation of membrane systems for integrated natural gas dehydration and sweetening. *Separation and Purification Technology, 247*, 116993. https://doi.org/10.1016/j.seppur.2020.116993

4 Membrane for Separation of Olefin/ Paraffin

Prarthana Bora and Swapnali Hazarika

4.1 INTRODUCTION

The sources of olefin are mainly the cracking processes of gasoline, such as steam, as well as the catalytic cracking and catalytic dehydrogenation processes of the paraffin molecule [1,2]. Usually, olefins are obtained in an impure form and are always contaminated with an amount of paraffin; this urges the separation of this mixture to obtain polymer-grade olefins [2]. The petroleum industries have been facing problems for years with this separation of olefin and paraffin. The main difficulties lie in their similar physicochemical properties, like boiling point, molecular size and solubility [3]. The undeniable demand for low molecular weight olefin and paraffin in the production of various chemicals and being the feedstock for polymers made this separation process a principal one [1,3]. The conventionally used separation technique at the industrial level is cryogenic distillation, which involves a high energy requirement, a large distillation column and a high reflux ratio, i.e., >10 [3–5]. It is a highly heat-sensitive process since the term cryogenic itself stands for low-temperature treatment [1]. All of this made cryogenic distillation disadvantageous and prompted the researchers to search for newer, more energy-efficient methods. Separations based on selective absorption are reported to be done based on the chemical complexation between the adsorbate and adsorbent molecules [6,7]. In this study, we are mainly focusing on the membrane-based separations of lightweight olefin and paraffin. Membrane-based separation has gained huge attention for olefin–paraffin separation in recent years as it can eliminate the requirement of large columns and has other additional benefits like being easy to set up and easy to maintain, as well as being an energy-efficient process. Because of their wide range of applications, membrane-based olefin and paraffin separation processes are mostly concerned with the separation of lightweight olefin and paraffin, i.e., ethane/ethylene and propane/ propylene gas mixtures. Thus, different types of membranes used for the separation of these two pairs of olefin/paraffin can be organised based on design, property and materials used, as shown in Figure 4.1.

DOI: 10.1201/9781003441359-4

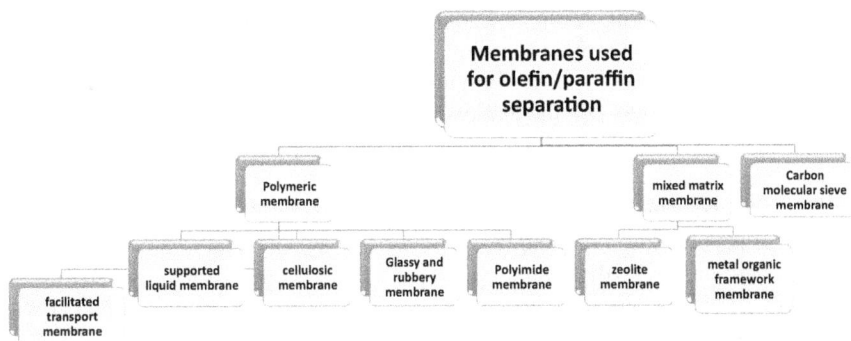

FIGURE 4.1   Different types of membranes used for olefin/paraffin separation.

4.2   OLEFIN/PARAFFIN SEPARATION BY POLYMERIC MEMBRANE

Over the last few decades, polymeric membranes have proven to be useful in a wide range of separation-based applications. This type of membrane is advantageous due to its low capital cost, easy fabrication, stability against high pressure, energy efficiency, mechanical stability and most importantly, the fact that it is easily scalable [8]. These advantages made the polymeric membrane a good candidate to be applied at the industrial level as a substitute for energy-intensive conventional processes. Mechanisms involved in traditional polymeric membranes can be explained through a solution-diffusion model where we can divide the permeation process into three steps: the first step involves the contact between molecules and the membrane, followed by adsorption and finally getting dissolved on the membrane surface. In the second step, due to the application of pressure, temperature, or the presence of a concentration gradient as a driving force, dissolved molecules get transported from the feed side to the permeate side of the membrane, and in the last step, desorption of the molecule at the permeate side of the membrane occurs [2]. Many works have been reported to use polymers alone for membrane fabrication without any carrier, such as cellulosic membranes [9], polyimide membranes and glassy and rubbery membranes [10]. Cellulosic membranes, being eco-friendly and renewable, have been used for different gas separations for a few decades, and cellulose acetate, cellulose triacetate, ethyl cellulose and cellulose acetate butylate (CAB) are a few examples of the cellulosic polymers used in most of the membranes. A report of their use in propane/propylene mixture separation was investigated, and ethyl cellulose (EC) was proven to show the highest performance among these cellulosic polymers, but the brittleness of EC often interferes with membrane fabrication [9]. EC membrane shows selectivity of 5–6. Glassy polymeric membranes made of different polymers, such as polyphenyleneoxide (PPO), copolymers based on PPO, polytetrafluoroethylene (PTFE), polyethylene (PE), polyethylene terephthalate (PETP) and polysiloxane (PS) were applied in the separation of a mixture of methane–ethane–ethylene, and PPO and its copolymer were proved to be compatible in permeation as well as the selectivity, which can be attributed to the higher diffusivity of olefin than paraffin through these polymer matrices [2]. Glassy membranes are responsible for high selectivity but show low permeance, which can be explained by the rigid nature of glassy polymers, which can stand as an obstacle in the path of gas permeation, resulting in low permeability. On the other hand, rubbery polymers show high permeability but low selectivity, which is the opposite of the glassy one, and this behaviour of high permeability is based on the presence of flexible chains in the polymeric structure, which allow the gases to pass through it easily [2]. The main reason that these cellulosic, polyimide, glassy and rubbery membranes are not widely used is that they have to compromise between selectivity and permeability, and they have poor resistance to plasticisation. To overcome these disadvantages, many steps are taken, such as transforming the polymeric membrane to a carbon molecular sieve (CMS) membrane [11] with better performance and, in some cases, introducing functionalisation to the polymeric precursor to produce a highly cross-linkable one. Another improvisation to the traditional polymer is the introduction of the carrier into the polymeric membrane with the support of a solvent or electrolyte, which increases the permeability as well as the selectivity of olefin significantly. Thus the supported liquid membrane (SLM) has come into the light and opens the gate for a bright future application in separation.

4.3   OLEFIN/PARAFFIN SEPARATION BY SLM

SLMs are porous supports where the pores are filled with liquid media driven by capillary forces, and these membranes are observed to be used in a wide range of gas separation experiments [12]. In liquid membranes, the preparation is done by simple impregnation of the membrane in a facilitation liquid solution, where the liquid is held in the pores through capillary forces [1]. SLMs involve the facilitated transport mechanism, where the presence of a carrier molecule preferentially mediates the transportation of one component of a mixture. This selective permeation arises due to the interaction or complex formation between the carrier molecule and the feed component. Depending on the mobility of the carrier molecule, facilitated transport-based membranes can be divided into three classes: (i) membranes with mobile carriers, where the carrier is allowed for free diffusion; (ii) membranes with semi-mobile carriers, in which the carrier is allowed to move only when the activation energy of diffusion is high; otherwise, the carrier molecule will not move and (iii) membranes with fixed-site carriers, where the carrier couldn't migrate but could vibrate in a nano range space only [13]. Most of the olefin–paraffin separation membranes contain mobile carriers. The most commonly used carriers for preferential olefin transport are transition metals since they are known to interact reversibly with olefins in solution [14]. Olefins having double bonds are capable of synergistically interacting with a transition metal carrier through π-bond complexation, which is responsible for their efficient separation, as shown in Figure 4.2 [15]. The carrier molecules mostly seen to be used in facilitating the transport of olefins are silver nanoparticles, silver ions (I) of silver salts, such as $AgBF_4$, $AgNO_3$, $AgNTf_2$ and $AgCF_3SO_3$ [14–23] and copper NPS as well as copper(I) ion of CuCl [3,12,24] as shown in Table 4.1. One of the best results was found with selectivity of 35.8 and 39.8 with ionic liquids [Emim][Me_2PO_4] and [Emim][Et_2PO_4] with $AgBF_4$ concentration of 2 mol/L, which is 15 times higher than that obtained with a pure aprotic ionic liquid (AIL) membrane [16]. The intensity of π-complexation is directly proportional to the electronegativity of the transition

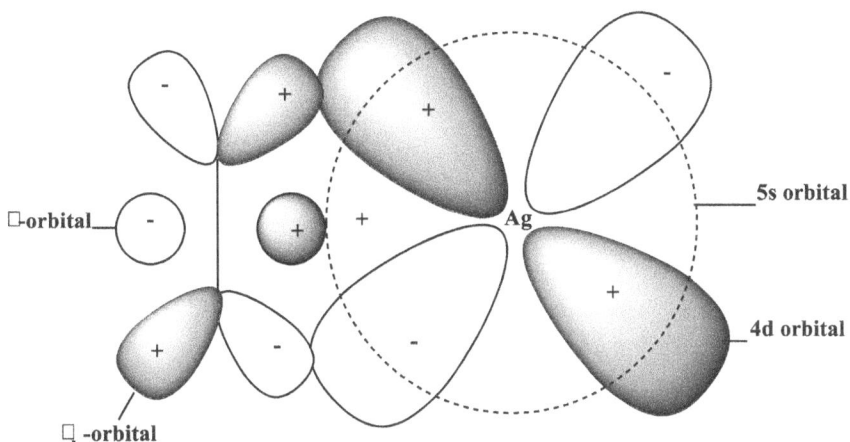

FIGURE 4.2 Schematic presentation of π-complexation between olefin and metal (Ag+).

TABLE 4.1
A List of Metal Carrier, Solvent Used and Corresponding Performances

Sl. No	Support Polymer	Carrier	Solvent(S)/ Additive (A)	Gas Mixture	Selectivity	References
1	PVDF	Ag$^+$ (AgNO$_3$)	–	C$_3$H$_6$/C$_3$H$_8$	–	[14]
2	PVDF	Ag$^+$ (AgBF$_4$)	RTIL	C$_3$H$_6$/C$_3$H$_8$	20	[17]
3	Polysulfone	Ag$^+$ (AgBF$_4$)	Ionic liquid	C$_3$H$_6$/C$_3$H$_8$	–	[18]
4	Nylon flat sheet	Ag$^+$ (AgNO$_3$)	Ionic liquid	C$_2$H$_4$/C$_2$H$_4$	–	[15]
5	PVDF	Ag$^+$ (AgBF$_4$)	Ionic liquid	C$_2$H$_6$/C$_2$H$_4$	40	[16]
6	Polysulfone	Ag$^+$ (AgBF$_4$)	Ionic liquid	C$_3$H$_6$/C$_3$H$_8$	4.9	[19]
7	Polysulfone	Ag$^+$ (AgBF$_4$)	Al(NO$_3$)$_3$ (A)	C$_3$H$_6$/C$_3$H$_8$	10	[20]
8	PVDF	Ag$^+$ (AgNTf$_2$)	Ionic liquid	C$_2$H$_6$/C$_2$H$_4$	–	[21]
9	Polysulfone	Ag$^+$ (AgCF$_3$SO$_3$)	Ionic liquid	C$_3$H$_6$/C$_3$H$_8$	9	[22]
10	Polysulfone	Ag nanoparticles	TCNQ (A)	C$_3$H$_6$/C$_3$H$_8$	17.5	[23]
11	PTFE	Ag$^+$(AgBF$_4$, AgNTf$_2$)	Ion gel	C$_3$H$_6$/C$_3$H$_8$	–	[27]
12	PVDF	Ag$^+$ (AgBF$_4$)	–	C$_2$H$_6$/C$_2$H$_4$	40	[26]
13	Polysulfone	Ag$^+$ (AgBF$_4$)	–	C$_3$H$_6$/C$_3$H$_8$	17	[28]
14	Polysulfone	Cu nanoparticle	Ionic liquid	C$_3$H$_6$/C$_3$H$_8$	2	[24]
15	Nylon	Cu$^+$ (CuCl)	DES	C$_2$H$_6$/C$_2$H$_4$	20	[3]
16	PVDF	Cu$^+$ (CuCl)	Ionic liquid	C$_2$H$_6$/C$_2$H$_4$	11.8	[12]
17	Polysulfone	–	Ionic liquid	C$_3$H$_6$/C$_3$H$_8$	2.8	[29]

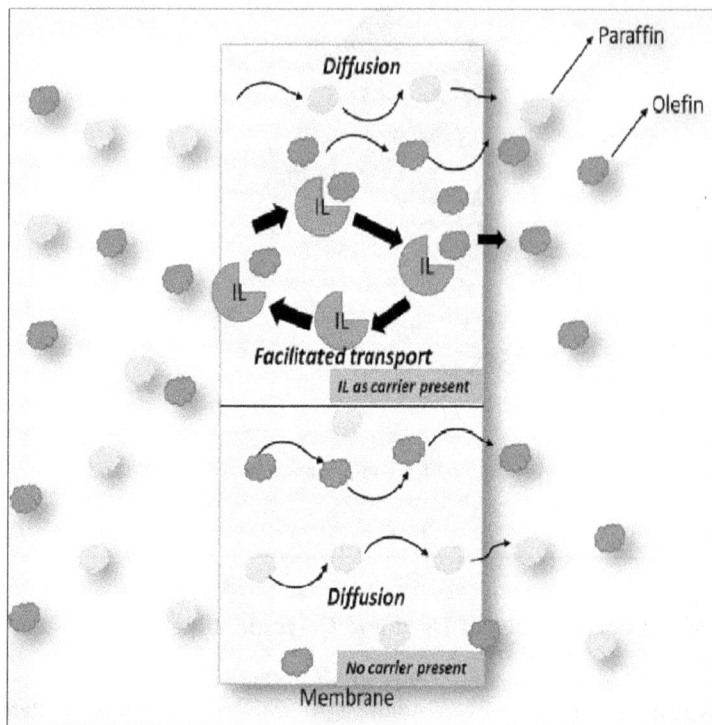

FIGURE 4.3 Schematic diagram representing facilitated transport.

metal. Silver and copper are more suitable than other transition metals with greater electronegativity as carriers since they can interact reversibly with alkene molecules, whereas the others with high electronegativity will get irreversibly bonded so that olefin molecules cannot be recovered, which implies that electronegativity should not be too high or too low, and a preferential range of electronegativity is reported to be 1.6–2.3 [14]. Figure 4.3 illustrates the facilitated transport mechanism that occurs inside a carrier-mediated, SLM. The performances of a facilitated transport membrane with a metal carrier are seen to be very efficient, but the only problem lies with its sustainability. The poisoning of metal carriers may occur due to their reduction due to the presence of impurities or other environmental factors that make the carrier unable to interact with olefin via complex formation [19]. To overcome this problem associated with metal reduction, many additives, such as oxidisers and stabilisers, are introduced in the literature that can retain satisfactory performance by preventing the reduction of metal ions. Additives, for example, $Al(NO_3)_3$ [18,20,22], ionic liquid [15–19], tetracyanoquinodimethane (TCNQ) [5,23], p-benzoquinone (p-BQ) [25] are reported to be used as a polariser, which can polarise the metal nanoparticles formed and thus prevent the performance degradation. Among this literature, many of them have used pure metal nanoparticles instead of metal salt as a carrier and observed their performances in the presence of the aforementioned additives [23–25]. The selectivity is found to be good but comparatively lower than that of metal salt. The advantages associated with metal nanoparticles, such as their high surface area, make them chemically more active; thus, they can be a good substitute for reduction-prone metal ions, but for better performance, an additive with a better polarising effect will be needed.

There is another important factor that plays an important role in supporting liquid membranes: the solvent. A drawback associated with SLM is the eventual evaporation of solvent over time, and thus solubilised carrier also got removed from the pore, which hampers the performance of membrane [5]. Deep eutectic solvents [3] and ionic liquids [12,15–19,21,22,24] having negligible vapour pressure and chemical, electrochemical, as well as thermal stability, are being used as solvents in many studies [3]. A process of oxidisation of reduced metal by using peroxide/acid liquid [26] is also reported, and the selectivity for mixed ethylene/ethane gas is found to be 40 at the initial stages, which afterwards decreases to 1.1 in 34 days. In this report, PEBAX 2533/$AgBF_4$ is coated on a polyvinylidene fluoride-supported membrane, and in order to increase the stabilisation factor, a solution of H_2O_2/HBF_4 is added to the coating solution and found to have a positive result. As the use of ionic liquid is very promising but may have some weaknesses since its binding in membrane pores is a result of capillary force alone, a study on the gelation of ionic liquid was done to form an ion–gel membrane to overcome this weakness [27]. A series of membrane electrolytes are introduced, in which polymers are used to dissolve the metal salt, and they allow the free ion movement even in the absence of solvent. This membrane electrolyte can resist the leaching of carriers from the pore site and increase sustainability [2]. Different silver salts are proven to successfully get solubilised in polymers, such as poly(ethylene oxide) (PEO) [20], poly(ether-block-amide)-5513 (PEBAX5513) [19], poly(vinyl alcohol) (PVA) [28] and poly(vinyl pyrrolidone) (PVP) [22] and have shown better performance. Few studies of facilitated transport in the absence of carriers and with solvent only, e.g., ionic liquid, which can form a reversible as well as specific interaction with olefin, are reported. Still the resultant selectivity and permeability are much lower than those with a carrier molecule [29]. Thus, SLMs are hoped to be very useful industrially if works on improvisation on solvent used and metal stability are done.

4.4 OLEFIN/PARAFFIN SEPARATION BY CMS MEMBRANE

Another type of membrane made of CMS is usually applied in the separation of small molecules, e.g., olefin paraffin mixture, based on its molecular sieving property. When amorphous or thermosetting polymer precursors are heated in an inert environment, they produce a new class of material called CMS [30]. After pyrolysis, the CMS membrane, due to the imperfect packing of the carbon sheet, provides pores of different diameters: ultramicropores (<7 Å) and micropores (7–20 Å) (shown in Figure 4.4) [30]. During pyrolysis, the small gaseous molecules channelling their way out of the solid polymer matrix initiate the ultramicropores. At the same time, the micropores are formed because of the reservation of the original structure of the precursor membrane [31]. Most of the time, the CMS membranes applied for the separation of gaseous molecules have an amorphous and isotropic sp^2 hybridised carbon sheet morphology [30].

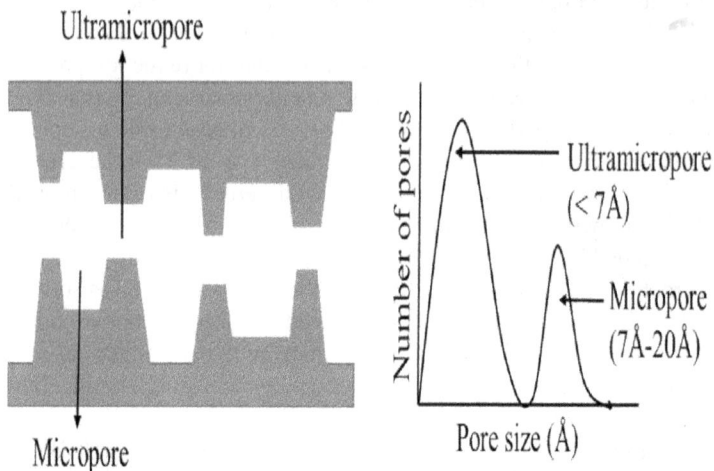

FIGURE 4.4   Pore size distribution in carbon molecular sieve membrane.

In literature, the separation strategy of olefin and paraffin through molecular sieving is divided into two types: (i) based on molecular size; as an example, we can idealise the separation case of C_3H_8 and C_3H_6 with diameters of 4.2 Å and 4.0 Å, respectively; if the membrane has a pore size in 4.0–4.2 Å range, obviously the smaller one, i.e., C_3H_6 will get permeated selectively; (ii) based on molecular shape; the membrane will allow the selective permeation of molecules based on its geometry [2]. The ultramicropores of CMS membrane perform gas separation by following the mentioned molecular sieving mechanism, and micropores offer diffusion-based permeation [30]. Thus, both types of pores generated can lead CMS membranes to show efficient separation and productivity. Although CMS membranes get aged over time, which effects their permselectivity, their higher permselectivity in addition to their thermal and chemical stability make them superior to other types of membranes, such as facilitated transport membranes and mixed matrix membranes [32].

Many studies have been done on this CMS, from its use as an adsorbent [33] to its combination with membrane. Polyimides, owing to their tunable properties and easier pyrolysis conversion to carbon with a high yield, can provide an excellent precursor for CMS. Hence, the use of 6FDA-based polyimides [30,32,34–39], spirobisindane-based polyimides [40] and Matrimid® polyimides [41–43] are mostly observed. Also, using cheap cellulosic material as a precursor has been proven to be an excellent adsorbent material, which shows 2.5 mol/kg propylene adsorption [33]. CMSM development became very attractive after its ability to form defect-less hollow fibre was discovered, and many works depending on CMS hollow fibre membrane were published [35,42–45]. In the literature, hybrid polymers are used as precursors in the formation of CMS membranes, such as the combination of poly-imide with a polysilsesquioxane (PQ) having a ladder-like structure, which results in a CMS membrane with very high productivity and shows a propylene/propane selectivity of 57 [38]. This production of high-performance CMSM is possible due to the suppression of thermal relaxation throughout the pyrolysis procedure, which is inspired by a rigid siloxane backbone in ladder-structured PQ, which is double-stranded [38]. Experiments were done on FDA polyimide with acid functionalities [30,32] and hydroxyl functionalities [37,40], and these functional groups allow cross-linking in the polyimide membrane and have been proven to show good separation performance. A very important parameter that can alter the performance of CMS membrane is the pyrolysed temperature; the most common temperature range used in the pyrolysis of CMS membrane is 500–800 °C, which is done at different stages under a controlled N_2 atmosphere [37]. The principal drawback associated with this CMS membrane is its ageing, which could be physical or chemical. The pores of the membrane tend to get blocked if molecules, such as water and other organics get absorbed physically and due to the chemical absorption of oxygen [46]. These absorptions cause the membrane to age and simultaneously degrade its transport properties. To eliminate this ageing problem, a few researchers have introduced many ageing-resistant precursors, for example, 6FDA-DABA, which is a polyimide-based acid functionalised precursor that is thermally cross-linkable and hence provides ageing resistance [32]. Super-hyperaging treatment is a recently coined term used for processing at high temperatures, which is used positively for the recovery of an aged CMS hollow fibre membrane and shows surprisingly good separation of propylene/propane [35]. This hyperaging process mainly involves the acceleration of CMS membrane ageing. To date, in spite of high performance, the commercialisation of CMS membranes has not become possible because they are sensitive to oxygen, air or water and also face difficulties in scaling up due to their brittle mechanical properties [47]. Improvising the precursors with age-resistance properties and tuning the mechanical properties to a suitable one may pave the way for CMS membranes in future applications. Table 4.2 shows the different CMS membrane precursors used in olefin paraffin separation and their performances.

4.5 OLEFIN/PARAFFIN SEPARATION BY POLYMERIC HYBRID MEMBRANES

Hybrid membranes are always getting attention from a researcher's point of view, as they have gained resistance properties as well as selectivity after incorporation of additive materials into polymer matrix. Mixed-matrix membranes (MMMs) are an example of hybrid membranes and are basically composed of two parts: the matrix part and the reinforced part. The matrix part is generally made up of polymers, while the reinforcing materials are usually microporous filler particles [49]. This type of membrane, with its high-performing filler and molecular sieving ability, can be expected to show very high performance in separation. MMMs are found to be great substitutes for other polymeric or inorganic membranes in the selective separation of gas molecules and also due to their plasticisation-suppressing capacity [50]. Recently, the metal–organic framework (MOF) of highly ordered porosity has gained great attention due to their compatibility with the matrix polymer, high surface area, tunable pore size, variable surface functionalities and most importantly, their potential to separate propane/propylene and ethane/ethylene gas pairs by molecular sieving [49,50]. MOFs are basically composed of metal ions and organic ligands forming a

TABLE 4.2
A Collection of CMS Membrane Precursors and Their Performances

Sl. No.	Precursor Material	Carrier Additives	Gas Mixture	Selectivity	References
1	6FDA-DAM:DABA	Fe^{+3}	C_2H_6/C_2H_4	11	[30]
2	6FDA-DABA	–	C_3H_6/C_3H_8	25	[32]
3	6FDA	γ-Alumina support	C_3H_6/C_3H_8	>30	[34]
4	6FDA1:BPDA1/DAM2	–	C_3H_6/C_3H_8	31.3	[35]
5	PIM-6FDA-OH	–	C_2H_6/C_2H_4	17.5	[37]
6	FDA-DAM:DABA/ LPSQ		C_3H_6/C_3H_8	57	[38]
7	6FDA-6FpDA:DABA	Zn^{+2}	C_2H_6/C_2H_4	24.1	[39]
8	6FDA-DAM, 6FDA/ BPDA-DAM	–	C_3H_6/C_3H_8	12–20	[44]
9	Matrimid® 5218	–	C_2H_6/C_2H_4	12	[41]
10	Matrimid® 5218 (BTDA-DAPI)	–	C_2H_6/C_2H_4	12	[42]
11	PIM-6FDA-OH	–	C_3H_6/C_3H_8	33	[40]
12	Matrimid® 5218	–	C_3H_6/C_3H_8	16.5	[43]
13	6FDA	γ-Alumina support	C_3H_6/C_3H_8	36	[48]
14	6FDA/BPDA-DDBT	–	C_3H_6/C_3H_8	22	[45]

self-assembled bridge structure and are prepared specifically with a precise pore size, which in combination with the ultra-permeability of polymers shows great stability as well as selectivity for gas permeance [49,51]. Two types of selectivity exist in the case of MMMs, i.e., selectivity of diffusion and selectivity of solubility; these two types need to be balanced to get a high selectivity. If the introduction of filler particles increases the flexibility of the polymer chain, then selectivity through diffusion will dominate, and on the other hand, solubility selectivity will dominate if there is an increase in membrane feed interaction after the incorporation of filler [52]. The mechanisms associated with MMM transport are actually driven by the combined effects of molecular sieving, Knudsen diffusion and surface diffusion [7]. Following different mechanisms for different materials, a lot of works using MMM are established. Firstly, we are going to discuss the most commonly used polymer matrices in the formation of MMM for olefin/paraffin separation. 6(4,4′-(Hexafluoroisopropylidene)-diphthalic anhydride)-based polyimides are reported to be the best choice in preparing MMM, and the 6FDA-DAM (2,4-diaminomesitylene)-based [44,45] and 6FDA-durene/DABA [53] polyimides are mostly used. The advantage lies in the presence of large CF_3 groups in its rigid structure, which helps in preventing the polymeric chains from getting proper packing, which is accompanied by other properties, such as thermal as well as chemical stability, spinability and mechanical strength [49] and shows a high potential in C_3H_6/C_3H_8 separation through diffusivity selectivity. Use of polysulfone [54], ethyl cellulose [55] as matrix materials is also found in the literature. Then the second most important factor in MMM is the filler particle, which has a dominant influence on the specific gas pair separation. A perfect combination between filler and polymer and its interfacial morphologies will be the main factors determining the performance of that MMM. The filler selection is the most important since the rate of discrimination-based separation is dependent on their capabilities. Filler particles can be pure inorganic or organic–inorganic, and they can be porous or non-porous. A comparison study has exposed that although organic–inorganic fillers (e.g., ZIFs, where zeolitic imidazolate frameworks (ZIFs) is one kind of MOF comprising of highly ordered porous frameworks with hydrothermal stability) are more expensive than inorganic ones hybrid membranes and are basically composed of two parts: the matrix part and the reinforced part. The matrix part is generally made up of polymers, while the reinforcing materials are usually microporous filler particles [49]. This type of membrane, with its high-performing filler and molecular sieving ability, can be expected to show very high performance in separation. MMMs are found to be great substitutes for other polymeric or inorganic membranes in the selective separation of gas molecules and also due to their plasticisation-suppressing capacity [50]. Recently, the MOF of highly ordered porosity has gained great attention due to their compatibility with the matrix polymer, high surface area, tunable pore size, variable surface functionalities and most importantly, their potential to separate propane/propylene and ethane/ethylene gas pairs by molecular sieving [49,50]. MOFs are basically composed of metal ions and organic ligands forming a self-assembled bridge structure and are prepared specifically with a precise pore size, which in combination with the ultra-permeability of polymers shows great stability as well as selectivity for gas permeance [49,51]. Two types of selectivity exist in the case of MMMs, i.e., selectivity of diffusion and selectivity of solubility; these two types need to be balanced to get a high selectivity. If the introduction of filler particles increases the flexibility of the polymer chain, then selectivity through diffusion will dominate, and on the other hand, solubility selectivity will dominate if there is an increase in membrane-feed interaction after the incorporation of filler [52]. The mechanisms associated with MMM transport are actually driven by the combined effects of molecular sieving, Knudsen diffusion and surface diffusion [7]. Following different mechanisms for different materials, a lot of works using MMM are established. Firstly, we are going to discuss the most commonly used polymer matrices in the formation of MMM for olefin/paraffin separation. 6(4,4'-(Hexafluoroisopropylidene)-diphthalic anhydride)-based polyimides are reported to be the best choice in preparing MMM, and the 6FDA-DAM (2,4-diaminomesitylene)-based [44,45] and 6FDA-durene/DABA [53] polyimides are mostly used. The advantage lies in the presence of large CF3 groups in its rigid structure, which helps in preventing the polymeric chains from getting proper packing, which is accompanied by other properties, such as thermal as well as chemical stability, spinability and mechanical strength [49] and shows a high potential in C_3H_6/C_3H_8 separation through diffusivity selectivity. Use of polysulfone [54] and ethyl cellulose [55] as matrix materials is also found in the literature. Then the second most important factor in MMM is the filler particle, which has a dominant influence on the specific gas pair separation. A perfect combination between filler and polymer and its interfacial morphologies will be the main factors determining the performance of that MMM. The filler selection is the most important since the rate of discrimination-based separation is dependent on their capabilities. Filler particles can be pure inorganic or organic–inorganic, and they can be porous or non-porous. A comparison study has exposed that although organic–inorganic fillers (e.g., ZIFs, where Zeolitic imidazolate frameworks are a kind of MOF comprising of highly ordered porous frameworks with hydrothermal stability) are more expensive than inorganic ones (e.g., zeolite), they are most likely to be used for separation due to their better compatibility with polymer matrix, functionality and thermal as well as chemical stability [52]. The polymer membrane alone, like 6FDA-based polyimides, could not avoid its eventual plasticisation and the impregnation of filler material is a good choice to prevent this plasticisation. And also, the dependency of the separation on the type of filler particle increases the effort to find new suitable materials, which results in the reports of using metal–gallate (M-gallate) [49], inorganic zeolite [56], zeolitic imidazolate framework-8 (ZIF-8) [47,50,51,53,57], ZIF-67-decorated porous graphene oxide nanosheets (ZPGO67) [50], MgO nanosheets [58], reduced graphene oxide nanoparticles (rGONPs) [59] and Cu@MIL-101(Cr) as filler material. The most widely used one among these filler materials is ZIF-based filler, which gains extra attention due to its highly ordered pore size distribution, thermochemical stability and the possibility of preparation using easily available solvents [47,51]. Metal ions, such as Zn(II) and Co(II) are tetrahedrally arranged in ZIF, which is linked together by the imidazolate moiety [47]. The pore size range exhibited by ZIFs is 3–5 Å and thus can allow the permeation of gas molecules having a molecular diameter within that range. Two types of membranes can be prepared by using ZIF: (i) one is MMM, where ZIFs are reinforced into the polymer matrix and (ii) another is a supported ZIF membrane, which involves the use of a macro- or mesoporous support on which a thin film membrane of ZIF is fabricated. When both are compared in terms of performance, the second one is found to be more capable, although ZIF-based membranes show good productivity and separation for propane/propylene. And in the case of scaling up, the ZIF-based MMM one is found to be more preferential [59]. In the case of M-gallate-derived MMM, the metals used are cobalt (Co), magnesium (Mg) and nickel (Ni); out of them, the Ni–gallate-containing membrane shows a higher permeability for C_2H_4 and a selectivity of 2.55 for C_2H_6/C_2H_4 (Figure 4.5) [49]. The significance of metal ions present in most of the MOF is to absorb the olefin molecule specifically via π-complexation and thus provide a preferable olefin transport. The use of ZPGO67 is reported, which show a high permeability as well as selectivity for C_3H_6 and C_3H_8, and these 2D nanofiller materials, being able to interact strongly with the polymer matrix, show plasticisation resistance [50]. Graphene oxides in their reduced form are proved to be good filler materials, and reportedly they show good selectivity for C_3H_6/C_3H_8 and the reason for this selective separation is due to the induced π–π interaction between olefin and nanosheet. Another study reported the synthesis of graphene nanosheets [55] having a high aspect ratio (>1000), offering an easily functionalisable surface and having a high surface area. After incorporation of this filler material, it can increase the twisted pathway for diffusion of gas, decrease the polymer chain mobility and allow the permeation of only propylene molecules with a lower kinetic diameter (450) and reject the larger molecules,

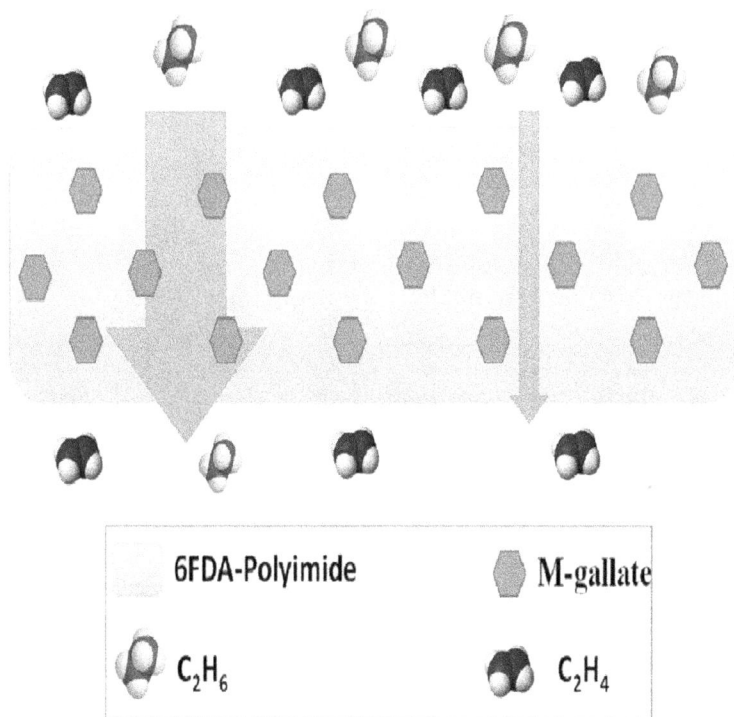

FIGURE 4.5 Schematic of 6FDA-polyimde MMMs incorporated with M-gallate MOF.

such as propane with a larger kinetic diameter (430 pm). An approach of using mesoporous magnesium oxide (MgO) nanosheets with a large surface area in combination with $AgBF_4$ as a filler material while the matrix part is a comb copolymer, has proved to be an effective membrane in terms of separation efficiency [58]. They have used two comb copolymers, poly(oxyethylene methacrylate)-g-poly(ethylene glycol) benzyl ether methacrylate (POEM-g-PEGBEM) and poly(2-hydroxypropyl 2-(methacryloyloxy)ethyl phthalate)-g-poly(ethylene glycol) biphenyl ether methacrylate (PHMEP-g-PEGBEM), and in a comparison study, the later one with PHMEP is observed to perform better in terms of olefin selectivity, which is justified by the presence of the phthalate moiety in PHMEP, which offers strong chelation, in the olefins. The overall enhanced performance of this reported membrane can be attributed to another reason, i.e., its dual functional nature: mesoporous MgO nanosheet is responsible for high diffusion while Ag^+ ions exhibit complex formation with olefin selectively. Another study of using copper salt in MMM is reported, here the salt is incorporated into a novel type of MOF, i.e., MIL-101(Cr), and the result of this combination is also based on dual functionality. Ag-exchanged X-type zeolite membrane [56] has been produced, and following dual functionality, it also shows higher selectivity and permeability. This membrane is proven to be specifically good for propane/propylene separation. Inorganic membranes mostly suffer in their compatibility with organic polymer matrixes, which can be modified by incorporation of organophilic

TABLE 4.3
A List of Mixed Matrix Membrane for Olefin/Paraffin Separation and Their Selectivity

Sl. No.	Filler Material	Gas Mixture	Selectivity	References
1	M-gallate	C_2H_6/C_2H_4	2.55	[49]
2	ZIF-8/PDA	C_3H_6/C_3H_8	94	[51]
3	ZPGO67	C_3H_6/C_3H_8	17.8	[50]
4	MgO Nano sheets and $AgBF_4$	C_3H_6/C_3H_8	12.9	[58]
5	rGO NPs	C_3H_6/C_3H_8	4.08	[54]
		C_2H_6/C_2H_4	2.8	
6	ZIF-8	C_3H_6/C_3H_8	55 (SF)	[57]
7	ZIF-8	C_3H_6/C_3H_8	-	[47]
8	ZIF-8	C_3H_6/C_3H_8	27.38	[53]
9	Cu@MIL-101(Cr)	C_3H_6/C_3H_8	2.0	[60]
10	Ag-X type zeolite membrane	C_3H_6/C_3H_8	55.4	[56]
		C_2H_6/C_2H_4	15.9	
11	Graphene nanosheet	C_3H_6/C_3H_8	10.42	[55]

materials. MOFs, although superior to zeolites and showing efficient and selective separation, can also be disadvantageous due to their chemical instability; for example, they get affected if they come in contact with water for a long time, even at room temperature. A lot of studies have been done to face this problem of chemical stability, and it could be a great field of research in the future. In Table 4.3, the reported studies on mixed matrix membrane-based olefin–paraffin separation are listed with their performances

4.6   CONCLUSION

Olefin/paraffin separation is a worthy field of study due to the great importance of pure-form olefin, which lies in its wide range of applications, ranging from the production of polymers to important chemicals. To overcome the limitations associated with energy-intensive conventional processes of olefin/paraffin separation, membranes are a promising alternative. Since membranes offer an advanced strategy with less energy needed for continuous processing, they are widely used in different separation-based applications along with olefin/paraffin. In membrane-based separation, the membrane material plays the leading role, which we can choose and develop with specific characteristics. The main criteria of a membrane are its high selectivity and its permeability. Thus, to fulfil these criteria, one must carefully select the membrane material. Many membranes have been developed for olefin/paraffin separation, which we can mainly categorise into polymeric, liquid, CMS and hybrid membranes. The separation that occurs through these types of membranes generally follows two mechanisms: one is facilitated transport and the other is molecular sieving. The facilitated transport is explained via the preferential chemical interaction of membrane material with one moiety, i.e., olefin and not paraffin, which allows the olefin to selectively pass through the membrane. This mechanism is mostly seen in dense membranes lacking permanent micropores. Selectivity obtained through such membranes is very high but compromises the permeability. Other mechanisms, i.e., sieving, can allow the tuning of membrane micropores' shape and size to sieve or permeate one molecule selectively. With the introduction of this family of porous materials, membranes attain more control over their pore size and topology. Membranes with dense structures and no permanent pores follow the solution diffusion model. Here, the olefins are first dissolved in the membrane matrix at the feed side, while diffusion occurs throughout the matrix to get permeated from the other side. However, newer materials, such as CMS and other hybrid membranes have been developed to overcome the trade-off between selectivity and permeability. Membrane technology-based olefin/paraffin separation is still growing day by day with the incorporation of different new advanced materials.

ABBREVIATIONS

RTIL	room temperature ionic liquid
DES	deep eutectic solvent
PVDF	poly (vinylidene) fluoride
PTFE	polytetrafluoroethylene
POZ	poly(2-ethyl-2-oxazoline)
SF	separation factor
ZIF	zeolitic imidazolate framework
PDA	polydopamine
ZPGO67	ZIF-67-decorated porous graphene oxide nanosheets
rGO NPs	reduced graphene oxide nanoparticles
MIL	materials Institute Lavoisier
LPSQ	ladder structured-poly (phenyl-co-pyridylethyl)silsesquioxane
6FDA	4,4'-(Hexafluoroisopropylidene)diphthalic anhydride
DAM	2,4,6-trimethyl-3,3-phenylenediamine
DABA	3,5-diaminobenzoic acid
PIM	polyimides with intrinsic microporosity
BPDA	3,3',4,4'-biphenyltetracarboxylic di-anhydride
DDBT	3,7-diamino-2,8(6)-dimethyldibenzothiophene sulfone

REFERENCES

1. Faiz, R., & Li, K. (2012). Polymeric membranes for light olefin/paraffin separation. *Desalination, 287*, 82–97. https://doi.org/10.1016/j.desal.2011.11.019
2. Hou, J., Liu, P., Jiang, M., Yu, L., Li, L., & Tang, Z. (2019). Olefin/paraffin separation through membranes: From mechanisms to critical materials. *Journal of Materials Chemistry A, 7*(41), 23489–23511. https://doi.org/10.1039/C9TA06329C
3. Jiang, B., Dou, H., Zhang, L., Wang, B., Sun, Y., Yang, H., & Bi, H. (2017). Novel supported liquid membranes based on deep eutectic solvents for olefin-paraffin separation via facilitated transport. *Journal of Membrane Science, 536*, 123–132. https://doi.org/10.1016/j.memsci.2017.05.004

4. Park, Y. S., Chun, S., Kang, Y. S., & Kang, S. W. (2017). Durable poly (vinyl alcohol)/AgBF4/Al (NO$_3$)3 complex membrane with high permeance for propylene/propane separation. *Separation and Purification Technology, 174*, 39–43. https://doi.org/10.1016/j.seppur.2016.09.050

5. Kim, M., & Kang, S. W. (2019). PEBAX-1657/Ag nanoparticles/7,7,8,8-tetracyanoquinodimethane complex for highly permeable composite membranes with long-term stability. *Scientific Reports, 9*(1), 4266. https://doi.org/10.1038/s41598-019-40185-6

6. Safarik, D. J., & Eldridge, R. B. (1998). Olefin/paraffin separations by reactive absorption: A review. *Industrial & Engineering Chemistry Research, 37*(7), 2571–2581. https://doi.org/10.1021/ie970897h

7. Jeazet, H. B. T., Staudt, C., & Janiak, C. (2012). Metal–organic frameworks in mixed-matrix membranes for gas separation. *Dalton Transactions, 41*(46), 14003–14027. https://doi.org/10.1039/C2DT31550E

8. Ito, A., & Hwang, S. T. (1989). Permeation of propane and propylene through cellulosic polymer membranes. *Journal of Applied Polymer Science, 38*(3), 483–490. https://doi.org/10.1002/app.1989.070380308

9. Iyer, G. M., Liu, L., & Zhang, C. (2020). Hydrocarbon separations by glassy polymer membranes. *Journal of Polymer Science, 58*(18), 2482–2517. https://doi.org/10.1002/pol.20200128

10. Karunaweera, C., Musselman, I. H., Balkus Jr, K. J., & Ferraris, J. P. (2019). Fabrication and characterization of aging resistant carbon molecular sieve membranes for C3 separation using high molecular weight crosslinkable polyimide, 6FDA-DABA. *Journal of Membrane Science, 581*, 430–438. https://doi.org/10.1016/j.memsci.2019.03.065

11. Sun, Y., Bi, H., Dou, H., Yang, H., Huang, Z., Wang, B., ... & Zhang, L. (2017). A novel copper (I)-based supported ionic liquid membrane with high permeability for ethylene/ethane separation. *Industrial & Engineering Chemistry Research, 56*(3), 741–749. https://doi.org/10.1021/acs.iecr.6b03364

12. Li, Y., Wang, S., He, G., Wu, H., Pan, F., & Jiang, Z. (2015). Facilitated transport of small molecules and ions for energy-efficient membranes. *Chemical Society Reviews, 44*(1), 103–118. https://doi.org/10.1039/C4CS00215F

13. Ravanchi, M. T., Kaghazchi, T., & Kargari, A. (2010). Supported liquid membrane separation of propylene–propane mixtures using a metal ion carrier. *Desalination, 250*(1), 130–135. http://dx.doi.org/10.1016/j.desal.2008.09.011

14. Dou, H., Jiang, B., Xiao, X., Xu, M., Wang, B., Hao, L., ... & Zhang, L. (2018). Ultra-stable and cost-efficient protic ionic liquid based facilitated transport membranes for highly selective olefin/paraffin separation. *Journal of Membrane Science, 557*, 76–86. http://dx.doi.org/10.1016/j.memsci.2018.04.015

15. Dou, H., Jiang, B., Xu, M., Zhou, J., Sun, Y., & Zhang, L. (2019). Supported ionic liquid membranes with high carrier efficiency via strong hydrogen-bond basicity for the sustainable and effective olefin/paraffin separation. *Chemical Engineering Science, 193*, 27–37. http://dx.doi.org/10.1016/j.ces.2018.08.060

16. Fallanza, M., Ortiz, A., Gorri, D., & Ortiz, I. (2012). Experimental study of the separation of propane/propylene mixtures by supported ionic liquid membranes containing Ag+–RTILs as carrier. *Separation and Purification Technology, 97*, 83–89. http://dx.doi.org/10.1016/j.seppur.2012.01.044

17. Kang, S. W., Kim, J. H., Won, J., & Kang, Y. S. (2013). Suppression of silver ion reduction by Al(NO$_3$)3 complex and its application to highly stabilized olefin transport membranes. *Journal of Membrane Science, 445*, 156–159. https://doi.org/10.1016/j.memsci.2013.06.010

18. Kim, S. Y., Cho, Y., & Kang, S. W. (2020). Preparation and characterization of PEBAX-5513/AgBF4/BMIMBF4 membranes for olefin/paraffin separation. *Polymers*, *12*(7), 1550. https://doi.org/10.3390/polym12071550

19. Song, D., Kang, Y. S., & Kang, S. W. (2015). Highly permeable and stabilized olefin transport membranes based on a poly (ethylene oxide) matrix and Al(NO$_3$)3. *Journal of Membrane Science*, *474*, 273–276. https://doi.org/10.1016/j.memsci.2014.09.050

20. Tomé, L. C., Mecerreyes, D., Freire, C. S., Rebelo, L. P. N., & Marrucho, I. M. (2014). Polymeric ionic liquid membranes containing IL–Ag+ for ethylene/ethane separation via olefin-facilitated transport. *Journal of Materials Chemistry A*, *2*(16), 5631–5639. https://doi.org/10.1039/C4TA00178H

21. Park, Y. S., & Kang, S. W. (2016). Role of ionic liquids in enhancing the performance of the polymer/AgCF3SO3/Al(NO3)3 complex for separation of propylene/propane mixture. *Chemical Engineering Journal*, *306*, 973–977. https://doi.org/10.1016/j.cej.2016.08.032

22. Jung, J. P., Park, C. H., Lee, J. H., Park, J. T., Kim, J. H., & Kim, J. H. (2018). Facilitated olefin transport through membranes consisting of partially polarized silver nanoparticles and PEMA-g-PPG graft copolymer. *Journal of Membrane Science*, *548*, 149–156. https://doi.org/10.1016/j.memsci.2017.11.020

23. Hong, G. H., Ji, D., & Kang, S. W. (2013). Highly permeable ionic liquid/Cu composite membrane for olefin/paraffin separation. *Chemical Engineering Journal*, *230*, 111–114. https://doi.org/10.1016/j.cej.2013.06.054

24. Kang, Y. S., Kang, S. W., Kim, H., Kim, J. H., Won, J., Kim, C. K., & Char, K. (2007). Interaction with olefins of the partially polarized surface of silver nanoparticles activated by p-benzoquinone and its implications for facilitated olefin transport. *Advanced Materials*, *19*(3), 475–479. https://doi.org/10.1002/adma.200601009

25. Merkel, T. C., Blanc, R., Ciobanu, I., Firat, B., Suwarlim, A., & Zeid, J. (2013). Silver salt facilitated transport membranes for olefin/paraffin separations: Carrier instability and a novel regeneration method. *Journal of Membrane Science*, *447*, 177–189. https://doi.org/10.1016/j.memsci.2013.07.010

26. Kasahara, S., Kamio, E., Minami, R., & Matsuyama, H. (2013). A facilitated transport ion-gel membrane for propylene/propane separation using silver ion as a carrier. *Journal of Membrane Science*, *431*, 121–130. https://doi.org/10.1016/j.memsci.2012.12.026

27. Park, Y. S., Chun, S., Kang, Y. S., & Kang, S. W. (2017). Durable poly (vinyl alcohol)/AgBF4/Al(NO3)3 complex membrane with high permeance for propylene/propane separation. *Separation and Purification Technology*, *174*, 39–43. http://dx.doi.org/10.1016/j.seppur.2016.09.050

28. Lee, J. H., Kang, S. W., Song, D., Won, J., & Kang, Y. S. (2012). Facilitated olefin transport through room temperature ionic liquids for separation of olefin/paraffin mixtures. *Journal of Membrane Science*, *423*, 159–164. https://doi.org/10.1016/j.memsci.2012.08.007

29. Chu, Y. H., Yancey, D., Xu, L., Martinez, M., Brayden, M., & Koros, W. (2018). Iron-containing carbon molecular sieve membranes for advanced olefin/paraffin separations. *Journal of Membrane Science*, *548*, 609–620. https://doi.org/10.1016/j.memsci.2017.11.052

30. Koresh, J. E., & Sofer, A. (1983). Molecular sieve carbon permselective membrane. Part I. Presentation of a new device for gas mixture separation. *Separation Science and Technology*, *18*(8), 723–734. https://doi.org/10.1080/01496398308068576

31. Karunaweera, C., Musselman, I. H., Balkus Jr, K. J., & Ferraris, J. P. (2019). Fabrication and characterization of aging resistant carbon molecular sieve membranes

for C3 separation using high molecular weight crosslinkable polyimide, 6FDA-DABA. *Journal of Membrane Science, 581*, 430–438. https://doi.org/10.1016/j.mem sci.2019.03.065

32. Andrade, M., Relvas, F., & Mendes, A. (2020). Highly propylene equilibrium selective carbon molecular sieve adsorbent. *Separation and Purification Technology, 245*, 116853. https://doi.org/10.1016/j.seppur.2020.116853

33. Ma, X., Williams, S., Wei, X., Kniep, J., & Lin, Y. S. (2015). Propylene/propane mixture separation characteristics and stability of carbon molecular sieve membranes. *Industrial & Engineering Chemistry Research, 54*(40), 9824–9831. https://doi.org/10.1021/acs.iecr.5b02721

34. Qiu, W., Xu, L., Liu, Z., Liu, Y., Arab, P., Brayden, M., ... & Koros, W. J. (2021). Surprising olefin/paraffin separation performance recovery of highly aged carbon molecular sieve hollow fiber membranes by a super-hyperaging treatment. *Journal of Membrane Science, 620*, 118701. http://doi.org/10.1016/J.Memsci.2020.118701

35. Rungta, M., Wenz, G. B., Zhang, C., Xu, L., Qiu, W., Adams, J. S., & Koros, W. J. (2017). Carbon molecular sieve structure development and membrane performance relationships. *Carbon, 115*, 237–248. http://dx.doi.org/10.1016%2Fj.carbon.2017.01.015

36. Salinas, O., Ma, X., Litwiller, E., & Pinnau, I. (2016). High-performance carbon molecular sieve membranes for ethylene/ethane separation derived from an intrinsically microporous polyimide. *Journal of Membrane Science, 500*, 115–123. http://dx.doi.org/10.1016/j.memsci.2015.11.013

37. Shin, J. H., Yu, H. J., Park, J., Lee, A. S., Hwang, S. S., Kim, S. J., ... & Lee, J. S. (2020). Fluorine-containing polyimide/polysilsesquioxane carbon molecular sieve membranes and techno-economic evaluation thereof for C3H6/C3H8 separation. *Journal of Membrane Science, 598*, 117660. https://doi.org/10.1016/j.mem sci.2019.117660

38. Wang, Q., Huang, F., Cornelius, C. J., & Fan, Y. (2021). Carbon molecular sieve membranes derived from crosslinkable polyimides for CO_2/CH_4 and C_2H_4/C_2H_6 separations. *Journal of Membrane Science, 621*, 118785. https://doi.org/10.1016/j.memsci.2020.118785

39. Swaidan, R. J., Ma, X., & Pinnau, I. (2016). Spirobisindane-based polyimide as efficient precursor of thermally-rearranged and carbon molecular sieve membranes for enhanced propylene/propane separation. *Journal of Membrane Science, 520*, 983–989. http://dx.doi.org/10.1016/j.memsci.2016.08.057

40. Rungta, M., Xu, L., & Koros, W. J. (2012). Carbon molecular sieve dense film membranes derived from Matrimid® for ethylene/ethane separation. *Carbon, 50*(4), 1488–1502. http://dx.doi.org/10.1016/j.carbon.2011.11.019

41. Xu, L., Rungta, M., & Koros, W. J. (2011). Matrimid® derived carbon molecular sieve hollow fiber membranes for ethylene/ethane separation. *Journal of Membrane Science, 380*(1–2), 138–147. https://doi.org/10.1016/j.memsci.2011.06.037

42. Kim, S. J., Lee, P. S., Chang, J. S., Nam, S. E., & Park, Y. I. (2018). Preparation of carbon molecular sieve membranes on low-cost alumina hollow fibers for use in C_3H_6/C_3H_8 separation. *Separation and Purification Technology, 194*, 443–450. https://doi.org/10.1016/j.seppur.2017.11.069

43. Xu, L., Rungta, M., Brayden, M. K., Martinez, M. V., Stears, B. A., Barbay, G. A., & Koros, W. J. (2012). Olefins-selective asymmetric carbon molecular sieve hollow fiber membranes for hybrid membrane-distillation processes for olefin/paraffin separations. *Journal of Membrane Science, 423*, 314–323. http://dx.doi.org/10.1016/j.mem sci.2012.08.028

44. Yoshino, M., Nakamura, S., Kita, H., Okamoto, K. I., Tanihara, N., & Kusuki, Y. (2003). Olefin/paraffin separation performance of carbonized membranes derived from an asymmetric hollow fiber membrane of 6FDA/BPDA–DDBT copolyimide. *Journal of Membrane Science*, *215*(1–2), 169–183. https://doi.org/10.1021/ie990209p

45. Xu, L., Rungta, M., Hessler, J., Qiu, W., Brayden, M., Martinez, M., ... & Koros, W. J. (2014). Physical aging in carbon molecular sieve membranes. *Carbon*, *80*, 155–166. https://doi.org/10.1016/j.carbon.2014.08.051

46. Alcheikhhamdon, Y., Pinnau, I., Hoorfar, M., & Chen, B. (2019). Propylene-propane separation using zeolitic-imidazolate framework (ZIF-8) membranes: Process techno-commercial evaluation. *Journal of Membrane Science*, *591*, 117252. http://dx.doi.org/10.1016/j.memsci.2019.117252

47. Ma, X., Lin, B. K., Wei, X., Kniep, J., & Lin, Y. S. (2013). Gamma-alumina supported carbon molecular sieve membrane for propylene/propane separation. *Industrial & Engineering Chemistry Research*, *52*(11), 4297–4305. https://doi.org/10.1021/ie303188c

48. Chen, G., Chen, X., Pan, Y., Ji, Y., Liu, G., & Jin, W. (2021). M-gallate MOF/6FDA-polyimide mixed-matrix membranes for C_2H_4/C_2H_6 separation. *Journal of Membrane Science*, *620*, 118852. https://doi.org/10.1016/j.memsci.2020.118852

49. Moghadam, F., Lee, T. H., Park, I., & Park, H. B. (2020). Thermally annealed polyimide-based mixed matrix membrane containing ZIF-67 decorated porous graphene oxide nanosheets with enhanced propylene/propane selectivity. *Journal of Membrane Science*, *603*, 118019. https://doi.org/10.1016/j.memsci.2020.118019

50. Jiang, X., Li, S., Bai, Y., & Shao, L. (2019). Ultra-facile aqueous synthesis of nanoporous zeolitic imidazolate framework membranes for hydrogen purification and olefin/paraffin separation. *Journal of Materials Chemistry A*, *7*(18), 10898–10904. https://doi.org/10.1039/C8TA11748A

51. Najari, S., Saeidi, S., Gallucci, F., & Drioli, E. (2021). Mixed matrix membranes for hydrocarbons separation and recovery: A critical review. *Reviews in Chemical Engineering*, *37*(3), 363–406. http://dx.doi.org/10.1515/revce-2018-0091

52. Askari, M., & Chung, T. S. (2013). Natural gas purification and olefin/paraffin separation using thermal cross-linkable co-polyimide/ZIF-8 mixed matrix membranes. *Journal of Membrane Science*, *444*, 173–183. http://dx.doi.org/10.1016/j.memsci.2013.05.016

53. Najafi, M., Sadeghi, M., Shamsabadi, A. A., Dinari, M., & Soroush, M. (2020). Polysulfone membranes incorporated with reduced graphene oxide nanoparticles for enhanced olefin/paraffin separation. *Chemistry Select*, *5*(12), 3675–3681. https://doi.org/10.1002/slct.202000240

54. Yuan, B., Sun, H., Wang, T., Xu, Y., Li, P., Kong, Y., & Niu, Q. J. (2016). Propylene/propane permeation properties of ethyl cellulose (EC) mixed matrix membranes fabricated by incorporation of nanoporous graphene nanosheets. *Scientific Reports*, *6*(1), 28509. https://doi.org/10.1038/srep28509

55. Sakai, M., Sasaki, Y., Tomono, T., Seshimo, M., & Matsukata, M. (2019). Olefin selective Ag-exchanged X-type zeolite membrane for propylene/propane and ethylene/ethane separation. *ACS Applied Materials & Interfaces*, *11*(4), 4145–4151. https://doi.org/10.1021/acsami.8b20151

56. Lee, M. J., Kwon, H. T., & Jeong, H. K. (2018). High-flux zeolitic imidazolate framework membranes for propylene/propane separation by postsynthetic linker exchange. *Angewandte Chemie International Edition*, *57*(1), 156–161. https://doi.org/10.1002/anie.201708924

57. Park, C. H., Lee, J. H., Jung, J. P., & Kim, J. H. (2017). Mixed matrix membranes based on dual-functional MgO nanosheets for olefin/paraffin separation. *Journal of Membrane Science, 533*, 48–56. http://dx.doi.org/10.1016%2Fj.memsci.2017.03.023

58. Ma, X., & Liu, D. (2018). Zeolitic imidazolate framework membranes for light olefin/paraffin separation. *Crystals, 9*(1), 14. https://doi.org/10.3390/cryst9010014

59. Jung, J. P., Kim, M. J., Bae, Y. S., & Kim, J. H. (2018). Facile preparation of Cu(I) impregnated MIL-101(Cr) and its use in a mixed matrix membrane for olefin/paraffin separation. *Journal of Applied Polymer Science, 135*(31), 46545. https://doi.org/10.1002/app.46545.

5 Membrane-Based Separation of Aromatics/Aliphatic Compounds

Prarthana Bora and Monti Gogoi

5.1 INTRODUCTION

One of the most important petroleum industry-based separations is the aromatic and aliphatic compound separation from their mixture, which has been the most significant source of research since 1960 [1]. The aromatic and aliphatic groups are the two main components of naphtha, which originates from the fractional distillation of crude oil. The aromatics are basically xylene, benzene, toluene and derivatives of benzene, which can be used as essential precursors for the production of chemicals in downstream processes, and they need to be separated from aliphatic hydrocarbons mostly in the range $C_4–C_{10}$. This separation has become very attractive, which can be attributed to two main reasons: firstly, in order to lower the aromatic concentration in the fuel mixture, and secondly, the removal of aliphatic can make the naphtha cracking product rich in BTX (benzene, toluene and xylene) [2]. The main problem associated with this separation is the close boiling point range of these hydrocarbons [3]. The conventional technologies available for aromatic/aliphatic separation are liquid extraction and extractive distillation, which are suitable for aromatics present in the range of 20–65wt.% and 65–90wt.%, respectively, for sample mixtures with an aromatic content greater than 90 wt.%, azeotropic distillation is normally used [3]. Among these procedures, the most commonly used one is liquid–liquid extraction, which involves containing an extraction column and adding polar solvent to the upper part of it, which will get mixed up with the aromatic component of the sample mixture due to the high aromatic affinity of the polar solvent. The mixture of aromatics and solvent accumulated inside the column, which led to separation. However, this conventional process suffers due to the need for a high flow rate [4]. The extraction of aromatics from its mixture with solvent and the involvement of extra purification steps make liquid extraction disadvantageous in terms of energy and cost. To overcome these problems, membrane-based separation called pervaporation is widely used since it offers many advantages over solvent-dependent liquid–liquid extraction and extractive distillation, such as energy efficiency, ease of operation and eco-friendliness [1,5]. However, recent studies have reported that the industrialisation of this technique is yet to be done due to the poor separation performance obtained to date. One of the main factors on which the pervaporation separation performance mainly depends is the membrane material [1]. This chapter covers an overview of

DOI: 10.1201/9781003441359-5

membrane-based separations of aromatic and aliphatic compounds reported to date, including the existing mechanism.

5.2   AROMATIC/ALIPHATIC SEPARATION BY PERVAPORATION (PV)

Pervaporation is the only technique based on membrane separation in which a phase transition occurs [2,6]. This membrane separation has been used for a few decades due to its energy efficiency, cost-effectiveness and environmental benignity [6]. It provides critical separations of close boiling point compounds and azeotropes. The main reason behind the superiority of this pervaporation method over distillation is the separation mechanism, which, in the case of PV, is dependent on the different properties of components, such as sorptivity and diffusivity, in addition to the selectivity existing due to the membrane. In contrast, distillation allows the separation only depending on the volatility difference. To understand pervaporation properly, one should thoroughly study the involved mechanisms in depth. The evaporation of permeate occurs after passing through the membrane; depending on their different rates of evaporation, they can be distinguished. This phase change from liquid to vapour during the separation is enough to make this technique a novel one [7]. The mechanism mainly involved in the pervaporation separation is believed to be the solution diffusion mechanism [1], which is performed in three steps: the first one is where absorption of the components on the membrane feed side occurs, some of which are dissolved; in the second step, diffusion of components from the feed side to the permeate side occurs; and in the last stage, the final desorption of the components occurs on the permeate side [1,8]. In short, from the mixture of ingredients, due to their variations in sorption and diffusion rates, one of them will separate preferentially through the membrane [9]. Figures 5.1 and 5.2 show the representative scheme of membrane-based pervaporation separation and membrane cell diagram, respectively.

Since the membrane plays a vital role in pervaporation, it should be selected very carefully. A few parameters are based on which membrane can be chosen for a specific sample mixture. The first one is the mass flux J, which is the permeation rate across the unit area of the membrane, while time is unity having unit its unit is kg μm/ (m^2 h) [7,8] and can be represented by the following equation (5.1),

$$J = m_p \cdot \delta_M / t \cdot A. \tag{5.1}$$

Secondly, the separation factor is the measure of separation performed by the membrane [9]. Several researchers have always faced problems because, between separation factors and membrane flux, one can only show better results at the expense of others. For measuring the ability of pervaporation membrane separation, an index named the pervaporation membrane index (PSI, kg/m^2 h) is commonly used by many researchers. But although this PSI factor seems useful in many similar types of membranes when applied to measure overall performance, it often fails since it is challenging to distinguish the PSI of a membrane having a low separation factor with high flux from a membrane where the separation factor is high, but the flux is low.

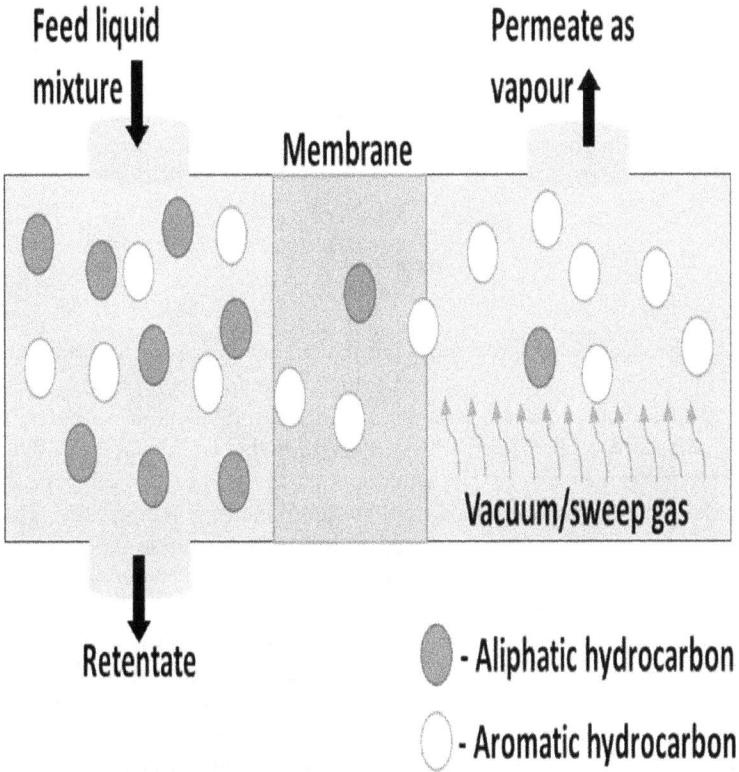

FIGURE 5.1 Representation of pervaporation membrane separation.

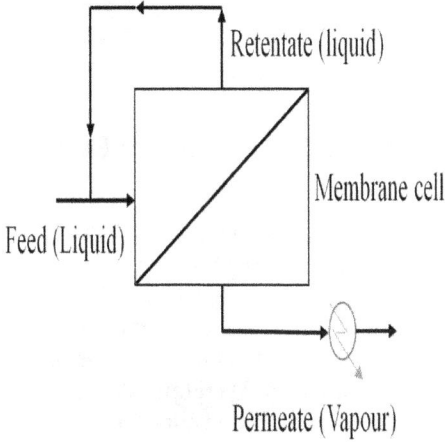

FIGURE 5.2 Schematic diagram of pervaporation membrane cell operation.

Another term, selectivity, can be used instead of the separation factor. Both imply the same meaning, and two types of selectivity exist: the ideal and the actual selectivity [7]. Ideal selectivity refers to the ratio of membrane flux values for pure components. At the same time, the actual one is calculated for a binary mixture and refers to the ratio between permeate concentration and feed concentration, as represented in equation (5.2).

$$\alpha = \frac{Ya/Yb}{Xa/Xb},$$

(5.2)

here, is the selectivity, and x and y are the concentrations of feed and permeate, respectively. And 'a' and 'b' are the experimental components. The resultant overall selectivity of a membrane is a combined effect of individual selectivities, such as diffusion selectivity and sorption selectivity [7]. Another additional factor that affects the overall performance of a pervaporation process is the operating parameters, such as downstream pressure, feed concentration and feed temperature. Therefore, these parameters are needed to be considered in a pervaporation separation process. There is a generalised belief that the effect of feed pressure on permeability and the pervaporation membrane is insignificant. Much research has been done to improve the material properties of membranes specifically for aromatic/aliphatic mixture separation. The membrane material selection has always been crucial in the separation process. The polymer precursor having a high affinity for one of the component mixtures has proven to be a good choice. Still, after a point, the membrane undergoes swelling since the affinity crosses the limit and thus makes the membrane lose its integrity and selectivity. Therefore, one must pay proper attention to preparing a membrane with controlled swelling to meet high-performance criteria. The most commonly used membrane materials are organic polymers, which have advantages, such as good membrane-forming ability, low price, easy fabrication and availability [1]. Here, we are trying to summarise the different membrane types used in pervaporation to separate aromatic and aliphatic compounds.

5.3   ALIPHATIC/AROMATIC SEPARATION BY A POLYMERIC MEMBRANE

Molecular structures and properties of a polymer are the main factors that affect the membrane material, followed by pervaporation separation. The membrane materials with selective aromatic affinity should be selected rather than the aliphatic ones. By using a polar group-containing polymer in the membrane matrix, there are possibilities for polarisation in aromatics due to delocalised π-bonds. At the same time, the aliphatic one will not be easily affected. Therefore, a membrane with polar groups can be a good choice for the preferential movement of aromatics through it. Correlating with the principle of like-to-like dissolution, a polymer having benzene rings in its main chain can also exhibit selective permeation of aromatics over aliphatic. A wide range of polymers are being used in aromatic/aliphatic hydrocarbon mixture separation, including polyimides [5,10–14], polyurethanes, poly(vinyl alcohol)s (PVA), polysiloxanes amides, poly(ether amides), polyacrylates, poly(methyl methacrylate) and alkyl esters of cellulose.

5.3.1 Polyimide-Based Polymeric Membrane

Polyimides have high thermal stability, excellent mechanical properties and other properties like dielectric and adhesive properties, which make them a good choice of polymer material for pervaporation membranes [12]. Although a few disadvantages, such as the poor solubility of polyimides in organic solvents, make the processing of such membranes difficult and limit the range of possible applications, this resistance property makes the polyimide polymer very stable [9]. The polyimides containing fluoride in their structure are ideal for pervaporation since they provide more stability, solubility in organics and free volume [12]. Polyimides are composed of two parts: the soft segment and hard segment where the former is responsible for selective interaction with aromatics and the latter makes the membrane mechanically stable [15]. Many copolyimides are introduced in the literature for pervaporation purposes in search of better performances. 6FDA-based polyimide in combination with 4MPD, 6FpDA and DABA monomer has been reported to improve cross-linking ability and reduce membrane swelling [5]. This study shows a separation factor of 5.8 for the benzene/cyclohexane mixture, whereas the flux was 8.9. Phosphorylated BPDA-TrMPD polyimides, when used for benzene/cyclohexane mixture pervaporation, result in a very high separation factor of 90 against the very low flux of 0.02; when they focus on increasing the flux value, the separation factor decreases to a value of 10 [10]. For toluene/n-heptane mixture separation, the use of various copolyimides, such as 4MPD-6FDA, FDA-DSDA, 6FDA-BDAF [11], 4MPD and APAF/4MPD [13], is reported, and the performances are mentioned in Table 5.1. The polyimide membranes tend to swell with increased aromatic content in the feed mixture [5]. This is why studies on improving the stability of polyimide membranes by incorporating cross-linking groups are being widely done. Since properties such as anti-swelling, structural integrity and chemical inertness are superior in the case of cross-linked membranes as compared to non-cross-linked ones [1]. Modifying molecular structures using crown ether is proven to show better permeation flux, which can be due to the cavity in its structure [16]. Further modifications are still ongoing to improve polyimide membrane-based separation efficiency. Figure 5.3 shows the chemical structures of a few widely used 6FDA-based non-cross-linked and cross-linked polyimide polymers.

Another type of polymeric material usually found to be used in aromatic/aliphatic hydrocarbon mixture separation is poly (ether amide), which can be modified by introducing different functional groups as a π electron acceptor to increase the sorption or diffusion rate of the corresponding membrane [1]. Examples of such groups having an affinity for π bond are the sulfone, phenyl, phosphorylate and acrylate groups, which also can interact with benzene, resulting in a charge transfer complex [17]. A literature review on industrially significant benzene/cyclohexane separation is reported where four poly(ether amides) soluble in organics are used. The modified polymers are obtained when terephthalic acid undergoes phosphorylation polyamidation with semi-fluorinated aromatic diamines. They have synthesised four

TABLE 5.1
Different Membrane Materials and Their Performances for Aliphatic Aromatic Separation

Membrane Type	Material Type	Hydrocarbon Mixture	Separation Factor	Flux g/ (m² h)	Normalized flux Kg μm/m² h	References
Polyimide	6FDA-based	Benzene/cyclohexane	5.8	—	8.9	[5]
	Phosphorylated BPDA-TrMPD	Benzene/cyclohexane	90	—	0.02	[10]
			10		24	
	4MPD-6FDA	Toluene/n-heptane	3.3	—	53.2 ± 1.7	[11]
	FDA-DSDA		11.5		0.585	
	6FDA-BDAF	toluene/n-heptane	6.49	—	0.66	[12]
	4MPD:APAF/4MPD (1:1)	Toluene/n-heptane	3.32	—	53.2 ± 1.7	[13]
			7.22		0.017 ± 0.002	
	6FDA–DABA-copolyimide	Naphthalene/n-decane	—	—	—	[14]
	DSDA–DABC/DDBT	Benzene/cyclohexane	30	—	—	[16]
	PBI-matrimid	Toluene/iso-octane	—	—	—	[50]
Poly(ether) amide	BTAPPI-TA BTAPPHI-TA BTAPPF-TA	Benzene/cyclohexane	6.9	—	23.40	[17]
	BTAPPPI-TA		6.5		29.12 (70 °C)	
			5.9		34.49	
			7.6		30.65	
	BTAPPI-IA BTAPPHI-IA BTAPPF-IA	Benzene/cyclohexane	6.9		23.66	[51]
	BTAPPPI-IA		6.5		29.89	
			5.9		31.42	
			7.6		30.85	
	PEBA-ceramic	Toluene/n-heptane	4.6	280 (80 °C)		[18]
	Pebax 4033	Benzene/cyclohexane	>3	—	—	[19]
Poly methyl methacrylate	Poly(methyl acrylate-co-acrylic acid)	Toluene/i-octane	13 (100 °C)	—	26 (100 °C)	[20]
	Polyacrylonitrile-block-poly(methyl acrylate) (40 mol %)	Benzene/cyclohexane	10.5 (30 °C)	—	4.20 (30 °C)	[22]

	NBR-PMMA-ethylene oxide/ epichlorohydrin/ allyl glycidyl ether (Hydrin)	Benzene/cyclohexane	7.9	–	28	[21]
Polydimethylsiloxane	(Glycidyl methacrylate) graft membrane	Benzene/cyclohexane	19–22	300–370	0.218	[24]
	PDMS/PEI	Benzene/cyclohexane	13.2	–	9.5	[29]
	PDSM/PAN-PMHS	Methanol/toluene	–	30	–	[30]
Cellulose	Cellulose alkyl esters	Benzene/cyclohexane	–	–	–	[31]
	CMC/SA	Benzene/cyclohexane	–	–	–	[32]
Cyclodextrin	β-CD/PVA/GA	Benzene/cyclohexane	27	51.41	–	[34]
	PEG-DMA; acrylated cyclodextrins	Toluene/cyclohexane	14	59.25	0.3	[25]
Inorganic filler	PVA/CS	Benzene/cyclohexane	49.9	61	–	[37]
	PVA/CMS	Benzene/cyclohexane	23.21	65.9	–	[44]
	β-CD-CNT/PVA	Benzene/cyclohexane	41.2	280 / 740	–	[41]
	PVA–CNT(CS)	Benzene/cyclohexane	53.4	90.7 / 40.2	–	[42]
	E-VOH	Benzene/cyclohexane	12.90	280	–	[43]
	E-VOH/CNT		9.03	740	–	[43]
	CG-PVA-06	Benzene/cyclohexane	100.1 / 344.5	–	90.7 / 40.2	[39]
	AgCl/PEO–PPO–PEO/PMMA	Benzene/cyclohexane	1.1	–	5000	[36]
	PAN/PVA	Toluene/n-heptane	4.5	–	42.4	[40]
	PAN/PVA-GO		12.9	–	27	[40]
	PVA-GPTMS	Benzene/cyclohexane	46.9	–	137.1	[46]
	$Cu_3(BTC)_2$/PVA	Toluene/n-heptane	17.9	–	133	[47]
	$Co(HCOO)_2$/PEBA	Toluene/n-heptane	5.1	–	771	[48]
Metal organic framework	Porous metal–organic molecule nanocages/W3000	Toluene/n-heptane	19	–	220.5	[49]
	MOP-SO_3NanHm/hyper-branched polymer	Benzene/cyclohexane	15.4	–	392.3	[52]
		Toluene/n-heptane	8.03	–	528	
	$[Cu_{24}(5\text{-}tBu\text{-}1,3\text{-}BDC)_{24}(S)_{24}]$ (MOP-tBu)	Benzene/cyclohexane	8.4	–	540	[53]
		Toluene/n-heptane	5.4	–	800	

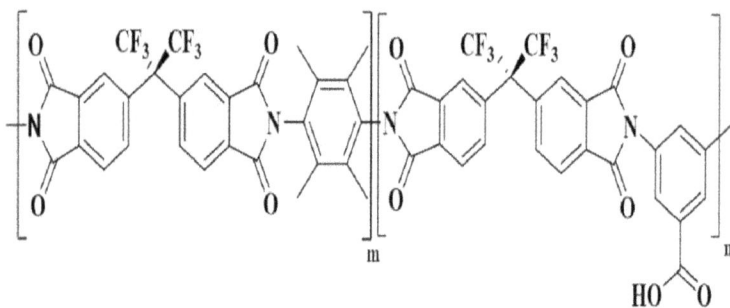

Non- crosslinked polyimide
6FDA-4MPD/DABA m:n

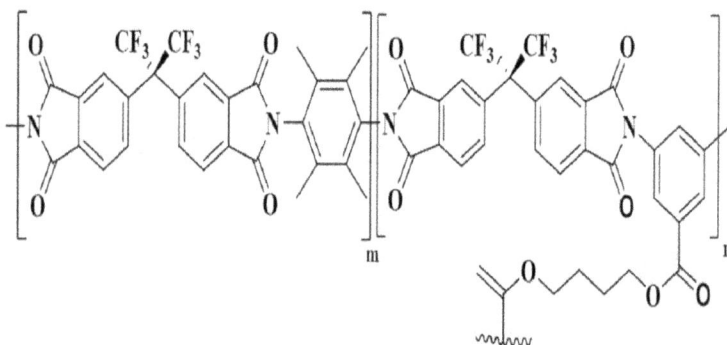

Crosslinked polyimide with 1,4-butananediol
6FDA-4MPD/DABA m:n

FIGURE 5.3 Chemical structures of 6FDA-based non-cross-linked and cross-linked polyimides.

different polyether-amides and named them PEAI, PEAII, PEAIII and PEAIV, which have the composition of BTAPPI-TA, BTAPPHI-TA, BTAPPF-TA and BTAPPPI-TA, respectively. After the performance experiment, the PEAIV polymer membrane, where the main polymer chain has a cardo phenolphthalein anilide group, exhibits a high separation factor of 7.6 at 50 °C temperature. While for the permeation flux value, the PEAIII membrane is found to have the highest value of 821 g/(m² h), whose normalised value is 34.49 kg μm/(m² h) at a temperature of 70 °C. This high permeation flux value can be explained by the increase in π–π interaction between benzene and PEAIII since the polymer has a cardo fluorine group in its structure [17]. The reaction between PEA and TA is represented in Figure 5.4.

PEA membranes are considered a good choice for high-permeation membranes since they show an affinity for aromatics. But in many cases, swelling still prevails. In order to remove this problem, the introduction of block copolymers was done, and by using commercial poly(ether-block-amide) (PEBA) with different filler particles [18], the performance and stability of the membrane were found to be better. PEBA has

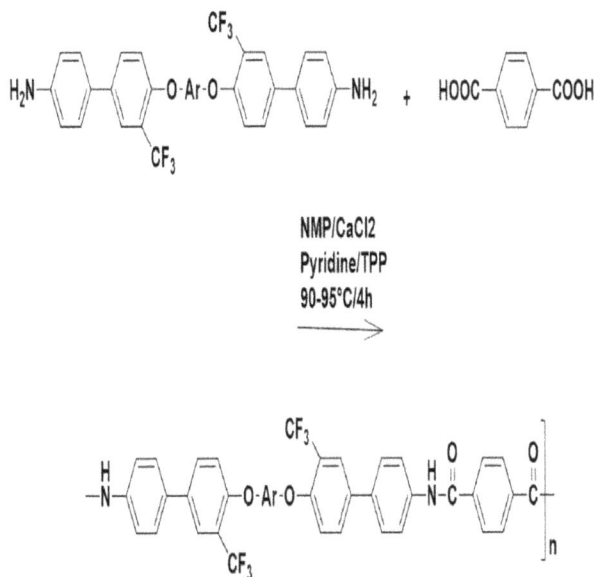

FIGURE 5.4   Scheme of reaction between PEA and TA.

two parts in its structure: a rigid polyamide part responsible for mechanical stability and a soft and flexible polyether chain part with very high mobility and high permeation. Incorporating filler such as ceramic tubular substrate will improve membrane stability by preventing swelling [18]. As this reported membrane is applied for separating a n-heptane/toluene mixture, the separation factor is found to be 4.6 at a PEBA concentration of 7%. Still, the flux value seems to have decreased from 96 g/(m² h) to 27 g/(m² h) with an increase in PEBA concentration. To increase the flux value, they do the permeation test at 80 °C and a high permeation value of 280 g/(m² h) is obtained, while at 40 °C, the flux value is only 65 g/(m² h). The fabrication of thermally crosslinked PEBA/ceramic membranes is shown in Figure 5.5.

Another method was reported for separating cyclohexane/benzene mixtures where PEBAX polymer (PEBAX 2533, 3533, 4033) is used. PEBAX reportedly shows good results due to its selectivity towards benzene [19]. As the hardness of the polymer increases, the separation factor value also increases, but the flux value decreases.

5.3.2   Poly(Methyl Methacrylate)-Based Polymeric Membrane

Poly(methyl methacrylate)s are another newly introduced branch of material that proved to be a good candidate for membrane matrix due to its high thermal, chemical as well as mechanical stability and wide compatibility range [1]. A reported analysis has shown that poly (methyl methacrylate-co-methacrylic acid) having a cross-linked structure results in a separation factor between 12 and 13, with a normalised flux value ranging from 24 to 28 kg μm/(m² h) [20]. Application of NBR/hydrin/PMMA membranes for benzene/cyclohexane separation has been reported [21], where NBR

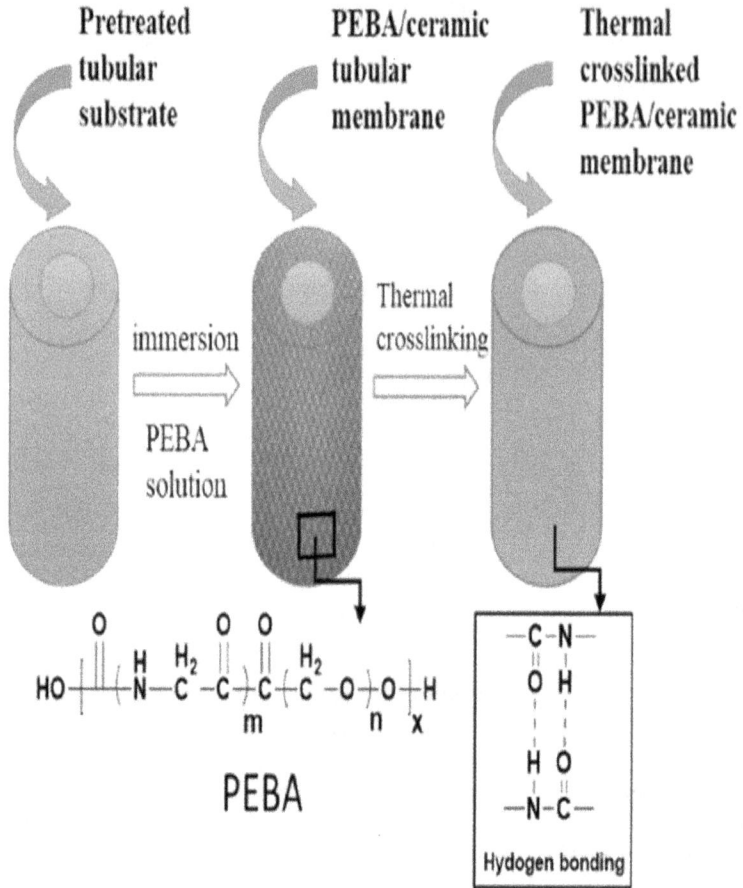

FIGURE 5.5   Schematic representation of PEBA/ceramic membrane.

and hydrin are used for controlling the solubility of the permeant, resulting in the prevention of membrane swelling. The prepared membrane in this report is more selective towards benzene since the polymer material PMMA has a similar solubility parameter to benzene than cyclohexane when the permeation experiment is done with a 50 wt.% mixture of benzene/cyclohexane through the membrane of composition NBR:hydrin:PMMA = 80:10:10 wt% permeation flux and separation factors are 160 g/m^2 h and 7.3, respectively. The permeation flux gradually increases with a decrease in the amount of NBR in the blend membrane, but the value of the separation factor remains almost the same. Another investigation is reported by changing the composition of a block copolymer, P(AN-b-MA) and its effect on membrane swelling properties and separation performance for a benzene/cyclohexane mixture [22]. As a result, they established that with the amount of benzene in the feed and increased in methacrylate content, benzene flux also increases while the separation factor subsequently decreases. A composite membrane obtained from plasma-induced polymerisation of GMA (glycidyl methacrylate) polymer grafted on a porous polyethylene substrate is prepared, and a thorough study of their morphological and pervaporation properties is done [23]. The GMA grafted membrane is applied for aromatic/non-aromatic separation, and the PV properties were investigated for the benzene/cyclohexane mixture [24]. In many literatures, the membrane often suffers from swelling at high temperatures. In order to control this temperature-mediated swelling, cross-linking membranes are introduced. Cross-linking between polymers containing compatible cross-linkable groups can result in the formation of covalent or other chemical bonds, depending on the reaction environment. Cross-linking is believed to be a counterfactor in membrane preparation that can connect the polymers to give a stable membrane form. Such an example of membrane preparation is reported by using a cyclodextrin molecule that has the ability to interact with the specific substrate through a host–guest mechanism [25]. Here, acrylate cyclodextrins are being used with polyethylene glycol dimethacrylate (PEG-DMA) polymer, which results in a cross-linked polymer; after performing toluene/cyclohexane separation at 60 °C, they get an outstanding selectivity of 14 for toluene, and permeability is found to be 0.3 kg/(mm^2 h) when toluene content in the feed is only 10 wt%.

5.3.3 POLYURETHANE-BASED POLYMERIC MEMBRANE

Polyurethanes (PUs) having different segments are also proven to be a perfect material for preparing pervaporation membranes. Polyurethane-based membranes are reported for separating aromatic/aliphatic component mixtures, although they can be a binary or multicomponent. A study of separation experiments reported that PU membranes show selectivity towards aromatics [26]. Another analysis is done for pervaporation-based benzene/cyclohexane separation, where a series of membranes are prepared with poly(oxytetramethylene) (PTMO)-based polyurethanes and utilised for the separation [27]. The chains of the polymer material are extended with chain extenders, such as aromatic diols of different types, for example, BHBP and HQE, which results in polyurethanes with varying morphologies of the domain. These synthesised membranes are applied for separating the benzene and cyclohexane, and the permselectivity for benzene is comparatively found to be better. A comparison study of the permselectivity and pervaporation flux of the synthesised TDI (2,4-tolylene diisocyanate)/PTMO membranes with chain extenders is done with neat copolymeric TDI/PTMO membranes, where the former one shows a better permselectivity. Still, a lower value of flux is obtained for the synthesised membrane than for the pure one. The hard segment of the membrane can control membrane swelling and is also responsible for variations in transport behaviour correlated with variations in composition in the feed mixture. It is generalised from the comparison study that in the case of both PU membranes, selectivity increases at the expense of permeability while the hard segment and soft segment lengths of the polymer chain increase and decrease, respectively [27]. The PTMO represents the soft segment in this reported membrane, while the TDI is responsible for the hard segment. Figure 5.6 depicts the chemical representation of the PU membrane.

According to another reported study, preparation of membranes with polyurethane (PU and polyurethaneimide (PUI) is done where they both have poly(ethylene adipate) diol as a soft segment but have varied hard segments; MDI-PMDA for PUI and

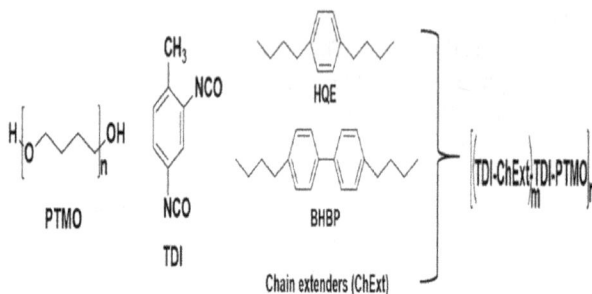

FIGURE 5.6 Chemical representation of PU membrane.

TDI-MDA for PU [28]. When these membranes are utilised in the separation of a 50 wt.% benzene/cyclohexane mixture, a separation factor of 6.29 is obtained with polyurethane membranes. At the same time, the permeation flux is found to be 264 g/ (m^2 h). In the case of PUI membrane, a separation factor of 8.25 is observed with a flux of 121 g/(m^2 h). When the performances and stability factors of both membranes were compared, the PUI membrane was found to be better, having a lesser tendency towards swelling than the PU membrane. In comparison to PUI, PU membranes showed a high flux vs. a low separation factor. This variation in swelling property can be explained by the weakening of hydrogen density between a hard and soft segment of the polymer, which led to the mobility of the softer segment composed of poly(ethylene adipate)diol.

5.3.4 POLYDIMETHYLSILOXANE-BASED POLYMERIC MEMBRANE

Polydimethylsiloxanes (PDMSs) are another material that is seen to be used in aromatic/non-aromatic separation, but a wide range of investigative reports are still needed [29]. For an increase in pervaporation performance, PDMS is often modified by cross-linking with suitable molecules. According to a study, microporous polyetherimide (PEI) was used as a support material for PDMS. It resulted in a composite membrane with a flat sheet structure used for the separation experiment with benzene/cyclohexane. The O-group in both PEI and PDMS tends to link together to form a tightly packed interface, and this cross-linking makes the membrane resistant to swelling. After cross-linking is introduced, an improvement in separation performance is also noticed. The average separation factor of 13.2 and flux of 218 g/(m^2 h) were found for the PDMS/PEI composite membrane. The stability of the membrane pervaporation performances remains intact until 180 h. Another approach for the modification of a PDMS-based composite membrane is done where polyacrylonitrile (PAN) is used with PDMS to prepare the composite [30]. Since structural stability is one of the main concerns in composite membranes due to the lack of compatibility between the monomers, poly(methylhydrosiloxane) (PMHS) is used in this report as a mediator for the improvement of membrane stability. PAN surface morphology and chemical structure were modified by plasma treatment with PMHS. After this treatment, an increase in interfacial adhesion was found. Due to this enhanced interfacial stability, a low value of swelling is obtained for the PDSM/PAN-PMHS membrane in toluene and n-heptane. Besides, this membrane is also proven to be a good choice for pervaporation performance since it gave a separation factor and flux value of 4.2 and 9.5 kg/(m^2 h) for methanol/toluene mixture, respectively. Researchers can work on stabilising the interfacial interaction of PDMS-based composite membranes in the near future with newer methodologies.

5.3.5 CELLULOSE-BASED POLYMERIC MEMBRANE

Cellulose is a good choice of material for controlling the swelling problem faced by most of the membranes in the pervaporation process due to its better durability towards various organic solvents [1]. Cellulose often fails in terms of high permeability value since it has a very low affinity towards both aromatics and aliphatic hydrocarbons; this made cellulose undergo modification through functionalisation to improve its affinity toward aromatic compounds. Such modified celluloses, such as benzoyl celluloses (BzCells), tosyl celluloses (TosCells) and butyryl celluloses (BuCells), are reported to be used for making cellulose membranes hydrophobic and applied for benzene/cyclohexane separation [31]. Specifically, herein, a series of cellulose alkyl esters are synthesised by varying the numbers of carbon atoms, and these membranes are utilised for benzene/cyclohexane separation where the feed mixture contains benzene of deficient concentration [31]. As a result, all these membranes show excellent selectivity for benzene/cyclohexane separation. The permeation rate is found to be enhanced as the number of carbons in the ester increases, but selectivity decreases. The high permeation value is due to the swelling of the membrane over time. There is a decrease in both solubility and diffusion selectivity, which leads to a total decrease in membrane selectivity. Another cellulose-based membrane is prepared by preparing blend membranes with varying compositions of sodium alginate (SA) and sodium carboxymethyl cellulose (CMC), where water is taken as solvent [32]. They found that increased CMC concentration in the blend solution also increased the flux value. An optimum selectivity and flux are obtained for a membrane having a composition of 25 wt.% SA and 75 wt.% CMC when applied to the benzene/cyclohexane mixture (19.6 wt.% benzene). Some polymers are structurally represented in Figure 5.7.

5.4 AROMATIC/ALIPHATIC SEPARATION BY HYBRID MEMBRANES

Polymers alone can be used for membrane preparation, but these membranes often seem to undergo swelling; furthermore, high selectivity and permeability cannot be achieved simultaneously through them. Therefore, additives are always needed to improve the shortcomings of a neat polymeric membrane. Hybrid membranes with a unique separation principle, which is the selective transport of the membrane through an absorptive separation, have become a perfect choice of material in separation studies [1,33]. Organic/inorganic hybrids are the most commonly used material in hybrid membrane preparation since they have various advantageous properties such as the flexibility and toughness of organic materials and anti-swelling; and the chemical,

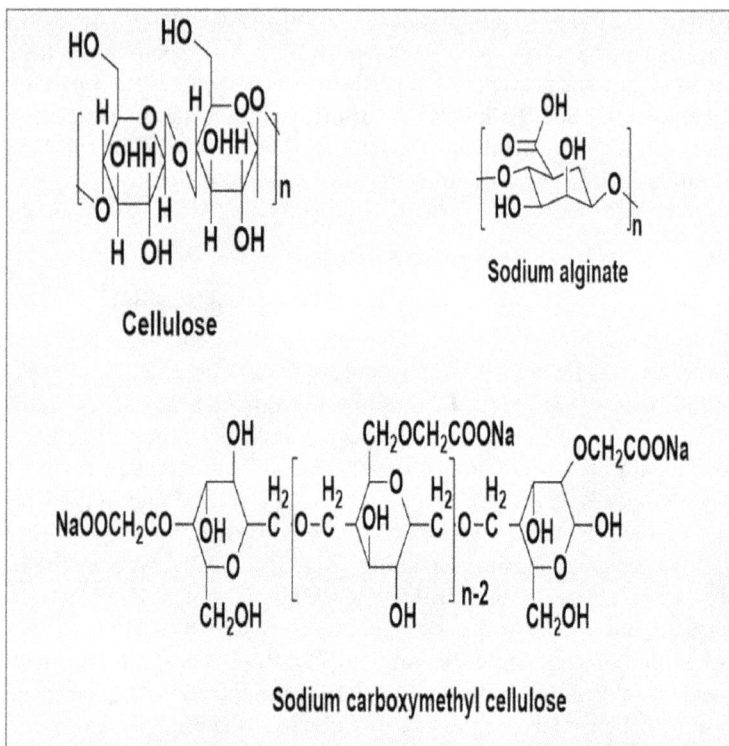

FIGURE 5.7 Structural representation of polymers.

mechanical and thermal stability of inorganic materials. Incorporating filler particles such as organic macromolecules, inorganic particles and metal–organic materials in a hybrid membrane is another novel approach to improving separation performances. Filler particles can interact specifically with one of the feed components in the mixture, mainly the aromatic one, resulting in facilitated transport. Introducing these particles is also responsible for increasing membrane properties, such as free volume and mass transfer channels, which improve the membrane's permeability. When the dispersion phase of hybrid membrane preparation is an organic macromolecule, it provides the advantages of compatibility with the polymeric matrix in addition to affinity for aromatics. PVA has proved to be an excellent membrane-forming material for many separation processes due to its polarity, hydrophilicity, cost-effectiveness and availability [34]. Cyclodextrin is a cyclic oligosaccharide with cavities of a hydrophobic nature. A study was reported where β-cyclodextrin (β-CD) is used with cross-linked PVA by incorporating it as a carrier and glutaraldehyde (GA) as a cross-linker, and an investigation on this prepared membrane on pervaporation-based benzene/cyclohexane separation was done. The result of this investigation is fruitful, as the β-CD/PVA/GA membrane showed high benzene permselectivity. They are incorporating β-CD into the PVA/GA membrane, which increases flux and separation factor values. Another study used acrylate cyclodextrin with PEG-DMA for the preparation of a

UV-mediated cross-linked copolymeric membrane to separate toluene/cyclohexane and found a highly selective separation of aromatics, i.e., toluene. Chitosan (CS) is another example of such a polysaccharide, which is cationic and used as a good membrane material. Due to its highly hydrophilic nature, chemical stability, availability, biodegradability, compatibility in membrane formation and, most importantly, high permeability because of the large free volume of its rigid structure, chitosan became one of the most common materials for preparing membranes [1]. Literature where the use of chitosan with PVA is reported to produce a blended membrane [35]. When applied to benzene/cyclohexane pervaporation separation, the resultant membrane showed a higher permeation value and a slight increase in benzene selectivity than the pure chitosan or pure PVA membrane.

On a similar note, the use of inorganic particles as a dispersion phase can also modify membrane properties in many ways, such as swelling prevention and mechanical, chemical and thermal stabilisations. Inorganic particles, including metal ions, zeolite molecular sieves, silicon dioxide and carbon materials, are used in literature to separate aromatic/aliphatic hydrocarbons. Due to their d-electrons availability, metal ions can interact with unsaturated compounds through d–π interaction, making metal ions a very useful mediator for separation. As reported, in situ polymerisation is done to prepare a hybrid membrane of AgCl/PMMA [36]. In the presence of surfactants dioctyl sodium succinate (AOT) and polyoxyethylene-polyoxypropylene-polyoxyethylene (F127), the size and morphological studies of AgCl are studied and a uniformly distributed nanoparticle of AgCl formation is found to have occurred in the PMMA membrane, which is due to the affinity of the surfactant towards PMMA. There is another reported study available where Ag nanoparticles (AgNPs) are being doped in a polymer matrix. Still, the hybrid membrane often faces problems with separation due to the ununiform distribution or aggregation of the nanoparticles on the membrane surface. To overcome this hurdle, many works have been done, one of which is the combination of Ag NPs with graphene oxide nanosheets (GONS), which can give uniformity in nanoparticle distribution [37]. Another work of GO-Ag NPS is reported, where composite membranes are fabricated through different reactant-induced impregnation reductions with a uniform Ag NPs distribution on the surface of GO and result in high separation efficiency [38]. Carbon-based materials are another commonly used inorganic material as a filler in hybrid membranes. Graphite, which has a similar structure to the benzene rings, can show preferential pervaporation separation of aromatics through σ and π bonds [1]. PVA blend membrane with the incorporation of carbon graphite crystalline flake is reported to produce CG-PVA membrane [39]. This membrane, in the benzene/cyclohexane separation experiment, showed a very high separation factor value and flux. With a benzene content of 50 wt.% in the feed, the highest separation factor and flux are 100.1 and 90.7 $g/(m^2 h)$ at 323 K. The reason may be due to the flexibility of graphite and the formation of defect-free voids at PVA and graphite interfaces, which somehow loosen the packing of the PVA chain and eventually, free volume in the membrane to increase. One of the main problems related to pervaporation membranes is stability. To improve this stability, many approaches are taken, one of which is the preparation of a pore-filling membrane where a PVA-GO nanohybrid layer is supported on the ultrafiltration support of asymmetric PAN [40]. With increased stability due to reduced membrane swelling, this

PVA-GO-PAN system also showed great affinity for aromatics, resulting in enhanced toluene/n-heptane separation. Besides graphite and its derivative, other carbon-based materials are available that are used for pervaporation separation of aliphatic/aromatic mixtures; they are carbon molecular sieves (CMS) and carbon nanotube-based hybrid membranes. Poly(vinyl alcohol)/carbon nanotube membrane (CNT) is prepared by dispersion of CNT with β-CD and applied for benzene/cyclohexane separation [41]. An excellent result with a separation factor of 41.2 is obtained, while the flux found is 61.0 g/(m^2 h) with the prepared membrane. Another modification to this PVA-CNT membrane is done by incorporating a multiwalled carbon nanotube (MWCNT) wrapped with chitosan, and separation factor and flux are found to be improved with values of 53.4 and 65.9 g/(m^2 h), respectively, for the benzene/cyclohexane mixture [42]. A recent study has reported that poly(ethylene-co-vinyl alcohol) (E-VOH) is used with carbon nanotube-filled poly (vinyl alcohol-co-ethylene) (E-VOH/CNT) as membrane material for pervaporation-based benzene/cyclohexane separation [43]. A comparison of the performances of both membranes was investigated, which proved that the incorporation of CNT filler made the membrane more capable in terms of permeability. However, the separation factor value did not improve. CMS is another carbon-based compound. It can be used as a filler material for the polymer matrix. An example is the incorporation of CMS into a PVA polymer matrix and the examination of its pervaporation performances for benzene/cyclohexane [44]. With an increase in CMS filling, the polymer chain's flexibility increases due to a decrease in hydrogen bonding interactions among polymer chains, thus producing a large free volume in the membrane structure. Due to relaxation in the polymer's chain packing, the sorption ability becomes higher, followed by an increase in permeation flux. The filling of the CMS should not be excessive. Otherwise, it may cause a retardation in the flux value. Other factors include the benzene content of the feed, the temperature at which the operation is done and the flow rate. While it increases, the flux value also increases, but the separation factor decreases. With and without filling of CMS, the membrane performances were studied, and a remarkable change in flux and separation factor values from 21.87 g/(m^2 h) and 16.7 to 59.25 g/(m^2 h) and 23.21 was obtained. Another filler material superior in terms of thermal, chemical and mechanical stability is a zeolite molecular sieve. Examples of using this filler as H-β-zeolite or Rh-loaded-H-β-zeolite were reported, which were incorporated into a polyvinyl chloride (PVP) polymer matrix and the prepared hybrid membrane was examined for benzene/cyclohexane mixture separation through pervaporation [45]. An improvement in pervaporation performances was found while comparing pure PVP with Rh-loaded-H-β-zeolite. With an enhanced Rh-loaded-H-β-zeolite concentration of 7%, the membrane showed optimum results with a separation factor of 26.44. However, a trade-off curve was observed with an increase in benzene content, where flux value increased while selectivity decreased. The decrease in separation factor value may be due to eventual membrane plasticisation or swelling. A lot of work is also reported with silane-based materials forming hybrid membranes. Since interaction between the interface of the polymer matrix and the inorganic particles is a major topic of concern, research is mostly directed at enhancing interfacial interaction. Introducing a silica network as a filler particle is only an outcome of this research. ɣ-(glycidyloxypropyl) trimethoxysilane (GPTMS) was reported to be used for hybrid membrane preparation where the polymer matrix is PVA, and the membrane is prepared through the sol–gel method [46]. With the enhancement of GPTMS content, membrane-free volume also increased; as a result, permeation flux, as well as separation factor value for the benzene/cyclohexane mixture rose from 20.3 g/(m² h) and 9.6 to 137.1 g/(m² h) and 46.9, respectively. Even if the impregnation of inorganic fillers gives a better separation performance, there is still room for improvement through research, such as control of particle size, agglomeration, compatibility with the polymer, leaching and work on increasing particle size loading in the polymer. Optimising all these factors may result in efficient separation through the membrane.

Another class of attractive porous material, a metal–organic framework (MOF) with both organic and inorganic fractions in its structure, has gained massive attention from researchers in the field of membrane preparation. MOFs are an excellent choice of material for the preparation of membranes in aromatic/aliphatic mixture separation because they contain metal ions in their structures that can interact with aromatics via d–π conjugation. In addition to that, there is a possibility of π–π conjugation between organic linkers of MOF and aromatics in feed [1]. In addition to these advantages, the compatibility of MOF with the polymer matrix made it more favourable to incorporate MOF in the membrane as filler, which can increase flux value as well as the separation factor of an aromatic/aliphatic mixture. As reported, a ceramic substrate is used on which the fabrication of a nanohybrid membrane of MOF-based $Cu_3(BTC)_2$/PVA is done by following a pressure-driven assembly method [47]. When this membrane is subjected to toluene/n-heptane mixture (50 wt.%) separation, an outstanding result is obtained with flux and separation factor values of 133 g/(m² h) and 17.9, respectively. Another MOF reported is Co(II)-formate ($Co(HCOO)_2$), which was used as a carrier for preferential aromatic hydrocarbon permeation [48]. A solvothermal method is used to synthesise ($Co(HCOO)_2$) MOFs of varied particle sizes, followed by doping these particles in a PEBA polymer matrix. Finally, after the deposition of this hybrid on a ceramic substrate of tubular shape, it was applied for toluene/n-heptane separation. For a feed solution with 10 wt.% toluene, the membrane's flux value and separation factor were found to be 771 g/(m² h) and 5.1, respectively.

Although these MOF particles are proven to be effective in terms of compatibility with polymer matrix and performance, they still suffer from problems related to particle dispersions, and often, agglomerations occur on the membrane surface. To face this problem of MOF dispersion, a novel material called metal–organic polyhedra (MOPs) can be used as a good substitute. In one of the reports on the use of MOPs, the hyperbranched amphiphilic dendritic polymer Boltorn W3000 was used as a polymer matrix. At the same time, the fillers are porous nanocages of metal–organic molecules. When this fabricated membrane was subjected to pervaporation separation of aromatic/aliphatic mixtures, an excellent result was obtained. Also, the membrane is well formed and has good properties, which can be attributed to the excellent dispersion of the metal–organic molecule-based filler due to its better compatibility with the polymer. Another method involved functionalising and incorporating a series of MOPs into a hybrid membrane of hyperbranched polymer. These MOPs are isostructural with uniform size/shape and are soluble in solvents. These characteristics allow for uniform dispersion in the polymer matrix. The pervaporation performances of MOP-SO$_3$NanHm incorporated membranes were studied for

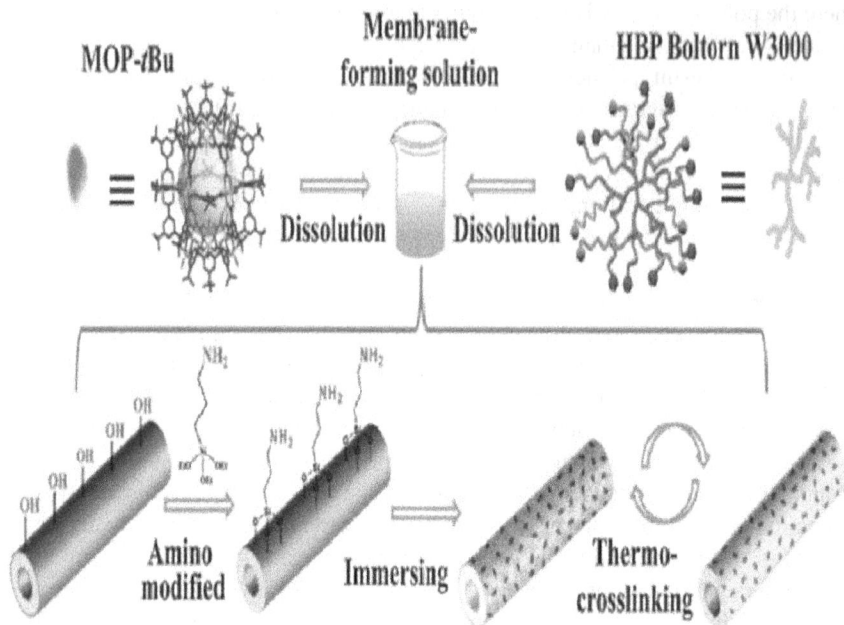

FIGURE 5.8   Fabrication of MOPs-tBu/W3000 membrane onto a tubular support.

Source: [49].

both benzene/cyclohexane and toluene/n-heptane mixtures. For toluene/n-heptane (50 wt.% mixture) separation, flux and separation factor were found to be 8.03 and 528 g/(m^2 h), while for benzene and cyclohexane mixture, these values were found to be 8.4 and 540 g/(m^2 h). The functional groups in MOPs induced polarity, which in turn affected the membrane's selectivity towards aromatics. However, it is challenging to get a pure MOP membrane that takes the utmost advantage of its nanocage. A very recent study has reported a hybrid membrane made of [Cu$_{24}$(5-tBu-1,3-BDC)$_{24}$(S)$_{24}$] (MOP-tBu) when supported on tubular ceramic support that is assisted by a nano-array of Co(OH)$_2$, a good pervaporation separation result was found for a toluene/n-heptane mixture [49]. MOPs-tBu molecules in the membrane are responsible for the formation of an ultrathin defect-free layer, while the presence of Cu^{2+} and benzene rings in the membrane matrix showed affinity towards aromatic compounds. When this hybrid membrane is applied for the separation of toluene/ n-heptane hydrocarbons (50 wt.% mixture), they find flux and separation factors of 800 g/(m^2 h) and 5.4, respectively. The Fabrication technique of MOPs-tBu/W3000 membrane onto a tubular support is shown in Figure 5.8.

5.5   CONCLUSION

The aromatic/aliphatic separation performances with the help of membranes have been studied in a wide range but still face a trade-off phenomenon between selectivity and permeability. To meet the requirements of the industrial sector, more modifications in membrane technology should be implemented. To overcome the trade-off barrier, one needs to choose a promising material for the membrane that exhibits high permselectivity for aromatics. In pervaporation, a solution diffusion mechanism is followed, where the separation performance is significantly dependent on the selective adsorption of aromatics by the membrane material. In the field of aromatic/aliphatic separation, the polymer is still a dominant membrane material. To improve the performance by facilitating transport, more efficient carrier molecules, such as MOFs, COFs, PAFs, MOPs, etc., can be designed and incorporated into a polymer matrix. These porous materials and their hybrids have inner cavities, which may help increase the permselectivity of the membrane. The membrane preparation method can also be modified to make it simple and effective while maintaining the membrane microstructure. In aromatic/aliphatic mixture separation, swelling of the membrane is frequently witnessed, which decreases the stability as well as the performance of the membranes. This problem of swelling can be decreased by controlling the microstructure of the membrane. Molecular modelling and simulation studies on the transport mechanism of aromatic and aliphatic compounds through various membranes can help properly understand flow properties. Most importantly, it will help screen the appropriate materials for the membrane.

ABBREVIATIONS

6FDA	4,4'-(Hexafluoroisopropylidene)diphthalic anhydride
DABA	3,5-Diaminobenzoic acid
BPDA	3,3',4,4'-Biphenyltetracarboxylic dianhydride
TrMPD	2,4,6-Trimethyl-1,3-phenylenediamine
4MPD	2,3,5,6-Tetramethyl-1,4-phenylene diamine
APAF	2,2-Bis(3-amino-4-hydroxyphenyl)-hexafluoropropane
BDAF	2,2-Bis[4-(4-aminophenoxy)phenyl]-hexafluoropropane
DSDA	Dipenyl sulphone tetracarboxylic dianhydride
NBR	Nitrile butadiene rubber
PMMA	Poly(methyl methacrylate)
BTAPPI	Bis-2,2'-[4-{2'-trifluoromethyl 4'-(4''-aminophenyl)phenoxy}phenyl] isopropylidene
BTAPPHI	Bis-2,2'-[4-{2'-trifluoromethyl4'-(4''-aminophenyl)phenoxy}phenyl] hexafluoro isopropylidene
BTAPPF	Bis-2,2'-[4-{2'-trifluo-romethyl-4'-(4''-amino phenyl)phenoxy}phenyl] fluorenylidene
BTAPPPI	3,3-Bis-[4-{2'-trifluoromethyl4'-(4''-aminophenyl)phenoxy} phenyl]-2-phenyl-2,3-dihydro-isoindole-1-one
TA	Terephthalic acid
IA	Isophthalic acid
P(AN-b-MA)	Polyacrylonitrile-block-poly(methyl acrylate)
PV	Pervaporation

REFERENCES

1. Liu, H. X., Wang, N., Zhao, C., Ji, S., & Li, J. R. (2018). Membrane materials in the pervaporation separation of aromatic/aliphatic hydrocarbon mixtures: A review. *Chinese Journal of Chemical Engineering, 26*(1), 1–16. https://doi.org/10.1016/S0376-7388(01)00560-9

2. Katarzynski, D., & Staudt-Bickel, C. (2006). Separation of multi component aromatic/aliphatic mixtures by pervaporation with copolyimide membranes. *Desalination, 189*(1–3), 81–86. https://doi.org/10.1016/j.desal.2005.06.015

3. Meindersma, G. W., Onink, S. A. F., Hansmeier, A. R., & de Haan, A. B. (2010). Ionic liquids as sustainable extractants in petrochemicals processing. In 2010 AIChE Annual Meeting.

4. Rölling, P., Lamers, M., & Staudt, C. (2010). Cross-linked membranes based on acrylated cyclodextrins and polyethylene glycol dimethacrylates for aromatic/aliphatic separation. *Journal of Membrane Science, 362*(1–2), 154–163. https://doi.org/10.1016/j.memsci.2010.06.036

5. Pithan, F., Staudt-Bickel, C., Hess, S., & Lichtenthaler, R. N. (2002). Polymeric membranes for aromatic/aliphatic separation processes. *ChemPhysChem, 3*(10), 856–862. https://doi.org/10.1002/1439-7641(20021018)3:10%3C856::AID-CPHC 856%3E3.0.CO;2-H

6. Liu, L., Wang, N., Liu, H. X., Shu, L., Xie, Y. B., Li, J. R., & An, Q. F. (2019). Nano-array assisted metal-organic polyhedra membranes for the pervaporation of aromatic/aliphatic mixtures. *Journal of Membrane Science, 575*, 1–8. https://doi.org/10.1016/j.memsci.2018.12.081

7. Villaluenga, J. G., & Tabe-Mohammadi, A. (2000). A review on the separation of benzene/cyclohexane mixtures by pervaporation processes. *Journal of Membrane Science, 169*(2), 159–174. https://doi.org/10.1016/S0376-7388(99)00337-3

8. Jonquieres, A., Clément, R., & Lochon, P. (2002). Permeability of block copolymers to vapors and liquids. *Progress in Polymer Science, 27*(9), 1803–1877. https://doi.org/10.1016/S0079-6700(02)00024-2

9. Chapman, P. D., Oliveira, T., Livingston, A. G., & Li, K. (2008). Membranes for the dehydration of solvents by pervaporation. *Journal of Membrane Science, 318*(1–2), 5–37. https://doi.org/10.1016/j.memsci.2008.02.061

10. Okamoto, K. I., Wang, H., Ijyuin, T., Fujiwara, S., Tanaka, K., & Kita, H. (1999). Pervaporation of aromatic/non-aromatic hydrocarbon mixtures through crosslinked membranes of polyimide with pendant phosphonate ester groups. *Journal of Membrane Science, 157*(1), 97–105. https://doi.org/10.1016/S0376-7388(98)00363-9

11. Ribeiro, C. P., Freeman, B. D., Kalika, D. S., & Kalakkunnath, S. (2012). Aromatic polyimide and polybenzoxazole membranes for the fractionation of aromatic/aliphatic hydrocarbons by pervaporation. *Journal of Membrane Science, 390*, 182–193. https://doi.org/10.1016/j.memsci.2011.11.042

12. Ye, H., Li, J., Lin, Y., Chen, J., & Chen, C. (2008). Synthesis of polyimides containing fluorine and their pervaporation performances to aromatic/aliphatic hydrocarbon mixtures. *Journal of Macromolecular Science, Part A: Pure and Applied Chemistry, 45*(2), 172–178. https://doi.org/10.1080/10601320701786984

13. Ribeiro Jr, C. P., Freeman, B. D., Kalika, D. S., & Kalakkunnath, S. (2013). Pervaporative separation of aromatic/aliphatic mixtures with poly (siloxane-co-imide) and poly (ether-co-imide) membranes. *Industrial & Engineering Chemistry Research, 52*(26), 8906–8916. https://doi.org/10.1021/ie302344z

14. Katarzynski, D., & Staudt, C. (2010). Temperature-dependent separation of naph-thalene/n-decane mixtures using 6FDA–DABA-copolyimide membranes. *Journal of Membrane Science*, *348*(1–2), 84–90. https://doi.org/10.1016/j.memsci.2009.10.043

15. Shao, P., & Huang, R. Y. M. (2007). Polymeric membrane pervaporation. *Journal of Membrane Science*, *287*(2), 162–179. https://doi.org/10.1016/j.memsci.2006.10.043

16. Yang, L., Kang, Y., Wang, Y., Xu, L., Kita, H., & Okamoto, K. I. (2005). Synthesis of crown ether-containing copolyimides and their pervaporation properties to benzene/cyclohexane mixtures. *Journal of Membrane Science*, *249*(1–2), 33–39. https://doi.org/10.1016/j.memsci.2004.08.029

17. Maji, S., & Banerjee, S. (2010). Preparation of new semifluorinated aromatic poly(ether amide)s and evaluation of pervaporation performance for benzene/cyclohexane 50/50 mixture. *Journal of Membrane Science*, *349*(1–2), 145–155. https://doi.org/10.1016/j.memsci.2009.11.042

18. Wu, T., Wang, N., Li, J., Wang, L., Zhang, W., Zhang, G., & Ji, S. (2015). Tubular thermal crosslinked-PEBA/ceramic membrane for aromatic/aliphatic pervaporation. *Journal of Membrane Science*, *486*, 1–9. https://doi.org/10.1016/j.memsci.2015.03.037

19. Yildirim, A. E., Hilmioglu, N. D., & Tulbentci, S. (2008). Separation of benzene/cyclohexane mixtures by pervaporation using PEBA membranes. *Desalination*, *219*(1–3), 14–25. https://doi.org/10.1016/j.desal.2007.02.031

20. Matsui, S., & Paul, D. R. (2002). Pervaporation separation of aromatic/aliphatic hydrocarbons by crosslinked poly (methyl acrylate-co-acrylic acid) membranes. *Journal of Membrane Science*, *195*(2), 229–245. https://doi.org/10.1016/S0376-7388(01)00560-9

21. Okeowo, O., Nam, S. Y., & Dorgan, J. R. (2008). Nonequilibrium nanoblend membranes for the pervaporation of benzene/cyclohexane mixtures. *Journal of Applied Polymer Science*, *108*(5), 2917–2922. https://doi.org/10.1002/app.27749

22. An, Q. F., Qian, J. W., Zhao, Q., & Gao, C. J. (2008). Polyacrylonitrile-block-poly(methyl acrylate) membranes 2: Swelling behavior and pervaporation performance for separating benzene/cyclohexane. *Journal of Membrane Science*, *313*(1–2), 60–67. https://doi.org/10.1016/j.memsci.2007.12.073

23. Wang, H., Lin, X., Tanaka, K., Kita, H., & Okamoto, K. I. (1998). Preparation of plasma-grafted polymer membranes and their morphology and pervaporation properties toward benzene/cyclohexane mixtures. *Journal of Polymer Science Part A: Polymer Chemistry*, *36*(13), 2247–2259. https://doi.org/10.1002/(SICI)1099-0518(19980930)36:13%3C2247::AID-POLA11%3E3.0.CO;2-A

24. Okamoto, K. I., Wang, H., Ijyuin, T., Fujiwara, S., Tanaka, K., & Kita, H. (1999). Pervaporation of aromatic/non-aromatic hydrocarbon mixtures through crosslinked membranes of polyimide with pendant phosphonate ester groups. *Journal of Membrane Science*, *157*(1), 97–105. https://doi.org/10.1016/S0376-7388(98)00363-9

25. Rölling, P., Lamers, M., & Staudt, C. (2010). Cross-linked membranes based on acrylated cyclodextrins and polyethylene glycol dimethacrylates for aromatic/aliphatic separation. *Journal of Membrane Science*, *362*(1–2), 154–163. https://doi.org/10.1016/j.memsci.2010.06.036

26. Cunha, V. S., Paredes, M. L. L., Borges, C. P., Habert, A. C., & Nobrega, R. (2002). Removal of aromatics from multicomponent organic mixtures by pervaporation using polyurethane membranes: Experimental and modeling. *Journal of Membrane Science*, *206*(1–2), 277–290. https://doi.org/10.1016/S0376-7388(01)00776-1

27. Wolińska-Grabczyk, A. (2006). Effect of the hard segment domains on the permeation and separation ability of the polyurethane-based membranes in benzene/cyclohexane separation by pervaporation. *Journal of Membrane Science, 282*(1–2), 225–236. https://doi.org/10.1016/j.memsci.2006.05.026

28. Ye, H., Li, J., Lin, Y., Chen, J., & Chen, C. (2008). Preparation and pervaporation performances of PEA-based polyurethaneurea and polyurethaneimide membranes to benzene/cyclohexane mixture. *Journal of Macromolecular Science, Part A: Pure and Applied Chemistry, 45*(7), 563–571. https://doi.org/10.1080/10601320802100697

29. Chen, J., Li, J., Lin, Y., & Chen, C. (2009). Pervaporation performance of polydimethylsiloxane membranes for separation of benzene/cyclohexane mixtures. *Journal of Applied Polymer Science, 112*(4), 2425–2433. https://doi.org/10.1002/app.29799

30. Zhou, H., Su, Y., Chen, X., Luo, J., Tan, S., & Wan, Y. (2016). Plasma modification of substrate with poly (methylhydrosiloxane) for enhancing the interfacial stability of PDMS/PAN composite membrane. *Journal of Membrane Science, 520*, 779–789. https://doi.org/10.1016/j.memsci.2016.08.039

31. Uragami, T., Tsukamoto, K., & Miyata, T. (2005). Permeation and separation characteristics of a mixture of benzene/cyclohexane through cellulose alkyl ester membranes during pervaporation. *Macromolecular Chemistry and Physics, 206*(6), 642–648. https://doi.org/10.1002/macp.200400488

32. Kuila, S. B., & Ray, S. K. (2014). Separation of benzene–cyclohexane mixtures by filled blend membranes of carboxymethyl cellulose and sodium alginate. *Separation and Purification Technology, 123*, 45–52. http://dx.doi.org/10.1016/j.seppur.2013.12.017

33. Souza, V. C., & Quadri, M. G. N. (2013). Organic-inorganic hybrid membranes in separation processes: A 10-year review. *Brazilian Journal of Chemical Engineering, 30*, 683–700. https://doi.org/10.1590/S0104-66322013000400001

34. Peng, F., Jiang, Z., Hu, C., Wang, Y., Lu, L., & Wu, H. (2006). Pervaporation of benzene/cyclohexane mixtures through poly(vinyl alcohol) membranes with and without β-cyclodextrin. *Desalination, 193*(1–3), 182–192. https://doi.org/10.1016/j.desal.2005.04.141

35. Lu, L., Peng, F., Jiang, Z., & Wang, J. (2006). Poly (vinyl alcohol)/chitosan blend membranes for pervaporation of benzene/cyclohexane mixtures. *Journal of Applied Polymer Science, 101*(1), 167–173. http://dx.doi.org/10.1002/1521-4125(200103)24:3%3C275::AID-CEAT275%3E3.0.CO;2-0

36. Wu, L. G., Wang, T., & Xiang, W. (2013). Regulation of AgCl in reverse microemulsion and its effect on the performance of AgCl/PEO–PPO–PEO/PMMA hybrid membranes. *Composites Science and Technology, 80*, 8–15. https://doi.org/10.1016/j.compscitech.2013.02.023

37. Mahmoudi, E., Ng, L. Y., Ba-Abbad, M. M., & Mohammad, A. W. (2015). Novel nanohybrid polysulfone membrane embedded with silver nanoparticles on graphene oxide nanoplates. *Chemical Engineering Journal, 277*, 1–10. https://doi.org/10.1016/j.cej.2015.04.107

38. Dai, S. Q., Jiang, Y. Y., Wang, T., Wu, L. G., Yu, X. Y., & Lin, J. Z. (2016). Enhanced performance of polyimide hybrid membranes for benzene separation by incorporating three-dimensional silver–graphene oxide. *Journal of Colloid and Interface Science, 478*, 145–154. https://doi.org/10.1016/j.jcis.2016.06.009

39. Peng, F., Lu, L., Hu, C., Wu, H., & Jiang, Z. (2005). Significant increase of permeation flux and selectivity of poly (vinyl alcohol) membranes by incorporation of

crystalline flake graphite. *Journal of Membrane Science*, *259*(1–2), 65–73. https://doi.org/10.1016/j.memsci.2005.03.014

40. Wang, N., Ji, S., Li, J., Zhang, R., & Zhang, G. (2014). Poly (vinyl alcohol)–graphene oxide nanohybrid "pore-filling" membrane for pervaporation of toluene/n-heptane mixtures. *Journal of Membrane Science*, *455*, 113–120. https://doi.org/10.1016/j.memsci.2013.12.023

41. Peng, F., Hu, C., & Jiang, Z. (2007). Novel ploy(vinyl alcohol)/carbon nanotube hybrid membranes for pervaporation separation of benzene/cyclohexane mixtures. *Journal of Membrane Science*, *297*(1–2), 236–242. https://doi.org/10.1016/j.memsci.2007.03.048

42. Peng, F., Pan, F., Sun, H., Lu, L., & Jiang, Z. (2007). Novel nanocomposite pervaporation membranes composed of poly (vinyl alcohol) and chitosan-wrapped carbon nanotube. *Journal of Membrane Science*, *300*(1–2), 13–19. https://doi.org/10.1016/j.memsci.2007.06.008

43. Zahlan, H., Saeed, W. S., Alqahtani, S., & Aouak, T. (2020). Separation of Benzene/cyclohexane mixtures by pervaporation using poly(ethylene-co-vinylalcohol) and carbon nanotube-filled poly(vinyl alcohol-co-ethylene) membranes. *Separations*, *7*(4), 68. https://doi.org/10.3390/separations7040068

44. Sun, H., Lu, L., Peng, F., Wu, H., & Jiang, Z. (2006). Pervaporation of benzene/cyclohexane mixtures through CMS-filled poly (vinyl alcohol) membranes. *Separation and Purification Technology*, *52*(2), 203–208. https://doi.org/10.1016/j.seppur.2006.04.002

45. Zhang, X., Qian, L., Wang, H., Zhong, W., & Du, Q. (2008). Pervaporation of benzene/cyclohexane mixtures through rhodium-loaded β-zeolite-filled polyvinyl chloride hybrid membranes. *Separation and Purification Technology*, *63*(2), 434–443. https://doi.org/10.1016/j.seppur.2008.05.028

46. Peng, F., Lu, L., Sun, H., Wang, Y., Wu, H., & Jiang, Z. (2006). Correlations between free volume characteristics and pervaporation permeability of novel PVA–GPTMS hybrid membranes. *Journal of Membrane Science*, *275*(1–2), 97–104. http://dx.doi.org/10.1016/j.memsci.2005.09.008

47. Zhang, Y., Wang, N., Ji, S., Zhang, R., Zhao, C., & Li, J. R. (2015). Metal–organic framework/poly(vinyl alcohol) nanohybrid membrane for the pervaporation of toluene/n-heptane mixtures. *Journal of Membrane Science*, *489*, 144–152. https://doi.org/10.1016/j.memsci.2015.04.012

48. Zhang, Y., Wang, N., Zhao, C., Wang, L., Ji, S., & Li, J. R. (2016). Co (HCOO) 2-based hybrid membranes for the pervaporation separation of aromatic/aliphatic hydrocarbon mixtures. *Journal of Membrane Science*, *520*, 646–656. https://doi.org/10.1016/j.memsci.2016.08.028

49. Zhao, C., Wang, N., Wang, L., Huang, H., Zhang, R., Yang, F., … & Li, J. R. (2014). Hybrid membranes of metal–organic molecule nanocages for aromatic/aliphatic hydrocarbon separation by pervaporation. *Chemical Communications*, *50*(90), 13921–13923. https://doi.org/10.1039/C4CC05279J

50. Kung, G., Jiang, L. Y., Wang, Y., & Chung, T. S. (2010). Asymmetric hollow fibers by polyimide and polybenzimidazole blends for toluene/iso-octane separation. *Journal of Membrane Science*, *360*(1–2), 303–314. https://doi.org/10.1016/j.memsci.2010.05.030

51. Maji, S., & Banerjee, S. (2010). Synthesis and characterization of new meta connecting semifluorinated poly(ether amide)s and their pervaporation properties for benzene/cyclohexane mixtures. *Journal of Membrane Science*, *360*(1–2), 380–388. https://doi.org/10.1016/j.memsci.2010.05.035

52. Zhao, C., Wang, N., Wang, L., Sheng, S., Fan, H., Yang, F., … & Yu, J. (2016). Functionalized metal–organic polyhedra hybrid membranes for aromatic hydrocarbons recovery. *AIChE Journal*, *62*(10), 3706–3716. https://doi.org/10.1002/aic.15263

53. Liu, L., Wang, N., Liu, H. X., Shu, L., Xie, Y. B., Li, J. R., & An, Q. F. (2019). Nano-array assisted metal-organic polyhedra membranes for the pervaporation of aromatic/aliphatic mixtures. *Journal of Membrane Science*, *575*, 1–8. https://doi.org/10.1016/j.memsci.2018.12.081

6 Produced Water Characteristics in Petroleum Industries

Sanjay Bhutani, Nikita Yadav, Ankush Thakur,
Amrendra Bhushan and Sudeepta Baruah

6.1 INTRODUCTION

The physicochemical characteristics of produced water are influenced by factors such as the geologic age, long-term geological transformation with which it has interacted for thousands of years, location of the field, depth, the type of organic chemicals made up of carbon and hydrogen being generated and geochemistry of the hydrocarbon-bearing formation, including heavy oil, medium oil, light oil, lean gas and rich gas [1,2]. Additionally, the qualities of the water produced are influenced by the chemical constitution of the gas and oil parts in the reservoir (Figures 6.1 and 6.2).

Produced water is a substantial wastewater stream generated during gas and oil extraction. This complex consists of dissolved and particulate inorganic and organic compounds, primarily salts, minerals and oils [3]. With the growth of oil and gas investigation and production, particularly from unconventional sources like shale oil and gas reservoirs, the volume of this effluent output is increasing globally, posing challenges for environmentally safe disposal [2]. Throughout a reservoir's projected lifetime, the characteristics and quantity of water that are produced is changed, with water production starting slowly and gradually increasing as the reservoir ages [2]. Region-specific studies are necessary to address the environmental concerns associated with produced water discharge, as the two samples are not identical [3].

The constitution of the produced water can differ even within the same field and between wells. Produced or generated water from the production or generation of oil and natural gas does not have a definite volume since it relies on the extraction method and the location. Around 667 million metric tonnes of generated or produced water were estimated to have been released the offshore worldwide (Figure 6.3). The source of the water (natural gas or crude oil) and the amount of water produced depend on various factors. In most instances, produced water is deoxygenated and contains both organic and inorganic substances, most of which comprise oleic acid salts and hydrocarbons that could contaminate the ecosystem upon discharge [4]. For example, total dissolved solids (TDS) concentrations in produced water may range from 500 to 600 ppm to over 100,000 ppm for natural gas from coal bed methane (CBM); also, the oil may be naturally dispersed in water [5].

DOI: 10.1201/9781003441359-6

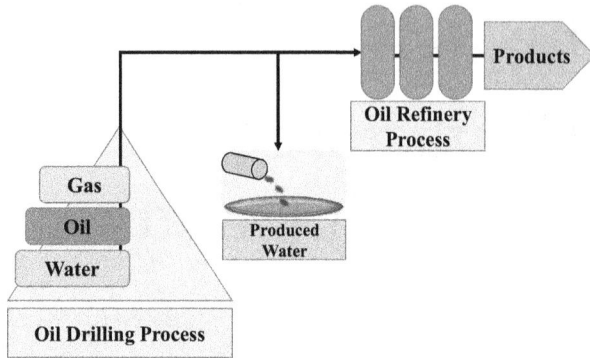

FIGURE 6.1   Process of formation of produced water.

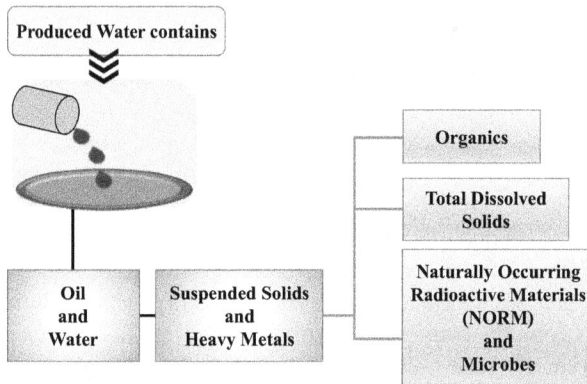

FIGURE 6.2   Constituents of produced water.

6.2   DISSOLVED AND DISPERSED OIL COMPOUNDS

6.2.1   DISSOLVED OR SOLUBLE ORGANIC COMPOUNDS

The number of dissolved hydrocarbons in produced or generated water rises due to the significant polar component of deep-sea crude oil. Factors such as pH and temperature can impact the soluble nature of organic molecules [6]. Naturally, hydrocarbons present in generated water are comprised of volatile hydrocarbons, polycyclic aromatic hydrocarbons (PAHs), organic acids and phenols. Although the individual toxicities of these hydrocarbons might not be significant, their combined toxicities can result in aquatic toxicity. These hydrocarbons likely add up to the toxic nature of produced water [7].

Produced water often contains soluble organics that are challenging to remove and discharged into the ocean or injected into the reservoir at onshore areas. Generally, when a compound's molecular weight falls, the concentration of organic chemicals in produced water rises. The effectiveness of the oil/water separation process has less

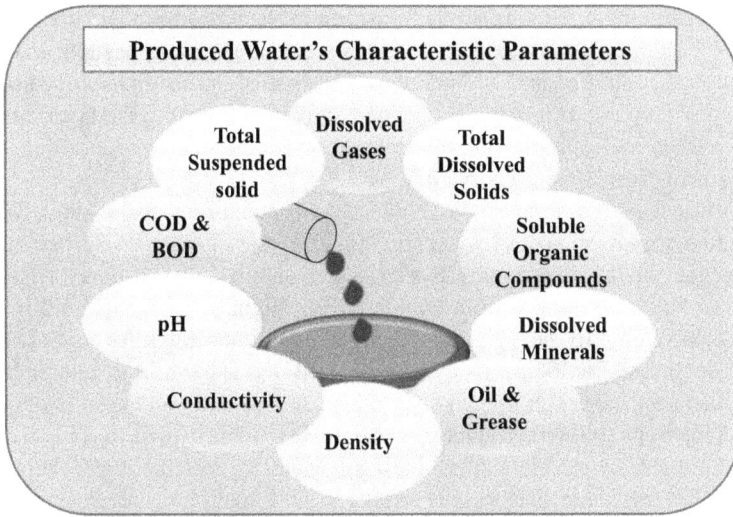

FIGURE 6.3   Characteristic parameters of produced water.

impact on lower molecular weight chemicals (naphthalene and BTEX) than on larger molecular weight PAHs [8], and they are not detected by the analytical method of oil and grease. Hydrocarbons that are volatile in nature are present in generated water, often found in greater amount in generated or produced water from gas condensate platforms than that in generated or produced water from the oil fields [8].

Hydroxybenzene or phenols, also known as phenolics, are aromatic organic compounds with an aromatic hydrocarbon group linked to one or more hydroxyl groups. There are different amounts of hydroxybenzene or phenols in the produced water from gas and oil wells [9], but the production of gas condensate was reported to have greater concentration. Gas field-produced water contains greater concentrations of phenol than oil field-produced water, based on the comparability of the amount of phenol in produced water from gas and oil fields. Besides, an investigation done by Neff and Johnsen and their team to evaluate the amount of phenol in produced water obtained from the Norwegian Region of the North Sea and the Louisiana Gulf Coast discovered that the maximum amount of phenols in produced water was in the range of 2.1–4.5 ppm and 0.36–16.8 ppm accordingly [10,11].

Methanol, acetic and propionic acids, as well as lesser molecular weight (C2–C5) carboxylic acids (fatty acids), alcohols, ketones, may be present in produced water. Certain highly dissolved organic compounds, such as acetone (up to 5000 ppm), are detected in produced waters. Despite their high solubility, they do not significantly contribute to grease and oil estimation, as the used organic solvent for grease and oil estimation cannot effectively extract them [12].

Slightly dissolved compounds include hydrocarbons of moderate to high-molecular weight (C6–C15). Although soluble in water at lesser concentrations, compared to hydrocarbons with lower molecular weights, they are less soluble. Since they are difficult to remove from produced water, they are often disposed of straight into the ocean. These compounds furnish to the sheen production, but their toxicity is the main concerning issue. Some of these constituents incorporate phenols, carboxylic acids, aliphatic and aromatic hydrocarbons. Aromatic hydrocarbons are compounds with C and H organised in a cyclic manner, similar to benzene. PAHs are remarked as hydrocarbon molecules having multiple cyclic rings and are produced under high pressure from organic material in crude oil.

Naphthalene is the simplest PAH having two connected cyclic rings (benzene rings), often appears in greater concentrations than that of other PAHs. For example, in Norwegian fields, offshore-produced water contains 95% or more of all PAHs. PAH water solubility ranges from comparatively 'light' compounds with moderate water-soluble nature to 'heavy' compounds with high liposolubility and less soluble in water. Increasing biological oxygen demand (BOD) is harmful to aquatic life, and it is a reason for cancer in humans and other animals. All can cause mutation and also be dangerous to the reproductive system. Heavy PAHs, like those found at the bottom, have strong organic matter-binding capabilities contributing to their persistence [13]. Greater molecular weight PAHs are more likely to be detected in or near dispersed oil due to their lower solubility in water. Alkylated phenols and aromatic hydrocarbons may be the two primary toxicants [14]. Alkyl groups bonded to phenols are believed to be endocrine-disrupting chemicals and may negatively impact fertility [14]. However, microorganisms and photo-oxidation can rapidly degrade phenols and alkyl phenols in seawater and marine sediments [15].

6.2.2 DISPERSED OIL

Oil is a significant ejected pollutant, since it may have harmful effects close to the discharge point. In the aqueous phase, tiny droplets of dispersed oil are suspended. Oil contamination and sediment deposition could happen if the dispersed oil interacts with the ocean floor, which could disrupt the benthic fauna. Dissipated oils can create sheen and increase BOD near the mixing zone as they rise to the surface and disperse [15]. The quantity of dissipated oil generated or produced by water relies on various factors, including the density of oil, friction between water and oil phases, variety and effectiveness of the use of chemicals, the size and type and also the efficacy of physical separative equipment [12]. Soluble organics and the use of chemicals in generated water fall down the interfacial tension between water and oil. The accumulation cycle may be impacted by water movement brought on by vertical mixing, currents, tides and waves. Moreover, since most recent approaches cannot eradicate particles smaller than 10 μm and precipitated droplets are typically between 4 and 6 μm in size, the small droplets can impede water treatment methods [16].

6.3 TOTAL SUSPENDED SOLID (TSS), TOTAL ORGANIC COUNT (TOC) AND TOTAL NITROGEN (TN)

Total suspended solids (TSS) in produced water can refer to any floating or drifting substances present in water, comprising silt, sand, sediment, plankton and algae. It is been reported that produced water has TSS concentrations between 8–5484 mg/

l (ppm) and 14–800 mg/l (ppm). Further research by Tibbetts revealed that the TSS concentration in oilfield-produced water ranged from 1.2 to 1000 ppm. As per the experiment of Rosenblum et al., the variation of time of TSS levels in produced water was almost 59% deduction in TSS concentration on 4 days and a 40% reduction on the 55th day.

For produced water samples obtained from diverse sources, TOC in the interval of 0–1500 ppm has been recorded. The TOC content in different types of natural sources of water is limited from less than 0.1 ppm to more than 11,000 ppm. Total nitrogen is the total sum of NO_3-N, NH_3-N, NO_2-N and also organically linked nitrogen, which are present in the water. The non-decaying portion of organic nitrogen is much more challenging to remove than the biodegradable portion, which is less hazardous to the ecosystem and easier to treat. Veil et al. and Bierman et al. looked into the existence of nitrite ion, nitrate ion, ammonia and ammonium ion in produced water at different sources like either oil and gas or mixed production wells and noticed maximum average amount limit of nitrate (NO_3^-) was found to be 2.71 ppm in produced water from gas wells, whereas largest concentration of ammonia and NH_4 was figured in produced water mostly from oil wells which is 92 ppm. On the contrary, all of the tested wells' produced water contained NO_2^- at a concentration of 0.05 ppm. However, it brings into consciousness that the soluble component is more difficult to isolate from the particulates.

6.4 BIOLOGICAL OXYGEN DEMAND (BOD) AND CHEMICAL OXYGEN DEMAND (COD)

For water produced or generated from a natural gas field, the BOD ranges from 75 to 2870 ppm [17]. Greater BOD amount in produced water taken straight from the well is caused by some reduced inorganic elements like Iron and Manganese, well drilling fluids and some added chemicals. Organic compounds high in amounts in drilling fluids have contributed to higher BOD levels in produced water. Besides that, water bodies are dumped by produced or generated water with a high amount of BOD concentration, which leads to a severe deficit of dissolved oxygen. Consequently, it is crucial to ensure that highly contented BOD water has undergone substantial oxidation in order to prevent its dumping into natural water bodies. In produced water, the calculated level of COD concentration ranging from 2600 to 120,000 ppm [17,18]. In Canadian oil fields, the estimated COD value was found to be 280 ppm. However, the produced water generated from a gas stream in a gas refinery in Iran was 270 ppm. Besides that, a higher range of COD concentration (27,000–35,000 ppm) is reported to be in the oil fields of the United States.

6.5 DISSOLVED FORMATION MINERALS

Produced water includes a variety of anions, cations, radioactive substances and heavy metals [19]. Common cations and anions include Na^+, Fe^{2+}, K^+, Mg^{2+}, Ca^{2+}, Ba^{2+}, Cl^-, Sr^{2+}, SO_4^{2-}, CO^{2-}, HCO^-. The two salt ions that are estimated to be most widespread in produced water are chloride and sodium, with phosphate having the lesser concentration. Sodium is the dominating cation in produced water from both conventional and unconventional wells, accounting for 81% in conventional wells and more than 90% in unconventional wells. So far, the composition of anions in conventional and unconventional wells differs [9]. While 97% of the anions in conventional wells are chloride anions, bicarbonate and chloride anions are present in 66% and 32% of unconventional wells, accordingly. High concentrations of insoluble sulphate and sulphide can occur due to the existence of sulphide and sulphate ions in produced water. Li^+ and Eu^{3+} among other ions might also be present [20,21]. Depending on the pH and other conditions, trace amounts of heavy metals like Hg, Cd, Pb, Cu, Ag, Zn, Cr and Ni might also be present in produced water [22]. Produced or generated water may have metals such as Zn, Fe, Ba, Cr, Ni and others. The geologic age, characteristics, volume of injected water and chemical composition all influence the type, chemical content and concentration of the metals [23]. Based on the gradually developed geology and duration of the well, the produced water from oilfield constitutes heavy metals like Hg, Pb as well as metalloids in varying amounts [8,24]. The concentration of metal in produced or generated water is typically greater than those in seawater [25,26]. The most frequently investigated metals are Cd, Cr, Hg, Ba, Pb, Cu, Ag, Ni and Zn [27]. Other trace elements found in produced water include Sr, B, Al, Fe, Mn, Li and Se. Due to their potential for bioaccumulation and/or toxicity, some metals raise special ecological concerns [27]. Sometimes produced water contains some naturally occurring radioactive materials (NORM). The two most common radioactive ions in produced water are ^{226}Ra and ^{228}Ra. The ions present influence the potential for scale formation, buffering capacity and the saline behaviour of produced water [22].

The saline behaviour of produced water is limited from a few to thousands of parts per million (ppm) [3,28]. Concentrations of rare and radioactive elements might fall below detection limits [29]. Produced water's salinity can range from a few parts per thousand (‰) to ~300 (saturated brine), which is significantly greater than seawater's salinity, ranging between 32 and 36‰. As a result, produced or generated water is frequently denser than that of seawater. As magnesium, calcium and potassium concentrations are frequently lower, dissolved chloride and sodium lead to higher salinity [24]. In Table 6.1, a list of the constituents in produced water from different oil fields is given.

6.6 PRODUCED CHEMICAL COMPOUNDS

Chemicals used in oil and gas production help manage operational challenges, such as facilitating the separation of gas, oil and H_2O, decreasing corrosion in the pipelines and preventing methane hydrate formation in the generation of gas systems. Three main categories of chemicals serve various production methods: stimulation and work-over chemicals, gas processing chemicals and production treatment chemicals like corrosion, scale, biocides and hydrate inhibitors. These categories can also classify water-treatment substances, including anti-foams and flocculants, breakers of emulsion, coagulants and reverse emulsion breakers treated in hydrocarbon restoration. These industrial chemicals dissolve in oil, eradicating the requirement for disposal methods. The application of environmentally harmful chemicals, like biocides

TABLE 6.1
A List of the Constituents in Produced Water from Different Oil Fields

Parameters	Values	Heavy Metal	Values (ppm)
Density (kg/m³)	1014–1140	Calcium	13–25,800
Conductivity (µS/cm)	4200–58,600	Sodium	132–97,000
Surface Tension (dynes/cm)	43–78	Potassium	24–4300
TOC (mg/L or ppm)	0–1500	Magnesium (Mg)	8–6000
COD (mg/L or ppm)	1220	Iron (Fe)	<0.1–100
TSS (ppm)	1.2–1000	Aluminium (Al)	310–410
pH	4.3–10	Boron (B)	5–95
Total dissolved Oil (ppm)	2–565	Barium (Ba)	1.3–650
Volatile (BTEX; ppm)	0.39–35	Cadmium (Cd)	<0.005–0.2
Base/Neutrals (ppm)	<140	Chromium (Cr)	0.02–1.1
Chlorides (ppm)	80–200,000	Copper (Cu)	<0.002–1.5
Bicarbonate (ppm)	77–3990	Lithium (Li)	3–50
Sulfate (ppm)	<2–1650	Manganese (Mn)	<0.004–175
Ammonia nitrogen (ppm)	10–300	Lead (Pb)	0.002–8.8
Sulphite (ppm)	10	Strontium (Sr)	0.02–1000
Total polar (ppm)	9.7–600	Titanium (Ti)	<0.01–0.7
Higher acids (ppm)	<1–63	Zinc (Zn)	0.01–35
Phenols (ppm)	0.009–23	Arsenic (As)	<0.005–0.3
Total non-volatile oil and grease base (µg/L)	275	Mercury (Hg)	<0.001–0.004

Sources: [24,27,30].

and corrosion inhibitors, has decreased as they were seen in generated or produced water in much smaller amounts [31,32].

- Corrosion inhibitors create a preventive coating on metal well constituents, protecting against corrosion caused by salts, gases and acids that cause corrosion [33–35]. Common examples include acetone, acetaldehyde, ethyl methyl derivatives, iso-propanol and formic acid [36].
- Scale inhibitors protect well pipes from clogging due to formation buildup. These inhibitors contain acrylate and polycarboxylate polymers [36].
- Citric acid, acetic acid, thioglycolic acid and sodium erythorbate prevent iron precipitation. Chelating agents that control iron prevent oxidation and subsequent Fe^{3+} precipitation by the formation of complexes with Fe^{2+} ions [37,38].
- Biocides prevent the growth of bacteria in well sites and boreholes because such expansion speed up the erosion of casings, equipment and degrades HF chemicals and well tubing [38,39]. Well-known biocides comprise of glutaraldehyde, quaternary ammonium compounds (QACs), tributyl tetradecyl phosphonium chloride (TTPC), tetrakis hydroxymethyl phosphonium sulfate

(THPS) and bromine-linked chemicals like 2,2-dibromo-3-nitrilopropionamide (DBNPA) [39,40,41].

- The most often utilised QACs as bioactive compounds are dialkonium and benzalkonium chlorides. Ammonium chloride (NH_4Cl) is found to be utilised [36,42].

6.7 DISSOLVED SOLIDS

Water produced may constitute silt, sand, carbonates, corrosion products, clays and other suspended particles originating from the generating formation and well-drilling activities, along with precipitated solids. Quantities can vary from negligible to a solid slurry, and excessive amounts can disrupt produced water treatment systems or damage a well. Solids can impact the future aspects and consequences of the generated/produced water, and fine-grained sediments can reduce oil/water separator efficiency, causing water discharge to exceed oil and grease limits [43]. Some solids create oil-contained sludge in manufacturing appliances, necessitating regular cleaning and disposal.

Guerra and his team found TDS in the range of 370–1940 ppm owing to high salt and bicarbonate concentrations [44]. Produced water management and reuse are affected by changes in water quality over time. The reuse and management of produced water impacted by factors such as the site of the well within the field, geologic variations among the basins and the source of produced or generated water can cause fluctuations in the count of TDS. Additionally, TDS concentrations were found to vary between conventional and unconventional wells, with conventional wells having a maximum of 400,000 mg/L (ppm) and CBM wells as low as 50,000 ppm or ppm in CBM wells.

Another characteristic that has been established is the conductivity measurement of the water produced in oil fields. It was reported that produced or generated water's conductivity from natural gas ranging from 4200 to 180,000 μS/cm, indicating that the degree of conducting electricity of generated or produced water might differ significantly. Another research reveals that the conductivity ranged from 136,000 to 586,000 μS/cm.

6.8 DISSOLVED GASES

Oilfield brines contain a considerable amount of dissolved gases, primarily hydrocarbons, but they also often include other gases like carbon dioxide, nitrogen and hydrogen sulphide. Generally, as the salinity of water and temperature increase and pressure rises, gas solubility in water decreases. Produced water frequently contains sulphides, with a significant portion being H_2S at neutral to low pH levels. This gas is flammable, toxic and corrosive, posing safety and air emission concerns. It can corrode downstream piping and foul downstream treatment equipment.

Several drill stem brine samples from water-bearing subsurface formations along the US Gulf Coast were analysed to identify the quantities and varieties of hydrocarbons [45]. Methane was usually the primary dissolved gas component, with appreciable amounts of ethane, propane and butane as well.

6.9 CONCLUSION

Produced or generated water is nothing but the huge waste stream related to gas and oil production in the researched oilfields. The qualities of generated water differ considerably from one place to another. Accurate knowledge of the origins, characteristics and management of produced water is crucial [9]. The petroleum industry relies heavily on water analysis for both upstream and downstream processes [46]. The findings mentioned in this chapter should be expanded in order to create a more comprehensive understanding of produced water and feasible management options. It is now necessary to encourage the reuse of produced water so as to fulfil the growing water needs and minimise the ecological consequences of the gas and oil industry. Reutilisation of produced water will minimise the negative environmental impacts as well as the stress on freshwater supplies [9]. This is crucial for civilisations where both residents and economic upswing are improving continuously.

REFERENCES

1. Duraisamy, R. T., Beni, A. H., & Henni, A. (2013). State of the art treatment of produced water. *Water Treatment*, *199*. http://dx.doi.org/10.5772/53478
2. Nasiri, M., & Jafari, I. (2017). Produced water from oil-gas plants: A short review on challenges and opportunities. *Periodica Polytechnica Chemical Engineering*, *61*(2), 73–81. https://doi.org/10.3311/PPch.8786
3. Neff, J., Lee, K., & DeBlois, E. M. (2011). Produced water: Overview of composition, fates, and effects. *Produced Water: Environmental Risks and Advances in Mitigation Technologies*, 3–54. https://doi.org/10.1007/978-1-4614-0046-2_1
4. Çakmakce, M., Kayaalp, N., & Koyuncu, I. (2008). Desalination of produced water from oil production fields by membrane processes. *Desalination*, *222*(1–3), 176–186. https://doi.org/10.1016/j.desal.2007.01.147
5. Harati, H. M. (2012). *Examination of produced water from the Al-Hamada oilfield, Libya*. Doctoral dissertation, Sheffield Hallam University, United Kingdom. http://shura.shu.ac.uk/19753/?
6. McFarlane, J., Bostick, D. T., & Luo, H. (2002). Characterization and modeling of produced water. In *Ground Water Protection Council Produced Water Conference, Colorado Springs, CO, Oct* (pp. 16–17). www.researchgate.net/publication/237809194
7. Glickman, A. H. (1998). Produced water toxicity: Steps you can take to ensure permit compliance. In *API Produced Water Management Technical Forum and Exhibition, Lafayette, LA* (pp. 17–18).
8. Utvik, T. I. R. (2003). Composition, characteristics of produced water in the North Sea. In *Produced Water Workshop, Aberdeen, Scotland* (pp. 26–27). https://doi.org/10.1051/e3sconf/20183103004
9. Mohammad, A. Al-G., Maryam, A. Al-K., Mohammad, Y. A. and Dana, A. D. (2019). Produced water characteristics, treatment and reuse: A review. *Journal of Water Process Engineering*, 28, 222–239. https://doi.org/10.1016/j.jwpe.2019.02.001
10. Neff, J., Sauer, T., & Hart, A. (2011). Bioaccumulation of hydrocarbons from produced water discharged to offshore waters of the US Gulf of Mexico, *Produced Water*, 441–477. https://doi.org/10.1007/978-1-4614-0046-2_24
11. Johnsen, S., Røe Utvik, Garland, T.I., de Vals, E. B., & Campbell, J. Environmental fate and effects of contaminants in produced water, SPE 86708. Paper presented at the Seventh SPE international conference on health, safety, and environment in oil and gas

exploration and production, Society of Petroleum Engineers, Richardson, TX, 2004, 9 pp. https://doi.org/10.2118/86708-MS

12. Ali, S. A., Henry, L. R., Darlington, J. W., & Occapinti, J. (1999). Novel filtration process removes dissolved organics from produced water and meets Federal oil and Grease guidelines. In *Produced Water Seminar* (pp. 21–22).

13. Danish, E. P. A. (2003). PAHs in the marine environment, ministry of environment and energy. *Danish Environmental Protection Agency, Faktuelt.* https://eng.mst.dk/

14. Frost, T. K., Johnsen, S., & Utvik, T. I. (1998). *Environmental Effects of Produced Water Discharges to the Marine Environment.* Norway: OLF.

15. Stephenson, M. T. (1992). A survey of produced water studies. *Produced Water: Technological/Environmental Issues and Solutions*, 1–11. https://doi.org/10.1007/978-1-4615-2902-6_1

16. Bansal, K. M., & Caudle, D. D. (1999). Interferences with processing production water for disposal. In *9th Produced Water Seminar*, Houston, TX 2p.

17. Fillo, J., Koraido, S., & Evans, J. (1992). Source characteristics, and management of produced waters from natural gas production and storage operations. *Produced Water*, 46,151–161.

18. Johnson, B. M., Kanagy Jr., L. E., Rodgers, J. H., & Castle, J. W. (2008). Feasibility of a pilotscale hybrid constructed wetland treatment system for simulated natural gas storage produced waters. *Environmental Geosciences*, 15, 91–104. https://doi.org/10.1016/j.biortech.2007.03.059

19. Fillo, J. P., Koraido, S. M., & Evans, J. M. (1992). Sources, characteristics, and management of produced waters from natural gas production and storage operations. *Environmental Science Research*, 46, 151–161.

20. Tibbetts, P. J. C., Buchanan, I. T., Gawel, L. J., & Large, R. (1992). A comprehensive determination of produced water composition. *Produced Water: Technological/Environmental Issues and Solutions*, 97–112. https://doi.org/10.1007/978-1-4615-2902-6_9

21. Tian, L., Chang, H., Tang, P., Li, T., Zhang, X., Liu, S., ... & Liu, B. (2020). Rare earth elements occurrence and economical recovery strategy from shale gas wastewater in the Sichuan Basin, China. *ACS Sustainable Chemistry & Engineering*, 8(32), 11914–11920. https://doi.org/10.1021/acssuschemeng.0c04971

22. Hansen, B. R., & Davies, S. R. (1994). Review of potential technologies for the removal of dissolved components from produced water. *Chemical Engineering Research & Design*, 72(2), 176–188.

23. Collins, A. (1975). *Geochemistry of oilfield waters.* Dallas: Elsevier.

24. Fakhru'l-Razi, A., Pendashteh, A., Abdullah, L. C., Biak, D. R. A., Madaeni, S. S., & Abidin, Z. Z. (2009). Review of technologies for oil and gas produced water treatment. *Journal of Hazardous Materials*, 170(2–3), 530–551. https://doi.org/10.1016/j.jhazmat.2009.05.044

25. Roach, R. W., Carr, R. S., & Howard, C. L. (1993). *An Assessment of Produced Water Impacts at Two Sites in the Galveston Bay System.* Galveston Bay National Estuary Program, Texas.

26. Igunnu, E. T., & Chen, G. Z. (2014). Produced water treatment technologies. *International Journal of Low-Carbon Technologies*, 9(3), 157–177. https://doi.org/10.1093/ijlct/cts049

27. Ray, J. P., & Engelhardt, F. R. (Eds.). (1992). *Produced Water: Technological/Environmental Issues and Solutions* (Vol. 46). Berlin: Springer Science & Business Media.

28. Pitre, R. L. (1984). Produced water discharges into marine ecosystems. In *Offshore Technology Conference*. OnePetro. https://doi.org/10.4043/4662-MS
29. Gäfvert, T., Færevik, I., & Rudjord, A. L. (2006). Assessment of the discharge of NORM to the North Sea from produced water by the Norwegian oil and gas industry. *Radioactivity in the Environment*, 8, 193–205. https://doi.org/10.1016/S1569-4860(05)08013-7
30. Lebas, R., Lord, P., Luna, D., & Shahan, T. (2013). Development and use of high-TDS recycled produced water for crosslinked-gel-based hydraulic fracturing. In *SPE Hydraulic Fracturing Technology Conference*. OnePetro. https://doi.org/10.2118/163824-MS
31. Karman, C. C., & Reerink, H. G. (1997). Dynamic assessment of the ecological risk of the discharge of produced water from oil and gas producing platforms. In *SPE/EPA Exploration and Production Environmental Conference*. OnePetro. https://doi.org/10.2118/37905-MS
32. Johnsen, S., Røe Utvik, T. I., Garland, E., de Vals, B., & Campbell, J. (2004). Environmental fate and effect of contaminants in produced water. In *SPE International Conference on Health, Safety, and Environment in Oil and Gas Exploration and Production*. OnePetro. https://doi.org/10.2118/86708-MS
33. Yang, J., Jovancicevic, V., Mancuso, S., & Mitchell, J. (2007). High performance batch treating corrosion inhibitor. In *CORROSION 2007*. OnePetro. https://onepetro.org/NACECORR/proceedings-abstract/CORR07/All-CORR07/118726
34. Rostami, A., & Nasr-El-Din, H. A. (2009). Review and evaluation of corrosion inhibitors used in well stimulation. In *SPE International Symposium on Oilfield Chemistry*. OnePetro. https://doi.org/10.2118/121726-MS
35. Al-Zahrani, A. A. (2013). Innovative method to mix corrosion inhibitor in emulsified acids. In *International Petroleum Technology Conference*. OnePetro. https://doi.org/10.2523/IPTC-16946-MS
36. Stringfellow, W. T., Domen, J. K., Camarillo, M. K., Sandelin, W. L., & Borglin, S. (2014). Physical, chemical, and biological characteristics of compounds used in hydraulic fracturing. *Journal of Hazardous Materials*, 275, 37–54. https://doi.org/10.1016/j.jhazmat.2014.04.040
37. Dill, W. R., & Fredette, G. (1983). Iron control in the Appalachian Basin. In *SPE Eastern Regional Meeting*. OnePetro. https://doi.org/10.2118/12319-MS
38. McCurdy, R. (2011). High-rate hydraulic fracturing additives in non-Marcellus unconventional shales. *For the Hydraulic Fracturing Study: Chemical & Analytical Methods*, 17. www.epa.gov/hfstudy/high-rate-hydraulic-fracturing-additives-non-marcellus-unconventional-shales
39. U.S. E.P.A. (2004). Hydraulic fracturing fluids. In *Evaluation of Impacts to Underground Sources of Drinking Water by Hydraulic Fracturing of Coalbed Methane Reservoirs*. https://nepis.epa.gov/Exe/ZyPURL.cgi?Dockey=P100A99N.TXT
40. Nobel, A. (2004). Macro-porous polymer extraction for offshore produced water removes dissolved and dispersed hydrocarbons. *Business Briefing, Exploration & Production, the Oil & Gas Review*.
41. FracFocus. (2015). *Hydraulic Fracturing*. Retrieved from www.fracfocus.org/learn/hydraulic-fracturing on 11 March 2023.
42. Aminto, A., & Olson, M. S. (2012). Four-compartment partition model of hazardous components in hydraulic fracturing fluid additives. *Journal of Natural Gas Science and Engineering*, 7, 16–21. https://doi.org/10.1016/j.jngse.2012.03.006

43. Cline, J. T. (1998). Treatment and discharge of produced water for deep offshore disposal. In *API Produced Water Management Technical Forum and Exhibition, Lafayette, LA* (pp. 17–18).

44. Guerra, K., Dahm, K., & Dundorf, S. (2011). *Oil and Gas Produced Water Management and Beneficial Use in the Western United States* (pp. 1–113). Washington, DC: US Department of the Interior, Bureau of Reclamation.

45. Buckley, S. E., Hocott, C. R., & Taggart Jr, M. S. (1958). Distribution of Dissolved Hydrocarbons in Subsurface Waters: Topical Papers.

46. Ibrahim, M. A. El L., Khaled, M. M., Nuri, M. T., & Somaya, A. A. (2021). Physiochemical characteristics of produced water in oilfields and its environmental impacts. *International Journal of Scientific Development and Research (IJSDR)*, 6(2), 288–300.

7 Produced Water Management

Ravi Kumar Lingam

7.1 INTRODUCTION

Produced water is wastewater coming out of oil and gas wells along with oil and gas. It is also called 'brine', 'salt water' or 'formation water'. It contains emulsified oil, dissolved solids, suspended solids and heavy metals. Oil droplets in produced water are emulsified and get stabilised in the water due to surfactants present in the produced water. Produced water is the largest volume of wastewater from the oil and gas industry. The cost of produced water volume management is vital for the industry for regulatory compliance as well as for environmental sustainability. Most of the produced water is injected underground to enhance oil production and disposal. This is the common practice from onshore wells. Produced water from off-shore wells is treated for disposal in oceans and some volume of it is used for enhanced oil recovery. The produced water characteristics vary from place to place, the way it is formed and the type of product that may be produced from the oil and gas wells. Produced water properties and volume from a particular reservoir varies throughout the lifetime of that particular reservoir. The oil and grease, salt content and total-dissolved solids (TDS) are the critical parameters for choosing the treatment method for recycling and reuse. Produced water from gas production contains condensed water which is not present in oil production wells [1,2].

7.2 PHYSICAL TREATMENT

7.2.1 Hydrocyclones

Hydrocyclone is a classifying device that uses centrifugal force to settle the particles, and it can be used in continuous operations. Hydrocyclone consists of a conical section and a cylindrical portion. The feed port is at one side of the equipment and has an opening at the top for overflow as well as preventing the short circuit of mixing the feed with overflow. The underflow is a thick slurry. The angle of the conical section is critical to the performance of the hydrocyclone. The feed enters the hydrocyclone through the tangential entry, and it imparts the swirling motion to the feed. There will be a vortex formation inside the hydrocyclone along the vertical axis, and the lighter particles, which do not settle transport along the vortex to the overflow line. The

DOI: 10.1201/9781003441359-7

hydrocyclones are made up of mild steel, stainless steel or ceramic. Hydrocyclones can be used for size separation in the range of 5–15 mm. Hydrocyclones are subjected to wear and tear due to particle abrasion [3].

7.2.2 GAS FLOTATION

Gas flotation can be used to separate the particles which cannot be removed by the sedimentation technique. In this process, finer air bubbles are generated in two methods: dissolved gas flotation and induced gas flotation. In dissolved gas flotation, the gas enters the chamber by pressure drop or vacuum. In induced gas flotation, mechanical stirrers are used to create bubbles. Gas flotation can handle the particle size up to 2.5 mm. This process will be highly efficient if the bubble sizes are less than the oil droplet or other particle sizes. When gas is entered and produced water in a flotation chamber and fine bubbles are generated. The bubbles carry the contaminants and bring them to the surface. These will be floated on the surface and form some kind of froth. This froth is collected by skimming the surface of the produced water. Sometimes, coagulants may be added to remove the particulates; otherwise, chemical use is not high. The sludge generated by this process has to be treated to manage the solid waste in the process.

7.2.3 ADSORPTION

Adsorption is used in wastewater treatment processes for removing phenols, heavy metals, BTEX and total organic carbon (TOC). There are different types of adsorbents available for various applications. The most common adsorbents in wastewater treatment are granular-activated carbon (GAC) and powdered-activated carbon (PAC). Activated alumina, organoclays, zeolites, amberlite, etc., are also used as adsorbents in various processes. Adsorption technology is used as a polishing step in some processes. The adsorbent requirement, column aspect ratio and feed stream details are calculated based on the treated water quality requirements. Based on the targeted impurities in the treated water, the operating parameters and design calculations are optimised. The adsorbent needs regeneration if it is saturated with impurities. The chemical treatment process is widely used for adsorbent regeneration. The exhausted adsorbent beyond the regeneration process is discarded and treated separately by solid waste management guidelines.

7.2.4 MEDIA FILTRATION

Media Filtration is the separation of suspended solid particles from liquid feed streams by a filter media. It can also be used for removing oil and grease. Various media for filtration technology, such as sand, walnut shell, gravel, anthracite, etc. Walnut shell is widely used in oilfield wastewater treatment. This process can be augmented by adding coagulants to the feed stream before filtration. Solid waste disposal is difficult to handle.

7.2.5 Ion Exchange Process

The ion exchange process uses resin for treating wastewater, and the resin has to be regenerated once the active sites of the resin material are saturated with contaminants. This process can be used for removing ions and metals from oil field wastewater. Reverse osmosis permeate of oil field wastewater can be treated in an ion exchange process for removing boron. Chemicals may be used for disinfection as well.

7.2 CHEMICAL TREATMENT

7.2.1 Chemical Oxidation

Chemical oxidation is a process in which oxidants oxidise or reduce the undesired contaminants, organic and inorganic compounds from produced water. This process can remove colour, odour, COD and BOD to desirable limits from the feed stream. The widely used oxidants are ozone, hydrogen peroxide, potassium permanganate, oxygen and chlorine. The oxidant breaks down the pollutants, thereby reducing their load in the treated water. Sludge handling is an issue in this process.

7.3 THERMAL TREATMENT

There are various technologies used for produced water treatment. They are multi-stage flash (MSF) distillation, multi-effect evaporation (MEE) and mechanical vapour compression (MVC) [3].

7.3.1 Multistage Flash

MSF is a process in which produced water is pumped to a flash tank at reduced pressure. The water flashes immediately and comes out through the outlet. It is condensed and can be reused. The major problem is the deposition of dissolved solids on the walls of the flash tank. Corrosion inhibitors and scale chemicals may be used for scale removal.

7.3.2 Multi-Effect Evaporator

A multi-effect evaporator (MEE) is an energy-efficient process where one kilogram of steam can produce three to four kilograms of water vapour. There are different types of configurations available for this system: feed forward and feed backward. Two common types of evaporators are available, i.e., falling film evaporator and forced circulation. The process requirements are the guiding principles for designing number of effects, pre-heater, vapour separators, condensers and plate type of heat exchangers for circulating water for cooling the pump seals. This process can be used for treating produced water.

7.3.3 MECHANICAL VAPOUR COMPRESSION

MVC is used in produced water treatment, where electrical energy is used to supply thermal energy for desalination. The feed stream is preheated by suitable heat exchangers using the heat from condensate. The feed is pumped to the feed distributor plate, where the feed is homogeneously spread across the distributor plate and thin layers are formed in the process are evaporated. Evaporated water vapour is suctioned to the vapour compressor, which discharges superheated steam to the heat transfer tubes. The tube bundles can be horizontal or vertical. The evaporation system contains non-condensable gases, and they will be vented off. MVC process is a high-energy-intensive process where high-grade electrical energy is needed to operate this process.

7.4 BIOLOGICAL TREATMENT

Biological treatment of produced water (PW) can be carried out in several ways, i.e., membrane bioreactors (MBR), activated sludge treatment and bio-electrochemical treatment [4]. The biological treatment of PW can reduce the concentration of unwanted constituents from oilfield wastewater [5]. Biological wastewater treatment is a classic method to treat and recycle or dispose the treated wastewater. This method removes the organic compounds from wastewater. These processes use aerobic pathway degradation. The heterotrophic biomass oxidises organic matter for its metabolism. The autotrophic biomass draws its energy from the oxidation of inorganic matter (nitrifying organisms). Biological treatment methods focussed on the salinity values and hydraulic retention time values. An increase in salinity changes the physicochemical properties of microorganisms. Hypersaline-produced water requires microorganisms suited for that environment. Halophilic microorganisms can survive under hypersaline conditions. These microorganisms are taken from hypersaline soils and enriched before inoculation.

The following processes are developed for biological treatment [6]:

7.4.1. Conventional-activated sludge
7.4.2. Biological-aerated filter
7.4.3. Membrane bioreactors
7.4.4. Hybrid processes

7.4.1 CONVENTIONAL ACTIVATED SLUDGE

Microorganisms are brought into contact with wastewater in the form of bioflocs. The size of bioflocs varies from 3 to 5 mm. The *Pseudomonas species* flavo bacteria and alcaligenes are widely found microorganisms in activated sludge. After the biological assimilation of organic compounds, a separation step (gravitational settling) is needed to obtain clarified water free from suspended matter.

Conventional activated sludge (CAS) processes are characterised by the organic loading rate (OLR; kg COD/m^3/d), the biological loading rate (kg COD/mixed liquor volatile suspended solids (MLVSS)/d), the hydraulic retention time, sludge retention time (days). Produced water with domestic sewage is treated using a CAS process in a batch reactor operated for 24 h [7]. Even if the gradual acclimation of the activated sludge was initially sampled from municipal water treatment, the COD efficiency reached 50%. With an OLR of 0.82 kg COD/m^3/d, they obtained a COD removal efficiency of 97%.

The microorganisms are acclimatised for ten days and they are collected from a contaminated site. These microorganisms removed the COD up to 97% at an OLR of 0.82 kg/COD/m^3/d [8,9]. CAS process is easy to operate and operation cost is less. However, the carbon footprint for this process is high and particle settling is an issue.

7.4.2 Biological-Aerated Filters

Biological operation is suitable for effluents containing complex and poorly bio-degradable compounds. In this process, microorganisms are fixed on the bed material on which wastewater flows through. The thickness of the biofilm (microbial) varies from a few microns to more than one centimetre [10]. The packing material consists of gravel, wood, coal, activated carbon, plastic carriers, etc. [11]. Biomass is fixed in the bedding material, thereby increasing microorganisms' concentration. Biomass concentration for BAF
could reach 100,000 mg/L (mixed liquor volatile suspended solids) compared to 2500 mg/L for CAS [10]. In the bench scale, BAF could maintain an oil/grease removal efficiency of 75–99%. The TOC removal rate is 78% even if the produced water is deficient in nitrogen and phosphorous [12]. BAFs loaded with activated carbon have two-fold advantages. The dissolved organic carbon could be adsorbed by activated carbon followed by organic matter removal by microorganisms. BAFs give high removal efficiency at less hydraulic retention time compared to the CAS process.

7.4.3 Membrane Bioreactors

MBR process consists of a suspended sludge aeration tank coupled with a physical separation. The physical separation may be microfiltration or ultrafiltration. Since the entire system is compact, the equipment footprint is less. Two configurations are generally used in MBR, i.e., immersed or submerged MBR where the membrane module is inside the aeration tank and side stream or external MBR where the membrane module is outside the aeration tank. Sometimes, the membrane module is set up outside the system to reduce the operating costs to counteract membrane fouling and maintenance. Two types of membrane filtration operations are used depending on the suitability of a particular process, i.e., dead-end filtration and crossflow filtration. Membranes are broadly classified as polymeric and inorganic or ceramic membranes. Ceramic membranes are highly resistant to chemicals and cleaning is easy compared to polymeric membranes. But inorganic membranes are expensive [13].

MBR's have advantages, viz., lower footprint, less sludge production or better effluent quality in terms of suspended solids. A major challenge in MBR is fouling. Cake layer formation on the surface of the membrane is the result of fouling. This layer consists of organic matter and inorganic salts. Irreversible fouling occurs at high transmembrane pressure [14]. Temperature influences membrane permeability. If the temperature is high, the viscosity of the liquor is decreased. Thereby, the membrane flux increases. However, high temperature may alter microorganisms. Membrane fouling is also dependent on water salinity. The cost associated with membrane replacement and maintenance is high, and in turn it affects the feasibility of implementing the MBR's in oil fields for treatment of produced water [15–17]. COD removal is less at a high salinity of wastewater.

7.5   MEMBRANE TREATMENT

7.5.1   PRESSURE-DRIVEN MEMBRANES

Produced water treatment for emulsified oil removal followed by salt and other metal removal using desalination process is vital for recycling the produced water. Membranes can be used for oilfield wastewater treatment to comply with the discharged or treated water physicochemical characteristics as per regulatory guidelines for safe disposal as well as recycling purposes. A membrane is a semipermeable barrier, which allows certain compounds to pass through, while retaining others. The removal of certain compounds, impurities, oil emulsions and other metals depends on the principles, viz., size, donnan exclusion and electric charge. The pore size of the membrane, surface charge on the membrane and hydrophilicity or hydrophobicity, contact angle of the membrane are the critical parameters to be accounted for removing the impurities or contaminants present in the produced water.

The emulsified oil in the produced water causes fouling while membrane treatment is being done. Fouling can inhibit the membrane efficiency or effectiveness. The membrane surface should have antifouling-coated layer to minimise fouling due to emulsified oil contact. The oil droplets flow through the pores of the membrane and it will get blocked and the membrane performance will be declined. The permeate flux will be decreasing due to pores blockage. The membrane surface needs to be sweep with inert gas. Backwashing with water or with chemical treatment of the surface, the pore blockage can be removed. The additional costs for membrane cleaning and other servicing factors are also accounted for viability of the membrane treatment.

The ultrafiltration (UF) membrane process operates at low pressure. It is used as a clarification or disinfection step. UF is used in combination with nanofiltration (NF) or reverse osmosis (RO) in desalination processes. Nanofiltration (NF) and reverse osmosis (RO) processes are pressure-driven membrane processes. Hydraulic pressure is applied to reverse the osmotic pressure of the feed solutions to get permeate with desired purity. The NF and RO membranes are very thin and supported by dense material to withstand high hydraulic pressure. RO membranes can remove the contaminants as small as $0.1 \, \mu m$ from the feed solution. Nanofiltration membranes can remove the contaminants as small as $10 \, \mu m$ from feed solution. But the stand-alone processes of NF or RO do not work. They have to be combined with pre-treatment steps for the effective removal of dissolved solids. If produced water has TDS of more than 50,000 ppm, then reverse osmosis (RO) is not effective.

7.5.2 FORWARD OSMOSIS

Forward osmosis is growing in the fields of wastewater treatment, desalination, food processing, etc. In this process, osmotic pressure is used as driving force between feed solution and draw solution. The selection of draw solute depends on the osmotic pressure of its solution at a particular concentration. The two most commonly used draw solutes are magnesium chloride and sodium chloride. Since the osmotic pressure is high for draw solution, it can draw clean water from feed solution using hollow fibre membrane or flat sheet membrane module. The permeate flux is very high in the case of hollow fibre membrane module and the flux is a little less in the flat sheet membrane module. The treatment time for wastewater using a hollow fibre membrane module is less compared to a flat sheet membrane module. The higher productivity for hollow fibre membrane module is due to higher packing density.

The clean water permeated through a permeable membrane gets mixed with the draw solution, thereby reducing the concentration of the draw solution. The draw solution may be made up of draw solute for desired osmotic pressure to draw the clean water. The resultant draw solution has to be treated in a reverse osmosis membrane process for separating the salts and water. Another option is to use thermal processes to evaporate the solution and condense the vapour to get clean water. The concentrated salt solution can be recycled in the forward osmosis process [18]. There is no high-pressure operating limit for FO, unlike reverse osmosis (RO).

7.5.3 THERMAL-DRIVEN MEMBRANES

Membrane distillation (MD) uses low-grade heat to drive separation. In this process, Products are separated from the feed solution by a semi-permeable membrane, which also prohibits the passage of liquid solution through it. The driving force for water vapour transport across the semi-permeable membrane is partial vapour pressure. Since this process uses the vapour pressure as a driving force, the permeate flux and salt rejection do not depend on feed water salinity. Various membrane configurations may be used for the application of membrane distillation. There are four common types of membrane configurations: air gap, vacuum membrane distillation, sweeping gas and direct contact distillation. In these methods, vapour pressure gradient and condensation methods vary from one configuration to another.

The driving force in direct contact MD is the temperature difference between the feed solution and permeate. The non-volatile compounds are retained in the feed, and pure water vapour passes through the membrane as permeation. The driving force is dependent on the feed solution temperature. If the temperature of the feed solution rises, then the water vapour flux is high. This is supported by the Clausius–Clapeyron relation in which the logarithmic of the vapour pressure and temperature are inversely proportional. The effect of salinity on water flux rate is not significant. If the TDS concentration of the feed increases from 35,000 to 75,000 mg/L, then the change in permeate flux is only 5%. In other words, membrane distillation is well-suited for high-salinity feed solutions. Another advantage of MD is less chance of fouling due to larger membrane pores and no applied hydraulic pressure in comparison with reverse osmosis (RO).

The critical challenge in MD is organic compounds and dissolved gases that exert partial vapour pressure equivalent to or slightly higher than water. These vapours from unwanted compounds are permeated along with water vapour and in turn causes contamination in the permeate. If the feed solution contains alcohols and surfactants, they reduce the surface tension of the feed solution. It causes wetting of membrane pores. The wetted membrane allows the feed solution to pass through the membrane pores, which also contaminates the permeate. Fouling is a major issue while handling high-salinity feed solutions. The high-salinity feed solutions form scale, inhibiting the vapour flow through membrane pores. This can be observed in flux decline behaviour. Fouling and scaling of membranes causes a decrease in operational productivity.

The MD process should accompany pre- and post-treatment processes for produced water. The pre-treatment removes the foulants and other chemicals that cause scale formation and membrane pore wetting. The post-treatment of the permeate stream removes the gases permeated to permeate stream as well as to remove unwanted volatile compounds. However, the pre and post treatment steps must be customised as produced water quality varies from geologic location and age of the well [18].

7.5.4 COMBINED TECHNOLOGIES (HYBRID MEMBRANE SYSTEM)

The hybrid system combines two or more processes for treating the produced water to remove contaminants, thereby meeting the targeted clean water specification. This system discusses biological treatment in BAFs followed by ultra-filtration membrane treatment and nanofiltration membrane treatment. Industrial wastewater contains organic matter and nutrients which can be removed effectively using biological processes. The critical limitation of biological treatment is high levels of TDS in the wastewater. Since the produced water TDS is high, biological treatment alone cannot fulfil the treated water specification. It has to be augmented by pre-treatment using an ultra-filtration membrane process followed by nanofiltration.

Biological processes for wastewater treatment are intended to remove biodegradable compounds. There are two types of these biological processes, i.e., suspended growth and attached growth processes. BAFs are categorised in the attached growth processes. In these processes, the medium hosts the microorganisms. The medium acts as a surface and enables the biomass to grow. The attached-growth system contains a higher concentration of biomass. The biomass grown on media is also called biofilm. These biofilms are adaptable to the variations in surrounding conditions. However, the contaminants are removed by filtering suspended solids, adsorption of dissolved solids on the media and biodegradation of organic carbon and other biodegradable compounds by oxidation of microorganisms. Sometimes, the microorganisms are acclimatised with the source waste water so that they can adapt to the conditions. GAC is used as a media in BAFs because of its highly specific surface area. The microporous structure and roughness of GAC favours biofilm attachment and protection, thereby enhancing adsorption capacity [6,19].

7.6 CONCLUSION

Produced water is a mixture of many constituents, and no single technology can provide a solution for the removal of contaminants from produced water. The combination of physical methods, chemical methods, thermal treatment, biological methods and membrane separations may be devised based on targeted contaminants removal efficiency and its final use. The existing technologies and advanced technologies which will be used commercially have to be thoroughly checked with the actual produced water of that particular geographical location for maximising the benefits.

REFERENCES

1. Clark, C. E., & Veil, J. A. (2009). *Produced water volumes and management practices in the United States* (No. ANL/EVS/R-09-1). Argonne National Lab. (ANL), Argonne, IL, United States. https://doi.org/10.2172/1007397
2. Oil Industry International Exploration, & Production Forum. (1997). Environmental management in oil and gas exploration and production: An overview of issues and management approaches. https://wedocs.unep.org/bitstream/handle/20.500.11822/8275/Environmental%20Management%20in%20Oil%20%26%20Gas%20Exploration%20%26%20Production-19972123.pdf?sequence=2%26isAllowed=y
3. Igunnu, E. T., & Chen, G. Z. (2014). Produced water treatment technologies. *International Journal of Low-Carbon Technologies*, 9(3), 157–177. https://doi.org/10.1093/ijlct/cts049
4. Abujayyab, M. A., Hamouda, M., & Hassan, A. A. (2022). Biological treatment of produced water: A comprehensive review and metadata analysis. *Journal of Petroleum Science and Engineering*, 209, 109914. https://doi.org/10.1016/j.petrol.2021.109914.
5. Katsoyiannis, I. A., & Zouboulis, A. I. (2004). Application of biological processes for the removal of arsenic from groundwaters. *Water Research*, 38(1), 17–26. https://doi.org/10.1016/j.watres.2003.09.011
6. Lusinier, N., Seyssiecq, I., Sambusiti, C., Jacob, M., Lesage, N., & Roche, N. (2019). Biological treatments of oilfield produced water: A comprehensive review. *SPE Journal*, 24(05), 2135–2147. https://doi.org/10.2118/195677-PA
7. Freire, D.D.C., Cammarota, M.C., & Sant'Anna, G.L. (2001). Biological treatment of oil field wastewater in a sequencing batch reactor. *Environmental Technology*, 22(10), 1125–1135. https://doi.org/10/1080/09593332208618203
8. Tellez, G. T., Nirmalakhandan, N., & Gardea-Torresdey, J. L. (2002). Performance evaluation of an activated sludge system for removing petroleum hydrocarbons from oilfield produced water. *Advances in Environmental Research*, 6(4), 455–470. https://doi.org/10.1016/S1093-0191(01)00073-9
9. Tellez, G. T., Nirmalakhandan, N., & Gardea-Torresdey, J. L. (2005). Kinetic evaluation of a field-scale activated sludge system for removing petroleum hydrocarbons from oilfield-produced water. *Environmental Progress*, 24(1), 96–104. https://doi.org/10.1002/ep.10042
10. Cohen, Y. (2001). Biofiltration – The treatment of fluids by microorganisms immobilized into the filter bedding material: A review. *Bioresource Technology*, 77(3), 257–274. https://doi.org/10.1016/S0960-8524(00)00074-2
11. Freedman, D. E., Riley, S. M., Jones, Z. L., Rosenblum, J. S., Sharp, J. O., Spear, J. R., & Cath, T. Y. (2017). Biologically active filtration for fracturing flowback and

produced water treatment. *Journal of Water Process Engineering*, *18*, 29–40. https://doi.org/10.1016/j.jwpe.2017.05.008

12. Zhao, X., Wang, Y., Ye, Z., Borthwick, A. G., & Ni, J. (2006). Oil field wastewater treatment in biological aerated filter by immobilized microorganisms. *Process Biochemistry*, *41*(7), 1475–1483. https://doi.org/10.1016/j.procbio.2006.02.006

13. Judd, S. J. (2016). The status of industrial and municipal effluent treatment with membrane bioreactor technology. *Chemical Engineering Journal*, *305*, 37–45. https://doi.org/10.1016/j.cej.2015.08.141

14. Barrios-Martinez, A., Barbot, E., Marrot, B., Moulin, P., & Roche, N. (2006). Degradation of synthetic phenol-containing wastewaters by MBR. *Journal of Membrane Science*, *281*(1–2), 288–296. https://doi.org/10.1016/j.memsci.2006.03.048

15. Fakhru'l-Razi, A., Pendashteh, A., Abdullah, L. C., Biak, D. R. A., Madaeni, S. S., & Abidin, Z. Z. (2009). Review of technologies for oil and gas produced water treatment. *Journal of Hazardous Materials*, *170*(2–3), 530–551.https://doi.org/10.1016/j.jhazmat.2009.05.044

16. Fakhru'l-Razi, A., Pendashteh, A., Abidin, Z. Z., Abdullah, L. C., Biak, D. R. A., & Madaeni, S. S. (2010). Application of membrane-coupled sequencing batch reactor for oilfield produced water recycle and beneficial re-use. *Bioresource Technology*, *101*(18), 6942–6949. https://doi.org/10.1016/j.biortech.2010.04.005

17. Pendashteh, A. R., Abdullah, L. C., Fakhru'l-Razi, A., Madaeni, S. S., Abidin, Z. Z., & Biak, D. R. A. (2012). Evaluation of membrane bioreactor for hypersaline oily wastewater treatment. *Process Safety and Environmental Protection*, *90*(1), 45–55. https://doi.org/10.1016/j.psep.2011.07.006

18. Shaffer, D. L., Arias Chavez, L. H., Ben-Sasson, M., Romero-Vargas Castrillón, S., Yip, N. Y., & Elimelech, M. (2013). Desalination and reuse of high-salinity shale gas produced water: Drivers, technologies, and future directions. *Environmental Science & Technology*, *47*(17), 9569–9583. https://doi.org/10.1021/es401966e

19. Riley, S. M., Oliveira, J. M., Regnery, J., & Cath, T. Y. (2016). Hybrid membrane bio-systems for sustainable treatment of oil and gas produced water and fracturing flowback water. *Separation and Purification Technology*, *171*, 297–311. https://doi.org/10.1016/j.seppur.2016.07.008

8 Advancement of Membrane Technology for the Separation of Oil from Oily Wastewater in Petroleum Industry

Chinmoy Bhuyan and Pratyashi Kondoli

8.1 INTRODUCTION

Rapid industrial development in the oil, gas industries, petrochemical industry, pharmaceutical industry, metallurgy and food industries, has resulted in a large volume of oily wastewater. Oil–water mixtures often occur because of various activities, like extraction, transportation, accidents in oil wells and also as waste from the petroleum industries. At the same time, water scarcity due to uncontrolled population growth is also a major concern [1]. Based on their chemical structure, the organic compounds that are present in oily sludge are categorised into four groups. These are aliphatic (alkanes and cycloalkanes), aromatics (benzene, toluene, phenols, xylenes, naphthalene or other polyaromatic hydrocarbons), nitrogen, sulphur, oxygen or NSOs and asphaltenes [2]. When oil-based wastewater with a high amount of organic matter is disposed of in nearby water reservoirs like rivers, wetlands, etc., the microorganisms that are present in the water consume an excess amount of oxygen, thereby decreasing the net oxygen level in the water reservoir or stream [3,4]. When oily wastewater is disposed of over soil, the fertility of the soil in the disposed area is badly affected. Moreover, oil contaminants generated from the petroleum industry have carcinogenic as well as genotoxic effects on human beings [5]. Therefore, it is of the utmost importance to treat oily water before disposing it in any body of water or soil. To address the water scarcity problem, it is also important to make the wastewater reusable after removing the contaminants. As the oil industry is ageing, the handling of produced wastewater becomes one of the main problems that adds additional costs to projects. It is crucial to treat this oily wastewater generated in various ways in order to avoid environmental and water scarcity issues. Several methods have been adopted for the treatment of oil-based wastewater, including both physical as well as chemical methods like adsorption (by carbon-based nanomaterials, zeolites, resins, organoclay and some copolymers)

DOI: 10.1201/9781003441359-8

and sand filtration. Chemical treatments include degradation or photodegradation by means of oxidation of oil contaminants, electrochemical processing, treatment by ozone and various ionic liquids at ambient conditions and the use of demulsifiers. These above-mentioned conventional methods often require tedious manual operation and are time-consuming. Moreover, the process of separation of an emulsion mixture of oil and water is pragmatic to a high extent, which results in incomplete separation with either water remaining in the oil or oil remaining in the water [6]. To fill the gaps, the membrane separation technique is an emerging technology that is being used in the 21st century because of its efficiency for the high removal of oil as well as due to the relatively simple operation process and other numerous additional advantages. This book chapter emphasises the recent advancement of membrane technology for oil industry-based wastewater treatment.

8.2 MEMBRANE FOR EFFICIENT SEPARATION OF OIL AND WATER FROM PETROLEUM WASTEWATER

The membrane technology can function without the involvement of chemical substances, and the energy requirement is low; handling is easy, and there are clear process controls. Membranes are now viable as compared with conventional water treatment techniques. Commonly used pressure-mediated methods for the separation of oil from oily wastewater include microfiltration (MF), ultrafiltration (UF), nanofiltration (NF) and reverse osmosis (RO) [7–9]. The mechanism or principle for oil–water separation through membranes is solely related to the size exclusion principle and the surface characteristics of the membrane.

8.2.1 SIZE EXCLUSION PRINCIPLE: SEPARATION OF OIL AND WATER

Oil–water separation through membranes is based on the pore size and capillarity of the membrane material. Oily wastewater is a mixture of various oil types, namely fat, grease, hydrocarbons and petroleum fractions, present in a variety of forms, including free oil ($d > 150$ m), dispersed oil ($d = 20$–150 m) and emulsified oil (20 m) at varying concentrations [10,11]. Therefore, the pore size tunability of the membrane is of great significance in order to get maximum rejection with excellent permeability. The pressure-mediated membrane-based separation consists mainly of MF, UF and NF, as well as FO and RO, among which UF membrane is found to be the most effective in oil–water separation. The pore size of UF membranes is in the range of 2–50 nm, which is favourable for the separation of all types of oil present in wastewater. UF polymeric membranes that are in the molecular weight cut-off range of 100,000–200,000 Da were successful in rejecting 96% of total hydrocarbons and several hazardous heavy metals like Cu and Zn by more than 95% [12]. Treatment of oil-based wastewaters, which have a very high salt content is usually done with reverse osmosis and nanofiltration membranes [13]. UF membranes (2–50 nm) and nanofiltration membranes (>1 nm) are suitable used for selectively separating oil from water. NF and UF membranes that have been developed by various research groups in recent years for efficient oil–water separation are discussed here.

Polysulfone (PSF), polyether sulfone (PES), polyvinylidene fluoride (PVDF), polyacrylonitrile (PAN) and cellulose acetate (CA) are the base polymers that have been used in porous polymeric membrane fabrication for oil–water separation [14–16]. Polyvinyl pyrrolidone (PVP), polyvinyl alcohol (PVA), polyethylene glycol (PEG), etc., are used as pore enhancing agents or pore-tuning additives. A blend of polymeric membranes with different weight percentages of PSF, PEG and PVP was developed, which gave rise to membrane surfaces with varying morphologies [14]. PVDF is a kind of polymer known for its exceptional properties, such as like excellent stability to heat and chemicals and high mechanical strength, hence its use in the design of MF and UF membranes. Another additive that is used for increasing the hydrophilicity of porous polymeric membranes is polymethyl methacrylate (PMMA). Upon addition of PMMA, the hydrophilicity as well as the surface porosity of the pristine PVDF membrane increase significantly, which ultimately results in enhanced water permeability. Also, membranes containing a higher percentage of PMMA are less prone to fouling [17].

The impregnation of nanomaterials on porous membrane structures for highly selective oil separation from oily wastewater is a novel method in the field of separation and purification chemistry. Metal nanoparticles, graphene-based layered materials, carbon nanotubes and inorganic layered materials are such types of nanomaterials that are impregnated over porous membranes that ensure an increase in oil rejection without hampering membrane flux [18–21]. Yuliwati et al. developed a PVDF-UF hollow fibre membrane using LiCl, H_2O and TiO_2 nanoparticles with varying concentrations. The membrane with 1.95 wt% of TiO_2 showed a maximum permeability of 82.50 L/m^2 h and separation efficiency of 98.83% [19]. A UF TiO_2/ Al_2O_3–PVDF mixed matrix membrane was also designed by Yi et al. to study the influence of some important factors like trans-membrane pressure (TMP), feed properties, oil concentration and total dissolve solids (TDS) on oil–water emulsion separation through the membrane developed by the group. Experiments suggest that, except TDS all other parameters have a substantial role in the relative flux of the membrane. A comparison study revealed that the modified PVDF membrane exhibits more anti-fouling properties than the neat PVDF membrane. Also, very adequate flux recovery was noted when the contaminated membranes were washed with water and NaClO solution [20]. C. Ong et al. created a similar type of hollow fibre membrane with 2 wt.% TiO_2 nanoparticles demonstrated maximum separation efficiency with a significant increase in permeability regardless of oil percentage compared to the bare PVDF membrane [21]. It is noted that the TiO_2 concentration in membrane fabrication should be such that no agglomeration of the nanoparticles may occur. The performance of the membrane fabricated by Ong et al. was further enhanced by accompanying the membrane permeation experiment with photocatalysis. Experimental analysis showed that under UV irradiation, the oil components present in wastewater could be efficiently degraded in the presence of TiO_2 nanoparticles [22]. PSF is another polymer that has been used as a base material for fabricating porous polymeric membranes for oil–water separation [23,25]. Through a simple phase inversion process, Zhang et al. created phosphorylated TiO_2–SiO_2, or simply PTS/PSF composite membrane. Phosphorylation of TiO_2–SiO_2 nanoparticles leads to the formation of more hydroxyl on PTS particles, which enhances the hydrophilicity of the resulting membrane significantly. The developed membrane demonstrated 92% retention for waste-containing oil, with significant improvements in antifouling and anti-compaction characteristics [23]. Hydrous manganese dioxide (HMO) nanoparticle–PES mixed matrix membrane was developed by incorporation of HMO into the PES membrane, which improved the wetting behaviours of the membrane with promising permeability (573.2 L m^{-2} h^{-1} bar^{-1}) and maximum oil rejection (nearly 100%) [24]. A novel hybrid mixed matrix membrane based on sulphated Y-doped zirconia particles ($SO_4^{2-}/ZrO_2–Y_2O_3$ or SZY particles) as the dopant on a PSF membrane was developed by Zhang et al. SZY particles were incorporated into the membrane to improve its hydrophilicity and antifouling properties [25]. Graphene oxide (GO), MgAl-layered double hydroxides (MgAl LDHs) and sepiolite-coated cellulose acetate membranes were developed through layer-by-layer assembly for oil separation and highly anti-oil-fouling ability. Among the different types, graphene oxide-coated membranes showed the best results in flux and anti-oil fouling ability. The group also explains the effect of charge on membrane permeability. As oil is of negative charge and LDHs are of positive charge, oil is usually trapped on the layers, causing the membrane to be polluted with oil. This decreases water permeability in the membrane. This decrease in water permeability was not observed in the case of a GO-based membrane [26]. In another work GO, reduced GO and polybenzimidazole nanocomposite membranes were developed via phase inversion followed by dip coating for oil–water separation from petrochemical industrial water. The as-developed membrane had the highest flux of 91 L m^{-2} h^{-1} bar^{-1} with 100% oil rejection [27]. In Figure 8.1, a schematic diagram of the membrane system for the separation of oily wastewater is shown.

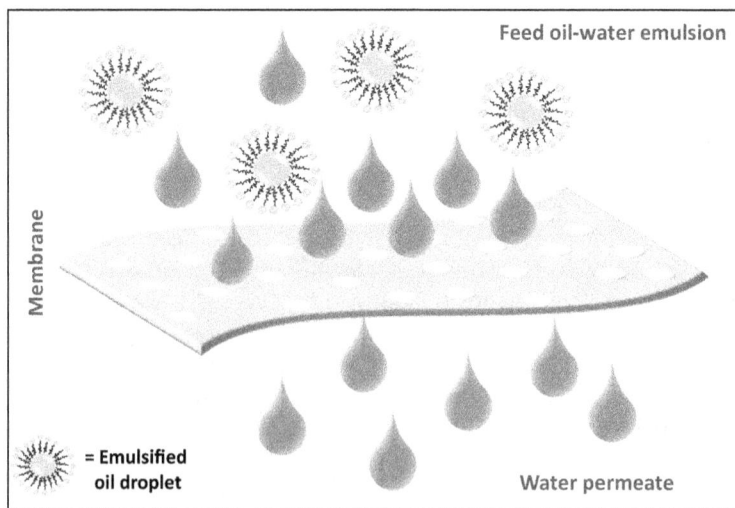

FIGURE 8.1   Schematic diagram of a membrane system for separation of oily wastewater.

8.2.2 OIL WATER SEPARATION BASED ON SURFACE PROPERTY AND WETTABILITY OF THE MEMBRANES

Another important factor that can control the oil–water separation through membranes is the membrane surface property, which is measured by hydrophilicity, hydrophobicity, oleophilicity, oleophobicity or more simply the surface wettability of the active membrane surface. Since hydrophobic membranes possess low surface energy therefore water droplets form a greater contact angle in the solid surface. In the case of hydrophilic membranes, water droplets can wet their surface. Designing a super-wetting surface is the key step to the fabrication of oil–water separating membranes.

8.2.3 PRINCIPLE OF OIL AND WATER THROUGH MEMBRANE

The contact angle is a standard measure for studying the surface wettability of a solid surface. It is the measure of wettability of an ideally considered hypothetical surface. In contact angle measurement, a droplet with a fixed volume of liquid falls on the surface of the membrane, and the angle made by the droplet is measured by microscopic imaging. Contact angle measurement is one of the most simple, common and consistent methods of determining the surface property of a solid surface. It is applicable when there is a distinction between intrinsic and apparent parameters. Apparent variations may be due to the porosity of the surface and surface roughness, alteration in surface morphology at the time of the measurement, surface heterogeneity as well as contamination by the solution(s) on the membrane surface [28,29].

8.2.3.1 Contact Angle Dependence on Surface Roughness

In a membrane surface, both the surface chemical composition and the microstructure surface control its wettability. The wettability of a smooth surface is increased or decreased by varying the surface roughness of the pristine membrane. Various studies have been done where the roughness of hydrophilic or hydrophobic polymer-based membranes was varied by introducing nanopatterns over the surface of the membrane [30].

8.2.3.2 Surface Wettability of Water in Air

Wettability can also be measured by analysing the surface topographical structure and chemical polar or non-polar interaction as well as composition.

For an ideal smooth surface, the air contact angle is represented by Young's equation [31].

$$Cos\Theta_Y = [(\Upsilon_{SV} - \Upsilon_{SL})/\Upsilon_{LV}] \tag{8.1}$$

where Υ_{SV}, Υ_{SL}, Υ_{LV} are respectively the interfacial energies between the interfaces of solid and vapor; solid and liquid; liquid and vapor and are determined by the chemical constituents of surface material.

But in the case of rough surfaces, Young's equation is replaced by two models: Wenzel and Cassie's model. In the Wenzel model, empty surface area of

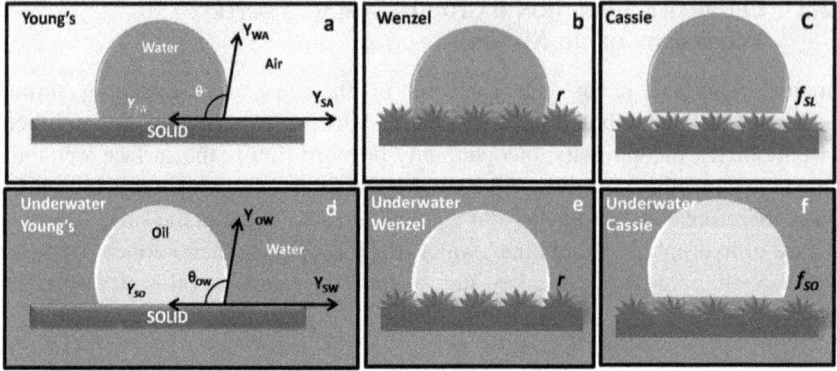

FIGURE 8.2 Schematic representation for water contact angle in an ideally smooth surface young's equation (a), and rough surface Wenzel explanation (b) and Cassie Explanation (c). Different explanation for Young's contact angle (d) Wenzel contact angle (e) and Cassie contact angle (f) on underwater wettability.

the membrane is completely filled with liquid, which is represented by the following equation,

$$\mathrm{Cos}\Theta_w = r \cdot \mathrm{Cos}\ \Theta_Y = r[(\Upsilon_{SV}-\Upsilon_{SL})/\Upsilon_{LV}] \tag{8.2}$$

The factor $r(r > 1)$ was introduced to describe the enlarged wettability on rough surfaces. For the rough surface, r is given as actual surface area of the surface divided by the apparent surface area. It is observed from the equation that the contact angle depends on both the surface morphology (reflected by r) and the chemical composition of the surface (reflected by Θ_o) of the membrane. A schematic representation of the water contact angle (WCA) for different conditions is shown in Figure 8.2.

The equation is further modified while there is occurrence of the composite wetting state, commonly when air traps under the droplet and forms a solid–water–air wetting. In that condition the apparent CA (Θ_{CB}) is given by the *Cassie equation*,

$$\mathrm{Cos}\Theta_{CB} = r f_{SL}\mathrm{Cos}\Theta_o + f_{SL}-1 \tag{8.3}$$

here, f_{SL} is used to express the solid water fraction under the area of contact, whereas the roughness ratio of the wet part is represented by r_r where $r_r < r$.

8.2.3.3 Surface Wettability of Oil in Under Water Environment

The theoretical explanation of the surface underwater wettability is explained by first assuming the surface to be an ideal surface. The underwater oil-contact angle (OCA) is determined by amalgamating Young's equation, Equation (8.1) of an ideal solid–air–water interface with the explanation of a solid–air–oil interface [32].

In an aqueous environment the apparent OCA in (Θ_{ow}) is expressed as,

$$\mathrm{Cos}\Theta_{ow} = (\Upsilon_{OA}\mathrm{Cos}\Theta_o - \Upsilon_{WA}\mathrm{Cos}\Theta_w)/\Upsilon_{ow} \tag{8.4}$$

where, Θ_w and Θ_o are respectively the WCA and in air OCA whereas, Υ_{OA}, Υ_{WA} and Υ_{OW} are respectively the interface surface energy of oil and air; water and air; oil and water interfaces.

It is observed from Equation (8.4) that a surface with hydrophilic properties also exhibits oleophilicity as in air environment oil possesses lower surface tension than that of water, i.e., in air environment $\Upsilon_{OA} < \Upsilon_{WA}$ ($\Theta_o < \Theta_w < 90°$) and hence $Cos\Theta_o$ and $Cos\Theta_w$ positive. As it is observed that the surface energy of oil or other organics in the liquid state is very less than that of water ($\Upsilon_{OA} \ll \Upsilon_{WA}$), therefore ($\Upsilon_{OA}Cos\Theta_o - \Upsilon_{WA}Cos\Theta_w$) will be negative; hence, it can be established that surfaces that are hydrophilic in air exhibit oleophobic properties when it is underwater at the interface of solid, water and oil. Again, it is noted that in an air environment, a membrane with oleophilic ($\Theta_o < 0°$) and hydrophobic ($\Theta_w > 90°$) surface exhibits oleophilicity in aqueous medium because the right-hand side numerator of Equation (8.1) is always positive. While constructing hydrophobic, oleophobic or underwater oleophobic surfaces, it should be noted that $\Upsilon_{OA}Cos\Theta_o$ value is greater than that of $\Upsilon_{WA}Cos\Theta_w$ value.

The Cassie and Wenzel equations in an underwater environment are attained by introducing roughness in the surface and contact phase fraction. The Wenzel equation for the underwater environment,

$$Cos\Theta_W = rCos\Theta_{OW} \tag{8.5}$$

Similarly, Cassie equation for underwater environment,

$$Cos\Theta_{CB} = r_f f_{SO} Cos\Theta_{OW} + f_{SO} - 1 = r_f Cos\Theta_{OW} - f_{SW}(r_f Cos\Theta_{OW} + 1) \tag{8.6}$$

where fraction at contact area of solid and oil is represented by f_{SO}.

8.2.3.4 Dependence of Contact Angle on Surface Polarity

While considering the wettability of a membrane surface, it is also important to study the possible interaction of the liquid to be separated from the molecules present on the membrane surface. Wettability will be high when the two molecules, i.e., the molecules on the membrane surface and the liquid that get in contact, contain groups of the same polarity and it will be low in the case of materials with different polarities. Cheng et al. explained the effect of surface polarity on wettability by assembling thiols in a copper substrate during the fabrication of surfaces with controlled oil wettability in an underwater environment. Figure 8.3 shows that increasing OH groups on the substrate increases hydrophilicity while making the surface oleophobic [33].

8.3 MODIFICATION IN POLYMERIC MEMBRANES FOR OIL–WATER SEPARATION

8.3.1 HYDROPHILIC MODIFICATION IN POLYMERIC MEMBRANES

Using hydrophilic additives, or more simply, molecules capable of participating in intermolecular hydrogen bonding with water, the membrane's water permeability is

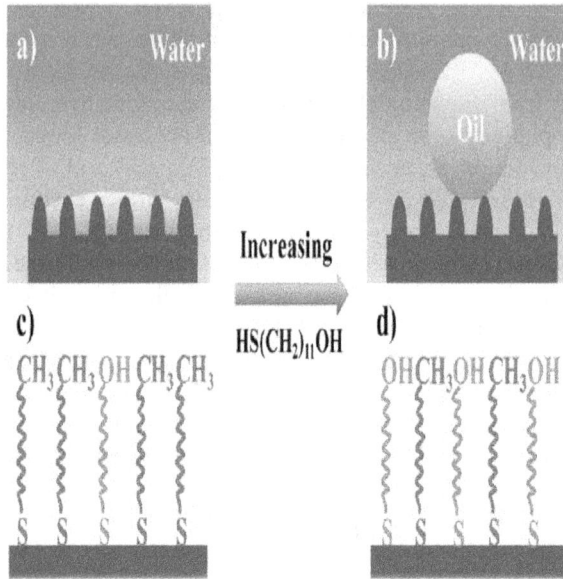

FIGURE 8.3   Change in wettability upon addition of more OH groups.

Source: [33].

increased, which in turn makes the membrane less prone to fouling. During the fabrication of a membrane, polyvinylpyrrolidone (PVP) and PEG are generally used as hydrophilic additives and pore-enhancing agents [14,17]. The development of blended polymeric membranes with the addition of hydrophilic polymers into the polymer matrix is another approach in the designing of membranes for the separation of oil and water. Some examples include PES/cellulose acetate/PEG asymmetric membrane [14], polyetherimide/PEG/polybenzimidazole asymmetric membrane [34], sulfonated polycarbonate/PVDF membrane by Masuelli et al. for emulsified oily water treatment [35] and polyetherimide/sulfonated poly(ether ether ketone) (PEI/SPEEK) membrane by Bowen et al. [36].

As mentioned above, based on wettability, oil–water separating membranes are mainly categorised in two ways, i.e., hydrophilic–oleophobic surfaces and hydrophobic–oleophilic surfaces. A surface having a water/oil contact angle above 90° is considered as hydrophobic/oleophobic surface, and one above 150° is deemed a superhydrophobic/superoleophobic surface. Again, when the water/oil contact angle lies between 10° and 90°, it is termed as hydrophilic/oleophilic surface. If it is less than 10 degrees then it is superhydrophilic/superoleophilic surface.

8.3.2   SUPERHYDROPHILIC AND SUPEROLEOPHOBIC MEMBRANES

As oils generally possess lower surface energy than water, hence a superhydrophilic as well as a hydrophilic surface generally shows oleophilicity while superoleophobic surfaces show superhydrophobicity and vice versa. Some surfaces can simultaneously exhibit hydrophilicity as well as oleophobicity depending upon the interaction of membrane materials with polar or non-polar liquids; these surfaces are named as stimuli-responsive-surfaces [37–39]. Tuteja et al. developed such hygroresponsive surfaces in the membrane which have both superhydrophilic as well as superoleophobic properties in air and under water, respectively. The dip coating method was employed for membrane fabrication where POSS°+°x-PEGDA complex was formed by crosslinking between fluorodecyl polyhedral oligomeric silsesquioxane (POSS) and poly(ethylene glycol) diacrylate (x-PEGDA). The membrane showed in air superoleophobic property with some fluorodecyl POSS aggregates. In contrast, in an aqueous environment, the aggregate disappears due to the action of water, which results in surface reconfiguration. The as-prepared membrane efficiently separated an oil–water emulsion in a gravity-driven separation system. The separation efficiency was greater than 99%, according to thermogravimetric analyses (TGA) [40].

8.3.3 SUPERHYDROPHOBIC AND SUPEROLEOPHILIC MEMBRANES

Membranes with superoleophilic and superhydrophobic surfaces have also been developed in recent days. These unique properties of superhydrophobicity and superoleophilicity will make it possible to spread, penetrate and easily absorb the oil phase on their surfaces while repelling the aqueous phase. As surface wettability depends on the collective effects of the chemical composition of the surface as well as the surface topology hence superhydrophobic and superoleophobic surfaces are developed by firstly constructing a hydrophobic rough structure followed by modifying it with low surface energy chemicals. The superhydrophobic and superoleophilic membranes are generally prepared by following three major steps: surface roughness modification, preparation of surface with activation and cleaning, surface chemistry modification.

A superhydrophobic-superoleophilic membrane based on PVDF was developed for the separation of oil and water emulsion systems. A modified-phase inversion method was employed where ammonia acts as the dehydrofluorinating agent to increase the mechanical strength of the PVDF-based membrane. A cluster form of PVDF was obtained, which finally resulted in the growth of spherical particles over the membrane, thereby making the membrane uniformly skinless. The as-prepared membrane was able to separate various types of oily water emulsions both surfactant-free and surfactant-stabilised emulsions, where oil droplet sizes range between micrometres to nanometres [41]. Zn-Al LDH and cellulose based membranes were fabricated by Yue et al. by employing a hydrothermal reaction between the metal precursors and the cellulose-based membrane followed by hydrophobic modification of the as-developed membrane. The superhydrophobic membrane was found to possess a good separation efficiency with excellent recyclability, which makes it a possible candidate for practical application in the field of oil-in-water separation [42].

8.3.4   Superhydrophilic and Underwater Superoleophobic Membranes

An ideal superhydrophilic surface possesses a WCA of 0°. Generally, a superhydrophilic surface exhibits superoleophobicity, in the air; oil repelling ability underwater. Moreover, extensive wettability in underwater environments or oil environments has been realised based on natural micro/nanoscale structures and their chemical composition. One such example is inspired from fish scale. Because fish scales are protected from oil contamination, hydrophilic and oleophobic surfaces of fish scales have received a tremendous response as an oil-water separating medium [43–47]. Designing membranes with hydrophilic and underwater superoleophobic properties will also overcome some problems related to the intrinsic properties of membranes, like membrane fouling and pore blocking due to intrinsic oleophilicity. Moreover, the deposition of wastes on the membrane and absorption of oil by the surface usually results in secondary pollution throughout the post-treatment of such materials. A hydrophilic membrane based on polyacrylamide (PAM) hydrogel in stainless-steel mesh was developed and applied to separate water from an oil-in-water emulsion system. The as-developed modified stainless steel mesh was prewetted before use, and it demonstrated a fascinating separation efficiency of more than 99% [48]. Later, the development of an oil–water separation barrier with the same property, i.e., hydrophilic and underwater oleophobic surfaces, was developed by various groups [49–52]. A superhydrophilic and superoleophobic ultrathin film composed of SWCNT and TiO_2 nanocomposite was developed by coating TiO_2 onto an ultrathin network film of SWCNT via sol–gel method. With a flux of more than 30,000 L m^{-2} h^{-1} bar^{-1} and a separation efficiency of more than 99%, nanocomposite film with pore-size in the nano regime can separate oil–water emulsion of both surfactant-containing or surfactant-free emulsions. The as-prepared membrane also has self-cleaning activity in the presence of UV radiation due to TiO_2 nanoparticles, which are known for their photocatalytic activity in the presence of light irradiation of specific wavelength [53]. A nanocomposite membrane was created by modifying a PVDF membrane and depositing SiO_2 on the surface with polydopamine, which was inspired by the hydrophilic–underwater oleophobic texture of fish scales. The as-developed membrane, which possessed a superhydrophilic/underwater superoleophobic surface exhibited excellent separation efficiency of more than 98% for a mixture of hexadecane-in-water emulsion [54]. LDHs, a type of inorganic layered material, have also gotten a lot of attention because of their unique structure for separation, purification and ion exchange capability [55,56]. Cui et al. developed a modified sisal-like NiCo LDH architecture on a polydopamine-impregnated highly porous PVDF membrane by the hydrothermal method. The superhydrophilic and underwater superoleophobic membranes demonstrated outstanding flux for various oil types as well as impressive antifouling performance with 99% oil rejection [57].

8.3.5 MEMBRANES WITH DUAL PROPERTY FOR SEPARATION OF OIL AND WATER

Another favourable way to tune the membrane surface characteristics is the addition of amphiphilic copolymers, which are polymers with both hydrophobic and hydrophilic moieties. W. Chen et al. used an amphiphilic copolymer, Pluronic F127, in varying concentrations to modify the surface of a neat PES membrane. The addition of amphiphilic polymers enhances the hydrophilicity of the membrane and thereby improves the permeate flux as compared to the pristine membrane. Also, fouling due to oil agglomeration over the membrane pores was minimised. With rising the amount of Pluronic F127 to 20%, the permeability of the designed membrane was improved to 82.98 L m^{-2} h^{-1} with a significant increase in selectivity of the membrane for rejection of oil [58]. In Table 8.1, some remarkable works on separation of oil and water by using polymeric membranes have been listed.

8.4 CONCLUSION

Rapid industrialisation and an excessive need for energy cause the generation of heavy amounts of oily wastewater in petrochemical industries, which has an adverse effect on the ecosystem. At the same time, water scarcity is also a concern. With the increasing population, the necessity of water for living beings has increased, and it has become a challenging task for various researchers to develop technology to purify wastewater so as to fulfil the demand for water. Various techniques have been established for the separation of oil and water from oil-based wastewater. Due to its various advantages, membrane technology and its development for the treatment of wastewater have become a trending topic of research. Membrane processes are commercially popular due to their cost effectiveness, low energy consumption and lack of risk of generating by-products. Membrane processes have also been tested at the bench scale and have been proven to be highly efficient for the treatment of large amounts of wastewater. Also, modifications in the membrane system and the development of a hybrid membrane system make the membrane technology a multipurpose one. Surface modification in membranes and various pore-tuning methods have been applied for better performance and separation efficiency. Operating parameters like pressure, temperature, permeate flow, etc., also have crucial roles in the performance of the membrane. In this book chapter, we have emphasised various approaches for the development of membranes for oil separation from oily wastewater. Various mechanisms for oil-water separation by membrane-based technology have also been briefly discussed. Researchers have made various efforts to advance high-performing membranes for environmental remediation as well as to address the demand for pure water. These endless and collective efforts in the advancement of proficient membranes provide an effective and economically viable technique for the remediation of oil-based wastewater.

TABLE 8.1
Some Recent Remarkable Works on Membrane Development for Oil-Based Wastewater Treatment

Membrane and Nature	Type of Membrane	Method of Fabrication	Type of Oil–Water Sample, Oil Content	Performance	References
Crossflow-UF membrane (PSf–PEG–PVP blend membrane)	Cross flow UF membrane	Immersion precipitation (i.e., phase inversion) method.	Synthetic oily water = 100 mg/L	Highest permeate flux = 128 L m^{-2} h^{-1} Oil recovery = more than 90%	[59]
PES/cellulose acetate (CA)/PEG asymmetric membrane	Hydrophilic UF membrane	Non-solvent-induced-phase inversion	Industrial oily water (produced) = 366 mg/L	Stable permeate flux = 27 Lm^{-2} h^{-1} Oil rejection = 88%	[15]
Pluronic F127 modified polyethersulfone ultrafiltration membranes	Hydrophilic UF membrane	Non-solvent-induced-phase inversion	Synthetic oil–water emulsion with 50 mg/L oil content	Permeate flux = 82.98 L m^{-2} h^{-1} Oil rejection = 100%	[58]
LiCl·H$_2$O–TiO$_2$·NPs–PVDF UF hollow fibre membrane	Hydrophilic UF membrane	Non-solvent-induced-phase inversion	Synthetic oil–water emulsion with 90 mg/L oil content	Permeate flux = 82.50 L m^{-2} h^{-1} separationefficiency=98.83%	[19]
PVDF–PVP–TiO$_2$ composite hollow fibre membranes	NF membrane	Dry-jet wet fibre spinning method	Refinery based synthetic wastewater	Permeate flux = 70.48 L m^{-2} h^{-1} oil rejection = 99.7%	[21]
Phosphorylated TiO$_2$–SiO$_2$ or simply PTS/ polysulfone composite membrane	Hydrophilic membrane	Non-solvent-induced-phase inversion	Synthetic oily solution	Permeate flux = 116 L m^{-1} h^{-1}, oil retention = 92%	[23]
Hydrous manganese dioxide nanoparticle-PES MMM	Hydrophilic mixed matrix membrane (MMM)	Non-solvent-induced-phase inversion	Synthetic oily wastewater with oil concentration of 46 mg/L	Maximum permeate flux = 573.2 L m^{-2} h^{-1} Oil rejection = 98.5%	[24]

Membrane	Wettability	Fabrication method	Feed	Performance	Ref.
Sulfated Y-doped zirconia particles-PS membrane	Hydrophilic membrane	Sol–gel process	Synthetic oily wastewater with 80 mg/L oil content	Permeate flux = 104.5 L m^{-2} h^{-1}; Oil rejection = 99.16%	[25]
Sepiolite-GO-cellulose acetate and Mg-Al LDH cellulose acetate membranes	Hydrophilic–oleophobic membrane	Non-solvent-induced-phase inversion followed by layer-by-layer assembly	Surfactant-free O/W emulsions with oil content 1 g/L	Permeate flux = 1910 L m^{-2} h^{-1}; Separation efficiency = more than 90%	[26]
ZnAl LDH on cellulose support	Superhydrophobic and super oleophilic	In situ growth of ZnAl LDH nanosheets followed by grafting with silane coupling agent	Oil water emulsion with 198mL of toluene followed by addition of 4 mL of water	Permeate flux = 500 L m^{-2} h^{-1} Without extra energy; Separation efficiency = 94%	[55]
NiAlFe LDH/PDA/PVDF (hydrophilic and UW oleophobic)	Hydrophilic–underwater oleophobic	In situ growth of LDH on the PDA modified PVDF membrane (hydrothermal treatment)	Oil:water = 1:99	Permeate flux = 1425 L m^{-2} h^{-1} bar^{-1}; Separation efficiency = 99%	[60]
pVA/PVDF/NiCo LDH	Hydrophilic/oleophobic	In situ growth of Ni-Co LDH microcrystal on the membrane surface	Oil:water = 1:99	Permeate flux = 705 L m^{-2} h^{-1}; Oil rejection = 99%	[57]
PET–cellulose–PEG membrane	Superhydrophilic	Phase inversion	Oil:water = 10:90	Permeate flux 86 L m^{-2} h^{-1}; Separation efficiency = 98%	[61]

REFERENCES

1. Shannon, M. A., Bohn, P. W., Elimelech, M., Georgiadis, J. G., Marinas, B. J., & Mayes, A. M. (2008). Science and technology for water purification in the coming decades. *Nature*, *452*(7185), 301–310. https://doi.org/10.1038/nature06599

2. Wang, B., Liang, W., Guo, Z., & Liu, W. (2015). Biomimetic super-lyophobic and super-lyophilic materials applied for oil/water separation: A new strategy beyond nature. *Chemical Society Reviews*, *44*(1), 336–361. https:// doi.org/10.1039/C4CS00220B

3. Mrayyan, B., & Battikhi, M. N. (2005). Biodegradation of total organic carbons (TOC) in Jordanian petroleum sludge. *Journal of Hazardous Materials*, *120*(1–3), 127–134. https://doi.org/10.1016/j.jhazmat.2004.12.033

4. Reddy, M. V., Devi, M. P., Chandrasekhar, K., Goud, R. K., & Mohan, S. V. (2011). Aerobic remediation of petroleum sludge through soil supplementation: Microbial community analysis. *Journal of Hazardous Materials*, *197*, 80–87. https://doi.org/10.1016/j.jhazmat.2011.09.061

5. Aguilera, F., Méndez, J., Pásaro, E., & Laffon, B. (2010). Review on the effects of exposure to spilled oils on human health. *Journal of Applied Toxicology: An International Journal*, *30*(4), 291–301. https://doi.org/10.1002/jat.1521

6. Robertson, S. J., McGill, W. B., Massicotte, H. B., & Rutherford, P. M. (2007). Petroleum hydrocarbon contamination in boreal forest soils: A mycorrhizal ecosystems perspective. *Biological Reviews*, *82*(2), 213–240. https://doi.org/10.1111/j.1469-185X.2007.00012.x

7. Pendergast, M. M., & Hoek, E. M. (2011). A review of water treatment membrane nanotechnologies. *Energy & Environmental Science*, *4*(6), 1946–1971. https://doi.org/10.1039/C0EE00541J

8. Kim, J., & Van der Bruggen, B. (2010). The use of nanoparticles in polymeric and ceramic membrane structures: Review of manufacturing procedures and performance improvement for water treatment. *Environmental Pollution*, *158*(7), 2335–2349. https://doi.org/10.1016/j.envpol.2010.03.024

9. Malaeb, L., & Ayoub, G. M. (2011). Reverse osmosis technology for water treatment: State of the art review. *Desalination*, *267*(1), 1–8. https://doi.org/10.1016/j.desal.2010.09.001

10. Anderson, G. K., Saw, C. B., & Le, M. S. (1987). Oil/water separation with surface modified membranes. *Environmental Technology*, *8*(1–12), 121–132. https://doi.org/10.1080/09593338709384470

11. Zhu, H., & Guo, Z. (2016). Understanding the separations of oil/water mixtures from immiscible to emulsions on super-wettable surfaces. *Journal of Bionic Engineering*, *13*(1), 1–29. https://doi.org/10.1016/S1672-6529(14)60156-6

12. Otitoju, T. A., Ahmad, A. L., & Ooi, B. S. (2016). Polyvinylidene fluoride (PVDF) membrane for oil rejection from oily wastewater: A performance review. *Journal of Water Process Engineering*, *14*, 41–59. https://doi.org/10.1016/j.jwpe.2016.10.011

13. Zhang, Y., Gao, B., Lu, L., Yue, Q., Wang, Q., & Jia, Y. (2010). Treatment of produced water from polymer flooding in oil production by the combined method of hydrolysis acidification-dynamic membrane bioreactor–coagulation process. *Journal of Petroleum Science and Engineering*, *74*(1–2), 14–19. https://doi.org/10.1016/j.petrol.2010.08.001

14. Cheryan, M., & Rajagopalan, N. (1998). Membrane processing of oily streams. Wastewater treatment and waste reduction. *Journal of Membrane Science*, *151*(1), 13–28. https://doi.org/10.1016/S0376-7388(98)00190-2

15. Mansourizadeh, A., & Javadi Azad, A. (2014) Preparation of blend polyethersulfone/cellulose acetate/polyethylene glycol asymmetric membranes for oil–water separation. *Journal of Polymer Research, 21*, 1–9. https://doi.org/10.1007/s10965-014-0375-x

16. Rahimpour, A., & Madaeni, S. S. (2007). Polyethersulfone (PES)/cellulose acetate phthalate (CAP) blend ultrafiltration membranes: Preparation, morphology, performance and antifouling properties. *Journal of Membrane Science, 305*(1–2), 299–312. https://doi.org/10.1016/j.memsci.2007.08.030

17. Ochoa, N. A., Masuelli, M., & Marchese, J. (2003). Effect of hydrophilicity on fouling of an emulsified oil wastewater with PVDF/PMMA membranes. *Journal of Membrane Science, 226*(1–2), 203–211. https://doi.org/10.1016/j.memsci.2003.09.004

18. Vatanpour, V., Madaeni, S. S., Moradian, R., Zinadini, S., & Astinchap, B. (2012). Novel antibifouling nanofiltration polyethersulfone membrane fabricated from embedding TiO_2 coated multiwalled carbon nanotubes. *Separation and Purification Technology, 90*, 69–82. https://doi.org/10.1016/j.seppur.2012.02.014

19. Yuliwati, E., & Ismail, A. F. (2011). Effect of additives concentration on the surface properties and performance of PVDF ultrafiltration membranes for refinery produced wastewater treatment. *Desalination, 273*(1), 226–234. https://doi.org/10.1016/j.desal.2010.11.023

20. Yi, X. S., Yu, S. L., Shi, W. X., Sun, N., Jin, L. M., Wang, S., … & Sun, L. P. (2011). The influence of important factors on ultrafiltration of oil/water emulsion using PVDF membrane modified by nano-sized TiO_2/Al_2O_3. *Desalination, 281*, 179–184. https://doi.org/10.1016/j.desal.2011.07.056

21. Ong, C. S., Lau, W. J., Goh, P. S., Ng, B. C., & Ismail, A. F. (2015). Preparation and characterization of PVDF–PVP–TiO_2 composite hollow fiber membranes for oily wastewater treatment using submerged membrane system. *Desalination and Water Treatment, 53*(5), 1213–1223. https://doi.org/10.1080/19443994.2013.855679

22. Ong, C. S., Lau, W. J., Goh, P. S., Ng, B. C., & Ismail, A. F. (2014). Investigation of submerged membrane photocatalytic reactor (sMPR) operating parameters during oily wastewater treatment process. *Desalination, 353*, 48–56. https://doi.org/10.1016/j.desal.2014.09.008

23. Zhang, Y., Liu, F., Lu, Y., Zhao, L., & Song, L. (2013). Investigation of phosphorylated TiO_2–SiO_2 particles/polysulfone composite membrane for wastewater treatment. *Desalination, 324*, 118–126. https://doi.org/10.1016/j.desal.2013.06.007

24. Gohari, R. J., Halakoo, E., Lau, W. J., Kassim, M. A., Matsuura, T., & Ismail, A. F. (2014). Novel polyethersulfone (PES)/hydrous manganese dioxide (HMO) mixed matrix membranes with improved anti-fouling properties for oily wastewater treatment process. *RSC Advances, 4*(34), 17587–17596. https://doi.org/10.1039/C4RA00032C

25. Zhang, Y., Shan, X., Jin, Z., & Wang, Y. (2011). Synthesis of sulfated Y-doped zirconia particles and effect on properties of polysulfone membranes for treatment of wastewater containing oil. *Journal of Hazardous Materials, 192*(2), 559–567. https://doi.org/10.1016/j.jhazmat.2011.05.058

26. Li, F., Gao, R., Wu, T., & Li, Y. (2017). Role of layered materials in emulsified oil/water separation and anti-fouling performance of modified cellulose acetate membranes with hierarchical structure. *Journal of Membrane Science, 543*, 163–171. https://doi.org/10.1016/j.memsci.2017.08.053

27. Alkhouzaam, A., & Qiblawey, H. (2021). Novel polysulfone ultrafiltration membranes incorporating polydopamine functionalized graphene oxide with enhanced flux and fouling resistance. *Journal of Membrane Science, 620*, 118900. https://doi.org/10.1016/j.memsci.2020.118900

28. Jung, Y. C., & Bhushan, B. (2006). Contact angle, adhesion and friction properties of micro-and nanopatterned polymers for superhydrophobicity. *Nanotechnology*, *17*(19), 4970. https://doi.org/10.1088/0957-4484/17/19/033

29. Chau, T. T., Bruckard, W. J., Koh, P. T. L., & Nguyen, A. V. (2009). A review of factors that affect contact angle and implications for flotation practice. *Advances in Colloid and Interface Science*, *150*(2), 106–115. https://doi.org/10.1016/j.cis.2009.07.003

30. Wenzel, R. N. (1936). Resistance of solid surfaces to wetting by water. *Industrial & Engineering Chemistry*, *28*(8), 988–994. https://doi.org/10.1021/ie50320a024

31. Young, T. (1805). III. An essay on the cohesion of fluids. *Philosophical Transactions of the Royal Society of London*, *95*, 65–87. https://doi.org/10.1098/rstl.1805.0005

32. Jung, Y. C., & Bhushan, B. (2009). Wetting behavior of water and oil droplets in three-phase interfaces for hydrophobicity/philicity and oleophobicity/philicity. *Langmuir*, *25*(24), 14165–14173. https://doi.org/10.1021/la901906h

33. Cheng, Z., Lai, H., Du, Y., Fu, K., Hou, R., Zhang, N., & Sun, K. (2013). Underwater superoleophilic to superoleophobic wetting control on the nanostructured copper substrates. *ACS Applied Materials & Interfaces*, *5*(21), 11363–11370. https://doi.org/10.1021/am403595z

34. Xu, Z. L., Chung, T. S., Loh, K. C., & Lim, B. C. (1999). Polymeric asymmetric membranes made from polyetherimide/polybenzimidazole/poly (ethylene glycol) (PEI/PBI/PEG) for oil–surfactant–water separation. *Journal of Membrane Science*, *158*(1–2), 41–53. https://doi.org/10.1016/S0376-7388(99)00030-7

35. Masuelli, M., Marchese, J., & Ochoa, N. A. (2009). SPC/PVDF membranes for emulsified oily wastewater treatment. *Journal of Membrane Science*, *326*(2), 688–693. https://doi.org/10.1016/j.memsci.2008.11.011

36. Bowen, W. R., Cheng, S. Y., Doneva, T. A., & Oatley, D. L. (2005). Manufacture and characterisation of polyetherimide/sulfonated poly (ether ether ketone) blend membranes. *Journal of Membrane Science*, *250*(1–2), 1–10. https://doi.org/10.1016/j.memsci.2004.07.004

37. Bowen, W. R., Cheng, S. Y., Doneva, T. A., & Oatley, D. L. (2005). Manufacture and characterisation of polyetherimide/sulfonated poly(ether ether ketone) blend membranes. *Journal of Membrane Science*, *250*(1–2), 1–10. https://doi.org/10.1016/j.memsci.2004.07.004

38. Howarter, J. A., Genson, K. L., & Youngblood, J. P. (2011). Wetting behavior of oleophobic polymer coatings synthesized from fluorosurfactant-macromers. *ACS Applied Materials & Interfaces*, *3*(6), 2022–2030. https://doi.org/10.1021/am200255v

39. Kota, A. K., Kwon, G., Choi, W., Mabry, J. M., & Tuteja, A. (2012). Hygro-responsive membranes for effective oil–water separation. *Nature Communications*, *3*(1), 1025. https://doi.org/ 10.1038/2027

40. Zhang, W., Shi, Z., Zhang, F., Liu, X., Jin, J., & Jiang, L. (2013). Superhydrophobic and superoleophilic PVDF membranes for effective separation of water-in-oil emulsions with high flux. *Advanced Materials*, *25*(14), 2071–2076. https://doi.org/10.1002/adma.201204520

41. Yue, X., Li, J., Zhang, T., Qiu, F., Yang, D., & Xue, M. (2017). In situ one-step fabrication of durable superhydrophobic-superoleophilic cellulose/LDH membrane with hierarchical structure for efficiency oil/water separation. *Chemical Engineering Journal*, *328*, 117–123. https://doi.org/10.1016/j.cej.2017.07.026

42. Chen, L., Liu, M., Lin, L., Zhang, T., Ma, J., Song, Y., & Jiang, L. (2010). Thermal-responsive hydrogel surface: Tunable wettability and adhesion to oil at the water/solid interface. *Soft Matter*, *6*(12), 2708–2712. https://doi.org/10.1039/C002543G

43. Wang, B., Liang, W., Guo, Z., & Liu, W. (2015). Biomimetic super-lyophobic and super-lyophilic materials applied for oil/water separation: A new strategy beyond nature. *Chemical Society Reviews*, *44*(1), 336–361. https://doi.org/10.1039/C4CS002 20B

44. Nosonovsky, M., & Bhushan, B. (2009). Multiscale effects and capillary interactions in functional biomimetic surfaces for energy conversion and green engineering. *Philosophical Transactions of the Royal Society A: Mathematical, Physical and Engineering Sciences*, *367*(1893), 1511–1539. https://doi.org/10.1098/rsta.2009.0008

45. Wang, B., Guo, Z., & Liu, W. (2014). pH-responsive smart fabrics with controllable wettability in different surroundings. *RSC Advances*, *4*(28), 14684–14690. https://doi.org/10.1039/C3RA48002J

46. Zhang, J., Wu, L., Zhang, Y., & Wang, A. (2015). Mussel and fish scale-inspired underwater superoleophobic kapok membranes for continuous and simultaneous removal of insoluble oils and soluble dyes in water. *Journal of Materials Chemistry A*, *3*(36), 18475–18482. https://doi.org/10.1039/C5TA04839G

47. Liu, X., Gao, J., Xue, Z., Chen, L., Lin, L., Jiang, L., & Wang, S. (2012). Bioinspired oil strider floating at the oil/water interface supported by huge superoleophobic force. *ACS Nano*, *6*(6), 5614–5620.https://doi.org/10.1021/nn301550v

48. Dong, Y., Li, J., Shi, L., Wang, X., Guo, Z., & Liu, W. (2014). Underwater superoleophobic graphene oxide coated meshes for the separation of oil and water. *Chemical Communications*, *50*(42), 5586–5589. https://doi.org/10.1039/C4CC014 08A

49. Wen, Q., Di, J., Jiang, L., Yu, J., & Xu, R. (2013). Zeolite-coated mesh film for efficient oil–water separation. *Chemical Science*, *4*(2), 591–595. http://doi.org/10.1039/CsSC21772D

50. Zhang, F., Zhang, W. B., Shi, Z., Wang, D., Jin, J., & Jiang, L. (2013). Nanowire-haired inorganic membranes with superhydrophilicity and underwater ultralow adhesive superoleophobicity for high-efficiency oil/water separation. *Advanced Materials*, *25*(30), 4192–4198. https://doi.org/10.1002/adma.201301480

51. Liu, N., Chen, Y., Lu, F., Cao, Y., Xue, Z., Li, K., ... & Wei, Y. (2013). Straightforward oxidation of a copper substrate produces an underwater superoleophobic mesh for oil/water separation. *ChemPhysChem*, *14*(15), 3489–3494. https://doi.org/10.1002/cphc.201300691

52. Gao, S. J., Shi, Z., Zhang, W. B., Zhang, F., & Jin, J. (2014). Photoinduced superwetting single-walled carbon nanotube/TiO_2 ultrathin network films for ultrafast separation of oil-in-water emulsions. *ACS Nano*, *8*(6), 6344–6352. https://doi.org/10.1002/cphc.201300691

53. Hossain, M. S., Ebrahimi, H., & Ghosh, R. (2022). Fish scale inspired structures: A review of materials, manufacturing and models. *Bioinspiration & Biomimetics*. https://doi.org/10.1088/1748-3190/ac7fd0

54. Yang, Y., Li, Y., Cao, L., Wang, Y., Li, L., & Li, W. (2021). Electrospun PVDF–SiO_2 nanofibrous membranes with enhanced surface roughness for oil–water coalescence separation. *Separation and Purification Technology*, *269*, 118726. https://doi.org/10.1016/j.seppur.2021.118726

55. Yue, X., Zhang, T., Yang, D., Qiu, F., Li, Z., Zhu, Y., & Yu, H. (2018). Oil removal from oily water by a low-cost and durable flexible membrane made of layered double hydroxide nanosheet on cellulose support. *Journal of Cleaner Production*, *180*, 307–315. https://doi.org/10.1016/j.jclepro.2018.01.160

56. Yang, Y., Yan, X., Hu, X., Feng, R., Zhou, M., & Cui, W. (2018). Development of zeolitic imidazolate framework-67 functionalized Co-Al LDH for CO_2 adsorption. *Colloids and Surfaces A: Physicochemical and Engineering Aspects, 552*, 16–23. https://doi.org/10.1016/j.colsurfa.2018.05.014

57. Cui, J., Wang, Q., Xie, A., Lang, J., Zhou, Z., & Yan, Y. (2019). Construction of superhydrophilic and underwater superoleophobic membranes via in situ oriented NiCo-LDH growth for gravity-driven oil/water emulsion separation. *Journal of the Taiwan Institute of Chemical Engineers, 104*, 240–249. https://doi.org/10.1016/j.jtice.2019.08.003

58. Chen, W., Peng, J., Su, Y., Zheng, L., Wang, L., & Jiang, Z. (2009). Separation of oil/water emulsion using Pluronic F127 modified polyethersulfone ultrafiltration membranes. *Separation and Purification Technology, 66*(3), 591–597. https://doi.org/10.1016/j.seppur.2009.01.009

59. Chakrabarty, B., Ghoshal, A. K., & Purkait, M. K. (2010). Cross-flow ultrafiltration of stable oil-in-water emulsion using polysulfone membranes. *Chemical Engineering Journal, 165*(2), 447–456. https://doi.org/10.1016/j.cej.2010.09.031

60. Zong, Y., Ma, S., Xue, J., Gu, J., & Wang, M. (2021). Bifunctional NiAlFe LDH-coated membrane for oil-in-water emulsion separation and photocatalytic degradation of antibiotic. *Science of the Total Environment, 751*, 141660. https://doi.org/10.1016/j.scitotenv.2020.141660

61. Bhuyan, C., Konwar, A., Bora, P., Rajguru, P., & Hazarika, S. (2023). Cellulose nanofiber-poly (ethylene terephthalate) nanocomposite membrane from waste materials for treatment of petroleum industry wastewater. *Journal of Hazardous Materials, 442*, 129955. https://doi.org/10.1016/j.jhazmat.2022.129955

9 Membranes for Removal of Phenolic Compounds from Petroleum Industry Wastewater

Ananya Saikia and Chinmoy Bhuyan

9.1 INTRODUCTION

One concerning issue facing the planet is water contamination, which endangers both humankind and the ecosystem. The amount of sewage being discharged into water resources has increased due to rising industrial and human activity; hence, the water has become polluted. Among the various pollutants in different industrial wastewater systems, phenol is considered a prior pollutant in wastewater. Some of the phenolic compounds are shown in Figure 9.1.

In refinery wastewater, phenols are typically regarded as one of the most dangerous organic contaminants since they are extremely harmful even in small concentrations. Phenol and phenolic compounds have adverse health effects on humans, and when disposed of in water streams, they severely affect aquatic organisms even at low concentrations. The wastewater of many different sectors, including coal treatment (9–6800 ppm), petroleum refineries (6–500 ppm), coking operations (28–3900 ppm) and petroleum distillate facilities (2.8–1220 ppm), contains phenols [1]. Phenolic compounds are in the 11th position among 126 hazardous chemicals as labelled by the United States Environmental Protection Agency [2,3]. Phenol and phenol-based compounds can enter the human body through contaminated surfaces and ground waterways. Additionally, the amount of phenol in the fluid sources can trigger oxidation and disinfection processes that generate other harmful substitute chemicals. Phenol exposure leads to high levels of irritation to the eye, skin and mucous membranes, as well as headaches and instability. Again, prolonged exposure may cause harm to the liver, kidneys, lungs and spleen. The ability of phenolic compounds to cause disease, particularly chlorophenols, has also been reported in different studies [4–7]. Therefore, considering the hazards due to phenol and its various derivatives, it is important to remove or neutralise them completely from the industrial wastes before disposing of them in soil or water bodies. Developing efficient abatement methods for their removal is particularly desirable given the substantial threat posed by contaminated industrial effluent containing phenol and its derivatives.

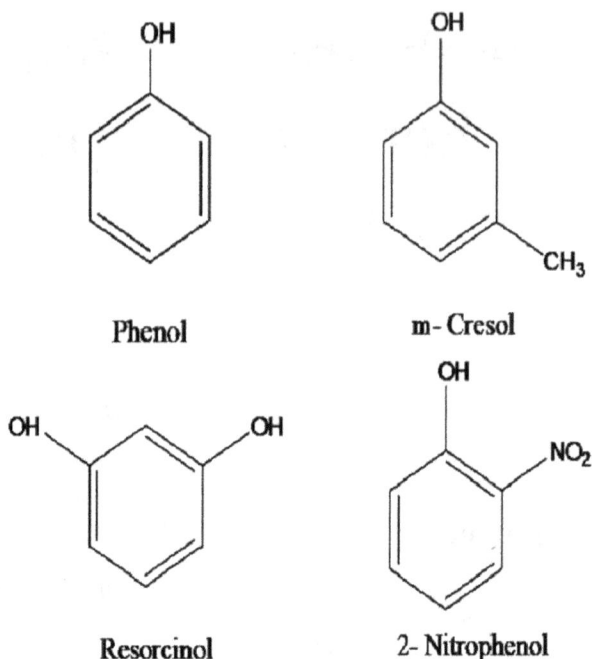

FIGURE 9.1 Chemical structure of some phenolic compounds.

Various techniques have been employed for the elimination of phenols from contaminated water, including solvent extraction, adsorption, chemical oxidation, biological treatment and distillation [8–12]. The conventional separation techniques were always considered unprofitable because of several problems like high energy consumption, the generation of secondary pollutants, the requirement of a large space and their high cost. Moreover, it is also important to ensure the nearly complete removal of phenol or phenolic compounds, as only a very small quantity of them is even harmful to human health, which the conventional techniques never achieve. Membrane, solvent extraction, pervaporation, liquid membrane and membrane distillation are the extensively employed membrane-based separation methods in industries for phenol removal [13]. Two promising techniques for removing phenol from wastewater include adsorption and pervaporation. Other attractive methods used with membranes to eliminate phenol in aqueous solutions include membrane bioreactors and photocatalytic membrane reactors, which efficiently remove phenol from wastewater [14,15].

9.2 MECHANISM FOR PHENOL REMOVAL THROUGH MEMBRANE

The mechanism of elimination of phenol present in a contaminated water system varies depending upon the separation process. In the case of selective removal of phenol through the membrane or the combined system with membrane, it mainly depends upon the membrane materials used for the fabrication of the membrane. Two main types of membrane material are available for the fabrication of membranes used primarily in membrane filtration owing to effluent water and water applications: polymers and ceramic materials. Ceramic membrane materials are inorganic materials that possess beneficial properties such as better separation and high flux, narrow and definite pore sizes, relatively high stiffness, improved thermal as well as chemical resistance, increased hydrophilicity, outstanding anti-bacterial property, convenience in washing the membrane and a prolonged operational lifetime. An example of an inorganic membrane is a ceramic membrane, which has three layers: a top layer that is microporous, an intermediate layer and a macroporous support layer. The most commonly used ceramic materials are titanium dioxide (TiO_2), alumina (Al_2O_3), zirconia oxide (ZrO_2) and silicon carbide (SiC). Extensively used processes for phenol separation are adsorption and separation by membrane. Phenol absorption occurs either through physisorption or chemisorption, though practically, it is physisorption that is responsible for phenol absorption via interaction of the active sites in the membrane with the phenol molecule through weak van der Waals interaction [16]. At the same time, the size exclusion principle also plays a vital role in phenol separation through membrane. The massive particles that first accumulate on the membrane surface cannot pass through the small pores. Depending on the pore sizes, the phenol-removing membranes are classified as microfiltration, ultrafiltration or nanofiltration. Therefore, phenol removal through membranes comprises a hybrid system of adsorption and separation due to size exclusion [17]. These two technologies are composed of either a mixed matrix membranes (MMMs) or a thin film composite membrane (TFCM). Another important factor in phenol removal by membrane is its interaction with chemicals and the membrane material, or more simply, the interaction between polar groups or oppositely charged molecules. Activated carbon is one such material that is often used in MMMs due to its unique properties for adsorbing almost anything, from pharmaceutical waste to heavy metals [18]. Depending on the charge of the membrane, phenolic compounds will be protonated or deprotonated over the membrane surface. Since reported literature revealed that functionalised activated carbon-based nanocomposite membranes effectively showed across the membrane surface, phenolic compounds in contact with this type of membrane compound will be deprotonated due to the negative part of the membrane surface. Therefore, the charge effect is considered a driving force for the effective reduction of phenolic compounds through an activated carbon-based membrane [19,20].

9.3 SIZE EXCLUSION AND ADSORPTION-BASED MEMBRANES

9.3.1 SIZE EXCLUSION

Size exclusion works on the basic principle of membrane filtration, in which pollutants are separated based on their sizes. Organic pollutants of greater size compared to the membrane pore's dimension are restored on the membrane due to the sieving effect. Membrane-based reduction of phenol present in phenolic wastewater is categorised in two ways: low-pressure membrane processes such as microfiltration (MF) and ultrafiltration (UF) and high-pressure membrane processes such as nanofiltration (NF) and reverse osmosis (RO) [2. Certain membrane types (reverse osmosis, nanofiltration or ultrafiltration) preferentially permit certain compounds while serving as a hindrance for others to separate the raw phenolic wastewater (feed) into clean water (permeate), also known as retentate [22]. This type of membrane mainly works on the principle of the driving force applied by the pressure on it. The fluid travelling across the membrane makes up the permeate, while the retentate stream is made up of particles and dissolved substances that are kept inside the membrane. The permeate stream is rich in solutes, with a nominal weight cutoff (NMWCO) below the porosity size of membranes. Among these nanofiltration membrane methods, separation is mostly employed in favour of the eradication of important organic and synthetic contaminants (e.g., phenol-based dyes) from effluent water generated from various industries [23,24]. Ghaemi et al. developed a membrane based on nanofiltration, and using this nanofiltration membrane, nitrophenols are separated from water-based solutions. They also investigated that by introducing surfactants that had a positive impact on membrane functionality (CTMB) and Triton X-100), there was a significant increase in the selectivity and flux of the membrane. As well it was shown that the pH of the solution, membrane properties and solute components all had an influence on the separation of polynitrophenols (PNP) and 3,5-dinitrosalicyclic acid (DNSA) [25]. According to research by Agenson et al., it was reported that the retaining effectiveness of NF membranes is influenced by the size and functional groups of the solutes [26]. The group developed a membrane for extracting alkyl phenols from aqueous solutions with more than 90% recovery. In another work, Arsuaga et al. [27] also developed a number of nanofiltration membranes to eliminate phenolic substances (phenol, 3-chlorophenol and 3-nitrophenol) from wastewater effluents. In the study, the group observed that the interactions between the membrane phase and solutes that are hydrophobic played a major role in membrane retention and also observed a significant increase in flux decline, which was due to the accumulation of the dissolved organic moieties onto the membrane's surface. Membranes including reverse osmosis and nanofiltration were employed by Bodalo et al. for the eradication of phenolic compounds from effluent water, where the reverse osmosis membrane with 40% selectivity performed lower as compared to the nanofiltration membrane (with 80% selectivity) [28]. RO membranes for the separation of 4-nitrophenol and 4-nitroaniline present in the effluents of factories were investigated and compared by Hidalgo et al. [29]. There was a greater rejection at pH 6 for 4-nitrophenol. Though full regeneration of the membrane wasn't possible, partial regeneration was achieved by ultrasound treatment. Similar work was done by Satto et al. using commercial membranes (reverse osmosis (BW-30) and nanofiltration (NF-90)) to study 2-nitrophenol and 2-chlorophenol regeneration present in effluent water. It was observed that the RO membrane demonstrated a better rejection percentage for 2-nitrophenol and 2-chlorophenol compared to the NF membrane [30]. MMMs are often referred to as a possible contender for wastewater treatment owing to their increased stability and varying mechanical properties with standard pH instances. A new carbon-based nanoparticle mixed polysulfone hollow fibre ultrafiltration composite membrane (UF membrane) was developed by Mukherjee et al. for their potential application in the elimination of phenol, toluene and benzene [31]. A permeation set-up of a dead-end ultrafiltration membrane process is shown in Figure 9.2. Nanoparticles included in MMMs often

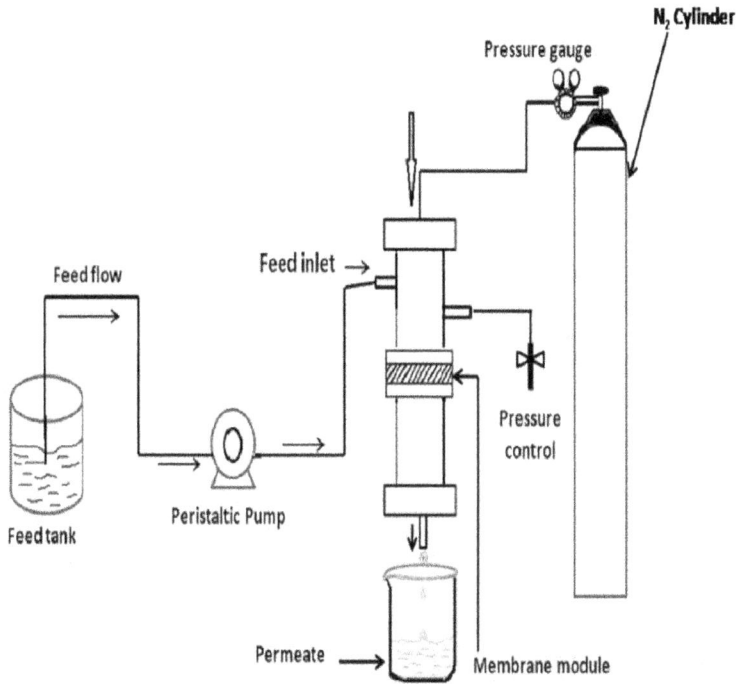

FIGURE 9.2   Permeation set-up of a dead-end ultrafiltration membrane process.

increase stabilities with greater durability than other commercial membranes like PEBA and PDMS. Gupta et al. blended cellulose acetate (CA) with choline chloride (ChCl),a novel composite membrane coated with a ceramic substrate material made of fly ash. Adding ChCl significantly increases the membrane wettability, chemical stability, pore size and permeability with a remarkable increase in phenol rejection. Increasing the pressure and concentration of the feed will lower the phenol rejection efficiency of the membrane, which is again increased with increasing pH [32].

9.3.2 ADSORPTION-BASED MEMBRANES

Recently, nanomaterial-based membranes, either coated on the base polymer membrane or as composites, have gained much importance due to their selectivity towards phenol removal. The most widely used techniques for treating wastewater are physical, chemical, biological and physical–chemical. The most widely employed approach to eliminating phenolic compounds present in an aquatic system is adsorption. Adsorption methods have the benefits of being easy, very efficient and non-polluting. The effectiveness of the adsorbent, which acts as a carrier for the passage of contaminants, is crucial to the treatment of wastewater. However, bigger pore sizes (MF, UF) are likely to exhibit better adsorption than smaller pore sizes (NF/RO), despite the fact that the adsorption site across the membrane surfaces is comparatively small. The well-known traditional adsorbent used for water filtration is activated charcoal (AC). Activated charcoal has two types: granular activated carbons (GACs) and powdered activated carbons (PACs). Generally, GACs are preferred over PACs for industries, as GACs are easily regenerated in industrial processes [33]. Activated carbon is one such material which considering its high performance, high adsorption capacity and cheaper operational cost, is considered to be one of the most commonly used membrane materials for the efficient removal of phenol. Moreover, activated carbon is regenerated and reused for several cycles [34]. Activated carbon has the inherent ability to absorb and removal of trace metal ions as well as toxic pollutants from aqueous solutions because of its high surface reactivity. To increase the efficiency of the method, new alternatives were developed, such as chemical modification of AC, infusion of nanoparticles with activated carbon, using the divergent activation method, etc. Rameshkumar et al. studied the functionalised activated carbon composite membrane with polyphenylsulfone (PPSU) and developed PPSU/FAC, applying a wet phase inversion approach. It was reported that adsorption effect improved pressure-driven permeation via pore diffusion [35]. Azizul et al. investigated elimination of bisphenol A (BPA) from drinking liquids using PVDF nanocomposite membranes were developed using the electrospun method with a variety of fillings. PVDF–MnO$_2$ nanoparticles and PVP as hydrophilic additive were incorporated into the PVDF membrane for efficient elimination of BPA present in the drinking water. They also accessed the possible interactions between BPA, PVP and MnO$_2$ nanoparticles by DFT study. Highly porous nanostructures in the membrane make it possible for the adsorption of BPA on the surface [36]. Also, DFT calculations confirm the hydrogen bonding interaction between MnO$_2$ and BPA. Nadavala et al. have done research on the adsorption activity of compounds containing phenol implementing pine bark, which is a cellulosic waste-carrying lignin that exists in forests. They have found a maximum pH of 6, stabilising bio-based absorption in 120 min, with a 143-ppm bio-based absorption capacity of phenol [37]. Table 9.1 represents various works on the development of different membranes for the removal of phenol from wastewater.

9.4 PERVAPORATION

To get rid of low volatile organic compounds, pervaporation technique is considered one of the most promising ways. In pervaporation, a non-porous membrane allows for the selective diffusion-vaporisation separation of two liquid mixtures. The feed pretreatment and distribution system, the permeate condensation/recovery system and a non-porous membrane are all components of the pervaporation [45]. Due to its simplicity, in recent days pervaporation has been extensively used as one of the most preferable separation methods in the removal of phenol or phenol-based compounds and in most cases the performance of recently modified membranes has been investigated with respect to membrane flux and efficiency of separation [46,47]. A schematic diagram of the pervaporation set up is shown in Figure 9.3. The advantages of pervaporation include low cost, easy installation of equipment, a lower chance of generating secondary pollutants and a lower energy consumption [48,49].

Polyurethane is one of the most commonly used polymeric materials for designing membranes for the pervaporation of phenol, as it has a greater affinity for phenol.

TABLE 9.1
Selected Works on Development of UF/RO/NF Membrane for Phenol Extraction from Wastewater

Membrane Operation/ Membrane Material	Source of Phenolic Compounds	Target Substance	Rejection	References
UF/100 kDa/Polysulfone/ Hollow fibre	Orange press liquor	Total polyphenols	58.30%	[38]
NF/200 Da/Polymeric/ Spiral Wound	Olive mill waste water	Total polyphenols	95%	[38]
UF/100 kDa/Polysulfone/ Flat sheet	Winery sludge from red grapes	Hydroxycinnamic acids, O-diphenols	81%	[39]
NF/DSS-HR98PP (polymeric membrane)	Waste water	Phenol	80%	[28]
UV/ROpolyamide thin film composite	Aqueous solution	Phenol	58%	[40]
UV and Ropolyamide thin film composite	Aqueous solution	Phenol	UV = 17% RO = 20%	[40]
Domestic low-pressure RO TFC polyamide low pressure reverse osmosis membrane	Aqueous solution	Phenol	71.7%	[41]
NF/Polysulfone	Pesticide waste water	2,4-dinitrophenol p-nitrophenol	>95% 90%	[42]
NF/Cellulose acetate	Pesticide waste water	p-Nitrophenol	90%	[25]
RO/Thin film polyamide composite membrane	Aqueous solution	Phenol Catechol Resorcinol 2-Chlorophenol 3-Chlorophenol 4-Chlorophenol 2-Nitrophenol 3-Nitrophenol 2-Nitrophenol	75% 88% 88% 71% 70% 63% 64% 68%	[43]
NF/Thin film polyamide composite membrane	Aqueous solution	Phenol Catechol Resorcinol 2-Chlorophenol 3-Chlorophenol 4-Chlorophenol 2-Nitrophenol 3-Nitrophenol 2-Nitrophenol	29% 46% 45% 36% 32% 25% 32% 26% 22%	[43]
NF/Polysulfone composite	Aqueous solution (100ppm)	2-Chlorophenol 2,4-Dimethylphenol Bisphenol-A	22.4% 12.4% 6.4%	[44]

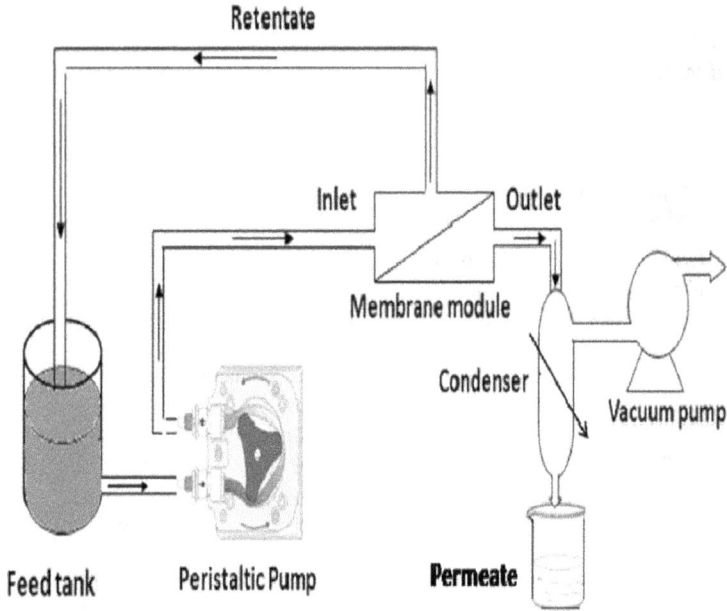

FIGURE 9.3 Schematic diagram of a pervaporation set up.

Moreover, it is versatile, biocompatible and biodegradable by nature [50,51]. A poly-urethane sandwich-type composite membrane was developed for the eradication of phenol present in an aqueous solution. The study record shows that phenol concentration in the permeate side was up to 7 from 0 wt.%, and at 68 °C the flux of the membrane reached 0.930 kg m² h⁻¹ [52]. Upon stimulation of the polyurethane membrane, the effectiveness of the eradication of phenol present in the phenol–water mixture was found to be dependent on the driving force, i.e., the transition from vacuum to low pressure, and also on the amount of phenol present in the supply stream [53]. In recent times, a blending of membranes with polyurethane (PU) and polyacrylonitrile (PAN) was developed which exhibits a pore size of 18 nm for treating wastewater. The membrane with $V_{PAN}:V_{PU} = 70:30$ exhibited a maximum flux recovery of 99% with good antifouling properties [54]. Experimental studies suggest that this modification is useful in the complete removal of organic contaminants, including phenol, present in industrial wastewater.

Another base polymer used for pervaporation is poly(ether block amide) or PEBA, which belongs to a group of copolymers having small biphasic structures that exhibit properties with both the constituting polymers, i.e., hard polyamide and soft polyether, which are not alone available in any of the polymers. In work by Hau et al., PEBA 2533 was used as the membrane material for studying the pervaporative extraction of phenolic compounds present in effluent streams. They studied the dependence of membrane performance on temperature and feed amount. The membrane showed good permeation and maximum separation efficiency in phenol/water separation, with an excellent preference for phenol sorption over water, and for a

2 wt.% phenol–water system, the selectivity for phenol was 80–90%. Also, they reported that at higher concentrations with selective temperatures, flux was increased at the initial stage of pervaporation [55]. In another work, a composite membrane was developed through adding a zeolitic imidazolate structure (ZIF), ZIF-8, into a PEBA 2533 polymer matrix for its application in phenol separation from wastewater. The role of ZIF-8 was to increase the membrane's hydrophobicity. The permeate flux regarding the developed membrane was notably increased compared to the pure PEBA-2533 membrane from 846 g m^2 h^{-1} to 1310 g m^2 h^{-1} also at 70 °C the separation factor increases from 39 to 53. The improvement in separation performance is justified by the ZIF-8 particles' capacity for developing phenol-permeable channels [56].

Composite membranes are another potentially employed membrane for pervaporation with regard to eliminating phenol from streams. In this regard, Wu et al. and their group developed the blend membranes of PVDF and polyurethane with varying PVDF content, and they reported that the membranes with 5 wt.% and 10 wt.% PVDF had good sorption capacities for phenol. These membranes showed a very good performance in the pervaporation of phenol compared to the neat PU membrane [57]. However, the performance of the materials developed was not satisfactory for industrial applications. Hence, researchers have developed some new techniques like integrating pervaporation with other methods, such as solvent extraction and distillation, which may be useful in industrial regarding the eradication of phenolic compounds present in effluent fluids. One such novel separation process, pertraction or a combination of pervaporation and membrane extraction, was studied by Xiao et al. [58] using poly (dimethylsiloxane) or membranes for membrane aromatic recovery made of PVDF and PDMS. The new membrane demonstrated mass transfer that was five times more powerful than silicone rubber membranes showing minimum permeability activation energy. The permeability of phenol through the pertraction and its diffusion coefficient were found to be lower compared to silicone-coated membranes. It was reported that in the phenol transport mechanism membrane mass transfer step was the rate-controlling step [58]. Hoshi and researchers examined how effectively a polyurethane membrane removed phenol present in an effluent stream. The group recorded the flux reached 0.930 kg m^2 h^{-1} at 68 °C and that the phenol content increased on the permeate side with 0–7 wt.% while feed side concentration decreased on the permeate side amount increased as the feed side raised from 0 to 7 wt.%. Later experimental results proved that better performance for the phenolic water separation was mainly dependent on the concentration and driving force of the feed solution. Also, they have concluded that as temperature increases the phenol flux also increases because of the solvability of phenol in the membrane composed of polyurethane [34]. Another study was done by Kujawskiet al. [59] using the PV technique. They studied the pervaporation of mixtures of water–phenol and water–acetone by the use of the zeolite-filled PDMS, PDMS and polyether-block-amide (PEBA) based composite membranes. The performance of the PEBA membrane gives good alertness regards to phenol. Using cellulose acetate phthalate–alumina nanoparticle membrane, Mukerjee et al. created a MMMs for its application in the separation of phenolic compounds (m-nitrophenol), catechol para-nitrophenol, ortho-nitrophenol and phenol present in wastewater using a pervaporation cell. Thess developed MMMs possessed outstanding efficiency in the rejection of chlorophenol (91%), which is considered as carcinogenic [60].

Another extensively employed membrane technology for phenol extraction of industrial wastes is membrane-based solvent extraction (MBSE). Due to its clear modular architecture, it is a continuous operational process with easy scalability and no risk of agitation or back mixing. The membrane resistance is one of the main drawbacks of traditional MBSE, which is reduced by employing membrane contactors comprising hollow fibres possessing a big surface area. The Liqui–Cel membrane contactor, a microporous propylene module, is presently the most widely used contactor module in industry. There are two steps in the MBSE procedure. One involves solvent extraction, and the other involves hollow fibre membrane contactors (HMCs), which are formed from a variety of polymers, including polypropylene (PP), polyethylene (PE), poly(tetrafluoroethylene-co-perfluorovinylether) (PFA), PVDF, polytetrafluoroethylene (PTFE), among others. The HMCs' numerous benefits include their large accessible surface area, lack of emulsion formation and simpler method [61,62]. To facilitate the separation of the aqueous solution, the procedure uses a membrane as a hindrance between phenol and water by applying a differential in transmembrane pressure. Widely used contactors are hollow fibre membranes with a variety of extractive agents like cumene, 1-decanol, MTBE, etc., among which because of the high polarity for phenol, MTBE was found to be the most suitable with the polypropylene (PP) membrane. MTBE was reported to be best appropriate for polypropylene (PP) membranes because of its high polarity with respect to phenol [63–65]. Using a silicone rubber membrane, Sawai et al. used the permeation and chemical desorption (PCD) approach to study the permeation efficiency with phenol and aniline [66]. As extractants, HCl and NaOH were utilised to separate phenol and aniline present in aqueous solutions. The penetration rates of aniline and phenol raise with the length of the alkyl chain. Because aniline has a lower diffusivity than phenol and a larger membrane distribution coefficient, it has a higher permeability. Pertraction is yet another MBSE method that is utilised commercially for the treatment of effluent water. The process is commercially important because the separation of liquid and extraction products is not required, or more simply, it is cost effective; hence, the technology is employed in treating industrial wastewater containing aromatic toxic chemicals generated from petrochemical industries. However, pertraction has several drawbacks to the eradication of hydrophilic pollutants like phenol and ethanol, as it is challenging to discover a good extraction agent [67]. Therefore, for effective removal of phenolic compounds, it is important to choose the appropriate extractive agents and contactors.

9.5 PHOTOCATALYTIC MEMBRANE FOR PHENOLIC COMPOUNDS DEGRADATION

Albeit membrane-based techniques are known to be advantageous methods for wastewater treatment, they possess some limitation phenomenon that inhibit various applications because of the hydrophobic nature of the membrane material. Foulants that build up on membrane surfaces can block membrane pores, thereby causing a significant decrease in membrane flux, shortening membrane life and consequently

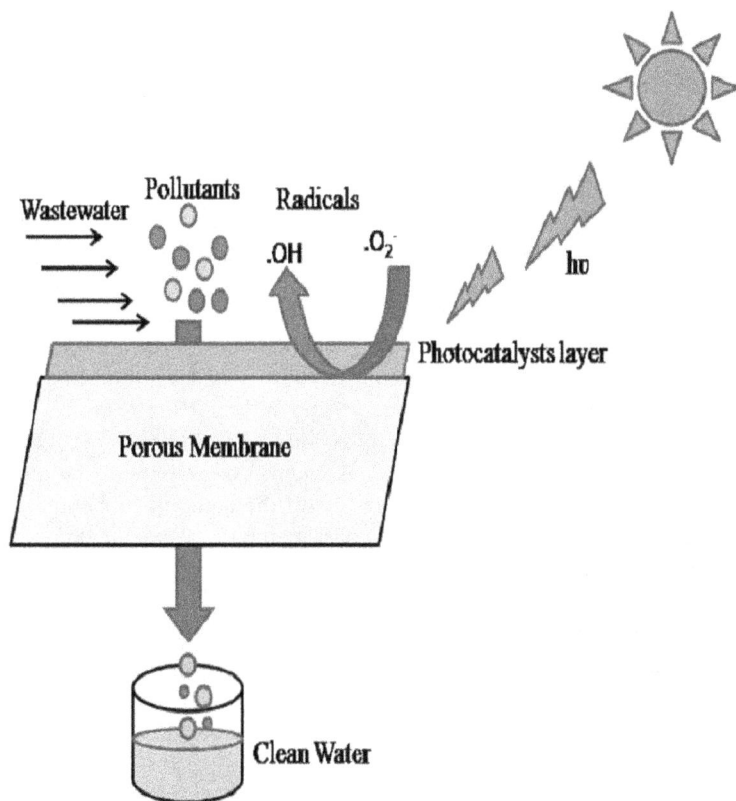

FIGURE 9.4   Schematic diagram of a photocatalytic membrane set up.

raising the cost of membrane technology. Environmentalists have been conducting research on the eradication of organic compounds through the photodegradation process for the past few decades. It is necessary and beneficial to create membranes with degraded foulant properties. Recent research has focused on using photocatalysis to remove contaminants, especially because the target contaminants are totally metabolised using this procedure. A schematic diagram of the photocatalytic membrane set up is shown in Figure 9.4. With applications in the conversion of organic and inorganic contaminants to acceptable chemicals, photocatalysis is regarded as a sustainable environmental technology. The inclusion of a photocatalyst into a membrane is known as a photocatalytic membrane. A few techniques exist for creating photocatalytic membranes, such as the liquid phase deposition (LPD) approach, the sol–gel method, anodization and simultaneous electrospinning and electrospraying [68–74]. When exposed to visible light or UV radiation, a supported layer of photocatalysts on a porous membrane engages in photocatalytic activity [73].

Photocatalysts work on an oxidation mechanism that generates hydroxyl radicals that remove and degrade organic pollutants present in water into simple, smaller and inorganic molecules without producing secondary waste. Among the several materials that exhibit some photocatalytic activity, TiO_2 photocatalyst has received extensive study for its potential use in the eradication of hazardous environmental contaminants. Titanium oxide (TiO_2) is the most efficient catalyst material for membrane fabrication owing to its high activity, outstanding thermal and chemical resistance with non-toxicity superhydrophilicity, longevity, lower price and transparency to visible light. In order to prevent being flushed out, TiO_2 photocatalysts are tightly attached to the carboxyl-containing membrane surface. In this section, we have presented different works and investigations related to photocatalysts in the photocatalytic degradation of phenolic compounds [69]. Many researcher have been performed to examine the immobilisation of TiO_2 upon the membrane surface. Song et al. performed the surface transformations of polypropylene macroporous membrane by grafting poly(acrylic acid) (PAAc) using the technique of photoinduced reversible addition-fragmentation. Eventually, they introduced TiO_2 photocatalysts upon the surface of the PAAc grafted membrane, afterwards carried out the decomposition of Phenol under UV irradiation. The result they have found that for the PAAc-modified membrane, the normalised membrane flow was 1.7 times greater than that of the unmodified membrane. They observed that the immobilisation of TiO_2 photocatalyst to the membrane surface slightly reduced the normalised membrane flux. With a decomposition percentage of 32.5% was obtained after 6 h UV light irradiation on the membrane surface [70]. Colon et al. discovered full conversion of phenol (50 ppm) with TiO_2 (1 g/l) after 90 minutes of illumination with a medium pressure 400 W Hg lamp (270 nm). These researchers have found that the decrease in activity by raising the calcination temperature and, as a result, the photo catalyst's surface area. Moreover, sulphuric acid treatment limits deactivation [71]. Lavinia et al. studied a closed-cycle method for the photocatalytic degradation of phenol using an exhausted Pd-based adsorbent. As Pd(II) is considered a precious metal with a significant economic interest, it is very much necessary to recover Pd. Therefore, the research team studied that Pd(II) is removed from an aqueous solution by adsorption onto Florisil (a solid inorganic support made of magnesium silicate) soaked in Cyphos IL 101. They found that the investigated materials' adsorption efficiency doubles when an ionic liquid (IL) is present in the adsorbent structure. Their study reported that the degradation of phenol from aqueous solutions using the investigated Pd-based photocatalyst and ionic liquid (IL) contributed to an enhancement in the phenol degradation process. These findings highlighted the sample F-Il-superior Pd's efficiency and stability as a photocatalyst for the process of breaking down phenol in aqueous solutions when compared to the sample F-IL [72]. Alireza et al. [74] investigated phenolic compounds photocatalytic degradation (phenol, resorcinol and cresol) using titanium oxide photocatalyst for mesoporous carbon (CMK-3) support under UV irradiation. In their observation, titanium isopropoxide was used as the titanium supply, and ordered mesoporous carbon served as the support. Researchers looked at the effects of many factors, including pH, catalyst content, phenolic content and irradiation period. High degradation efficiency of phenolic compounds was reported at the optimum conditions were 96% for phenol, 91% for m-cresol and 98% for resorcinol and total organic carbon (TOC) rejection efficiencies were 74%, 62% and 78%, respectively. As a consequence TiO_2/CMK-3 was found to be a promising photocatalyst that is utilised to degrade phenolic compounds by photocatalysis [74].

9.6 CONCLUSION

With the rapid growth of different chemical and petrochemical industries, the generation of phenolic wastewater has been increasing day by day. Phenolic compounds are nowadays considered as a priority pollutants that occur in water bodies because polluted effluent from the petroleum sector, agriculture and home activities is released into the water. As phenol has become an issue for the environment, its treatment is necessary before discharging to the water systems, and the worldwide standard for phenol discharge has become stricter. Extensive research has described methods, such as distillation, adsorption, liquid extraction, pervaporation, advanced oxidation processes, biological approaches, membrane technology and treatment of phenolic compounds present in industrial wastewater streams. Out of all these technologies, membrane technology is considered the most effective method for reducing phenolic compounds from wastewater due to its low startup cost, ease of scaling up, low energy requirement, environment friendly. In this chapter, we have discussed about the advantageous effect of some membrane processes with respect to membrane pore size for treating phenolic wastewater. Some reported works with their detailed performance were analysed, illustrating the membrane technology as green and sustainable method for eradicating phenol from industrial effluents. This chapter also reviewed two convenient methods of separation, i.e., adsorption and pervaporation, for the eradication of phenolic compounds in industrial effluents. Considering the harmful effects of phenol and phenol-based wastewater as well as the water scarcity issue, various efforts have been made to make industrial wastewater phenol free. The advancement of a new approach for the effective elimination of phenol and phenolic compounds in industrial wastewater will provide a new route in the separation and purification technology.

REFERENCES

1. Ahmed, S., Rasul, M. G., Martens, W. N., Brown, R., & Hashib, M. A. (2010). Heterogeneous photocatalytic degradation of phenols in wastewater: A review on current status and developments. *Desalination*, *261*(1–2), 3–18. https://doi.org/10.1016/j.desal.2010.04.062
2. EPA, Toxic and Priority Pollutants under the Clean Water Act. (1976). www.epa.gov/eg/toxic-and-priority-pollutants-under-clean-water-act#priority
3. NPRI, Substance list: National Pollutant Release Inventory. (2017). www.canada.ca/en/environment-climate-change/services/national-pollutant-release-inventory/substances-list.html
4. Czaplicka,M. (2004). Sources and transformations of chlorophenols in the natural environment. *Science of the Total Environment*, *322*(1–3), 21–39. https://doi.org/10.1016/j.scitotenv.2003.09.015
5. Tuomisto,J., Airaksinen, R., Pekkanen, J., Tukiainen, E., Kiviranta, H., &Tuomisto, J. T. (2017). Comparison of questionnaire data and analyzed dioxin concentrations as a measure of exposure in soft-tissue sarcoma studies. *Toxicology Letters*, *270*, 8–11. https://doi.org/10.1016/j.toxlet.2017.02.011

6. Michałowicz, J., &Duda, W. (2007). Phenols—Sources and Toxicity. *Polish Journal of Environmental Studies*, 16(3). www.pjoes.com/Phenols-Sources-and-Toxicity,87995,0,2.html

7. Igbinosa, E. O., Odjadjare, E. E., Chigor, V. N., Igbinosa, I. H., Emoghene, A. O., Ekhaise, F. O., & Idemudia, O. G. (2013). Toxicological profile of chlorophenols and their derivatives in the environment: The public health perspective. *The Scientific World Journal*, 2013. https://doi.org/10.1155/2013/460215

8. Rengaraj, S., Moon, S. H., Sivabalan, R., Arabindoo, B., &Murugesan, V. (2002). Removal of phenol from aqueous solution and resin manufacturing industry wastewater using an agricultural waste: Rubber seed coat. *Journal of Hazardous Materials*, 89(2–3), 185–196. https://doi.org/10.1016/S0304-3894(01)00308-9

9. Carmona, M., De Lucas, A., Valverde, J. L., Velasco, B., &Rodriguez, J. F. (2006). Combined adsorption and ion exchange equilibrium of phenol on Amberlite IRA-420. *Chemical Engineering Journal*, 117(2), 155–160. https://doi.org/10.1016/j.cej.2005.12.013

10. Miland, E., Smyth, M. R., & Fágáin, C. Ó. (1996). Phenol removal by modified peroxidases. *Journal of Chemical Technology & Biotechnology: International Research in Process, Environmental AND Clean Technology*, 67(3), 227–236. https://doi.org/10.1002/(SICI)1097-4660(199611)67:3%3C227::AID-JCTB563% 3E3.0.CO;2-J

11. Buchanan, I. D., &Han, Y. S. (2000). Assessment of the potential of *Arthromyces ramosus* peroxidase to remove phenol from industrial wastewaters. *Environmental Technology*, 21(5), 545–552. https://doi.org/10.1080/09593332108618091

12. Wagner, M., & Nicell, J. A. (2001). Peroxidase-catalyzed removal of phenols from a petroleum refinery wastewater. *Water Science and Technology*, 43(2), 253–260. https://doi.org/10.2166/wst.2001.0097

13. Kislik, V. S. (Ed.). (2009). *Liquid Membranes: Principles and Applications in Chemical Separations and Wastewater Treatment*. Elsevier. https://doi.org/10.1016/C2009-0-18491-X

14. Mohammadi, S., Kargari, A., Sanaeepur, H., Abbassian, K., Najafi, A., & Mofarrah, E. (2015). Phenol removal from industrial wastewaters: A short review. *Desalination and Water Treatment*, 53(8), 2215–2234. https://doi.org/10.1080/19443994.2014.883327

15. Mozia, S. (2010). Photocatalytic membrane reactors (PMRs) in water and wastewater treatment. A review. *Separation and Purification Technology*, 73(2), 71–91. https://doi.org/10.1016/j.seppur.2010.03.021

16. Nelson, N. C., Manzano, J. S., Sadow, A. D., Overbury, S. H., &Slowing, I. I. (2015). Selective hydrogenation of phenol catalyzed by palladium on high-surface-area ceria at room temperature and ambient pressure. *ACS Catalysis*, 5(4), 2051–2061. https://doi.org/10.1021/cs502000j

17. Mulder, M., & Mulder, J. (1996). *Basic Principles of Membrane Technology*. Springer Science & Business Media. http://dx.doi.org/10.1007/978-94-009-1766-8

18. Said, K. A. M., Ismail, A. F., Zulhairun, A. K., Abdullah, M. S., Usman, J., Azali, M. A., &Azali, M. A. (2021). Zinc ferrite migration dependence on magnetic induce membrane for phenol removal: Adsorption reaction and diffusion study. *Journal of Environmental Chemical Engineering*, 9(1), 105036. https://doi.org/10.1016/j.jece.2021.105036

19. Saranya, R., Kumar, M., Tamilarasan, R., Ismail, A. F., & Arthanareeswaran, G. (2016). Functionalised activated carbon modified polyphenylsulfone composite membranes for adsorption enhanced phenol filtration. *Journal of Chemical Technology & Biotechnology*, 91(3), 748–761. https://doi.org/10.1002/jctb.4641

20. Mohammadi, S. Z., Darijani, Z., &Karimi, M. A. (2020). Fast and efficient removal of phenol by magnetic activated carbon-cobalt nanoparticles. *Journal of Alloys and Compounds*, *832*, 154942.https://doi.org/10.1016/j.jallcom.2020.154942

21. Racar, M., Dolar, D., Špehar, A., &Košutić, K. (2017). Application of UF/NF/RO membranes for treatment and reuse of rendering plant wastewater. *Process Safety and Environmental Protection*, *105*, 386–392. https://doi.org/10.1016/j.psep.2016.11.015

22. Kaghazchi, T., Mehri, M., Ravanchi, M. T., &Kargari, A. (2010). A mathematical modeling of two industrial seawater desalination plants in the Persian Gulf region. *Desalination*, *252*(1–3), 135–142.https://doi.org/10.1016/j.desal.2009.10.012

23. Kim, S., Chu, K. H., Al-Hamadani, Y. A., Park, C. M., Jang, M., Kim, D. H., …&Yoon, Y. (2018). Removal of contaminants of emerging concern by membranes in water and wastewater: A review. *Chemical Engineering Journal*, *335*, 896–914. https://doi.org/10.1016/j.cej.2017.11.044

24. Gilron, J., Linder, C., &Wiesman, Z. (2017). *U.S. Patent Application No. 15/313,743.* https://cris.bgu.ac.il/en/publications/selective-phenol-removal-membranes-and valorization-of-olive-oil-

25. Ghaemi, N., Madaeni, S. S., Alizadeh, A., Daraei, P., Zinatizadeh, A. A., & Rahimpour, F. (2012). Separation of nitrophenols using cellulose acetate nanofiltration membrane: Influence of surfactant additives. *Separation and Purification Technology*, *85*, 147–156. https://doi.org/10.1016/j.seppur.2011.10.003

26. Agenson, K. O., Oh, J. I., & Urase, T. (2003). Retention of a wide variety of organic pollutants by different nanofiltration/reverse osmosis membranes: Controlling parameters of process. *Journal of Membrane Science*, *225*(1–2), 91–103. https://doi.org/10.1016/j.memsci.2003.08.006

27. Arsuaga, J. M., López-Muñoz, M. J., & Sotto, A. (2010). Correlation between retention and adsorption of phenolic compounds in nanofiltration membranes. *Desalination*, *250*(2), 829–832. https://doi.org/10.1016/j.desal.2008.11.051

28. Bódalo, A., Gómez, E., Hidalgo, A. M., Gómez, M., Murcia, M. D., & López, I. (2009). Nanofiltration membranes to reduce phenol concentration in wastewater. *Desalination*, *245*(1–3), 680–686. https://doi.org/10.1016/j.desal.2009.02.037

29. Hidalgo, A. M., León, G., Gómez, M., Murcia, M. D., Gómez, E., & Gómez, J. L. (2013). Application of the Spiegler–Kedem–Kachalsky model to the removal of 4-chlorophenol by different nanofiltration membranes. *Desalination*, *315*, 70–75. https://doi.org/10.1016/j.desal.2012.10.008

30. Sotto, A., Arsuaga, J. M., & Van der Bruggen, B. (2013). Sorption of phenolic compounds on NF/RO membrane surfaces: Influence on membrane performance. *Desalination*, *309*, 64–73. https://doi.org/10.1016/j.desal.2012.09.023

31. Mukherjee, R., & De, S. (2016). Novel carbon-nanoparticle polysulfone hollow fiber mixed matrix ultrafiltration membrane: Adsorptive removal of benzene, phenol and toluene from aqueous solution. *Separation and Purification Technology*, *157*, 229–240. https://doi.org/10.1016/j.seppur.2015.11.015

32. Gupta, V., Raja, C., &Anandkumar, J. (2020). Phenol removal by novel choline chloride blended cellulose acetate-fly ash composite membrane. *Periodica Polytechnica Chemical Engineering*, *64*(1), 116–123. https://doi.org/10.3311/PPch.14126

33. Busca, G., Berardinelli, S., Resini, C., & Arrighi, L. (2008). Technologies for the removal of phenol from fluid streams: A short review of recent developments. *Journal of Hazardous Materials*, *160*(2–3), 265–288. https://doi.org/10.1016/j.jhazmat.2008.03.045

34. Hoshi, M., Kogure, M., Saitoh, T., & Nakagawa, T. (1997). Separation of aqueous phenol through polyurethane membranes by pervaporation. *Journal of Applied*

Polymer Science, *65*(3), 469–479. https://doi.org/10.1002/(SICI)1097-4628(19970
718)65:3%3C469::AID-APP6%3E3.0.CO ;2-F

35. Saranya, R., Kumar, M., Tamilarasan, R., Ismail, A. F., & Arthanareeswaran, G.
(2016). Functionalised activated carbon modified polyphenylsulfone composite
membranes for adsorption enhanced phenol filtration. *Journal of Chemical Technology
& Biotechnology*, *91*(3), 748–761. https://doi.org/10.1002/jctb.4641

36. Zahari, A. M., Shuo, C. W., Sathishkumar, P., Yusoff, A. R. M., Gu, F. L.,
Buang, N. A., …&Yusop, Z. (2018). A reusable electrospun PVDF-PVP-MnO$_2$
nanocomposite membrane for bisphenol A removal from drinking water. *Journal
of Environmental Chemical Engineering*, *6*(5), 5801–5811. https://doi.org/10.1016/
j.jece.2018.08.073

37. Nadavala, S. K., Man, H. C., &Woo, H. S. (2014). Biosorption of phenolic compounds
from aqueous solutions using pine (*Pinus densiflora*Sieb) bark powder. *BioResources*,
9(3), 5155–5174. https://10.15376/biores.9.3.5155-5174

38. Castro-Muñoz, R., Yáñez-Fernández, J., & Fíla, V. (2016). Phenolic compounds
recovered from agro-food by-products using membrane technologies: An overview.
Food Chemistry, *213*, 753–762. https://doi.org/10.1016/j.foodchem.2016.07.030

39. Castro-Muñoz, R., Barragán-Huerta, B. E., Fíla, V., Denis, P. C., &Ruby-Figueroa,
R. (2018). Current role of membrane technology: From the treatment of agroindustrial by-products up to the valorization of valuable compounds. *Waste and
Biomass Valorization*, *9*, 513–529. https://link.springer.com/article/10.1007/s12
649-017-0003-1

40. Kargari, A., &Mohammadi, S. (2015). Evaluation of phenol removal from aqueous
solutions by UV, RO, and UV/RO hybrid systems. *Desalination and Water Treatment*,
54(6), 1612–1620. https://doi.org/10.1080/19443994.2014.891077

41. Khazaali, F., &Kargari, A. (2017). Treatment of phenolic wastewaters by a domestic
low-pressure reverse osmosis system. *Journal of Membrane Science and Research*,
3(1), 22–28. https://doi.org/10.22079/jmsr.2017.23344

42. Ghaemi, N., Madaeni, S. S., Alizadeh, A., Daraei, P., Badieh, M. M. S., Falsafi, M., &
Vatanpour, V. (2012). Fabrication and modification of polysulfone nanofiltration membrane using organic acids: Morphology, characterization and performance in removal
of xenobiotics. *Separation and Purification Technology*, *96*, 214–228. https://doi.org/
10.1016/j.seppur.2012.06.008

43. Arsuaga, J. M., Sotto, A., López-Muñoz, M. J., & Braeken, L. (2011). Influence of
type and position of functional groups of phenolic compounds on NF/RO performance. *Journal of Membrane Science*, *372*(1–2), 380–386. https://doi.org/10.1016/
j.memsci.2011.02.020

44. Muppalla, R., Jewrajka, S. K., & Reddy, A. V. R. (2015). Fouling resistant nanofiltration
membranes for the separation of oil–water emulsion and micropollutants from water.
Separation and Purification Technology, *143*, 125–134. https://doi.org/10.1016/j.sep
pur.2015.01.031

45. Peng, M., Vane, L. M., &Liu, S. X. (2003). Recent advances in VOCs removal from
water by pervaporation. *Journal of Hazardous Materials*, *98*(1–3), 69–90. https://doi.
org/10.1016/S0304-3894(02)00360-6

46. Crespo, J. G.,& Brazinha, C. (2015). Fundamentals of pervaporation. In *Pervaporation,
Vapour Permeation and Membrane Distillation* (pp. 3–17). Sawston, Cambridge:
Woodhead Publishing. https://doi.org/10.1016/B978-1-78242-246-4.00001-5

47. Drioli, E., &Romano, M. (2001). Progress and new perspectives on integrated membrane operations for sustainable industrial growth. *Industrial &Engineering Chemistry
Research*, *40*(5), 1277–1300. https://doi.org/10.1021/ie0006209

48. Bera, S. P., Godhaniya, M., & Kothari, C. (2022). Emerging and advanced membrane technology for wastewater treatment: A review. *Journal of Basic Microbiology*, 62(3–4), 245–259. https://doi.org/10.1002/jobm.202100259

49. Ong, Y. K., Shi, G. M., Le, N. L., Tang, Y. P., Zuo, J., Nunes, S. P., & Chung, T. S. (2016). Recent membrane development for pervaporation processes. *Progress in Polymer Science*, 57, 1–31. https://doi.org/10.1016/j.progpolymsci.2016.02.003

50. Das, S., Banthia, A. K., & Adhikari, B. (2008). Porous polyurethane urea membranes for pervaporation separation of phenol and chlorophenols from water. *Chemical Engineering Journal*, 138(1–3), 215–223. https://doi.org/10.1016/j.cej.2007.06.030

51. Yilgör, I., & Yilgör, E. (1999). Hydrophilic polyurethaneurea membranes: Influence of soft block composition on the water vapor permeation rates. *Polymer*, 40(20), 5575–5581.https://doi.org/10.1016/S0032-3861(98)00766-6

52. Hoshi, M., Kogure, M., Saitoh, T., & Nakagawa, T. (1997). Separation of aqueous phenol through polyurethane membranes by pervaporation. *Journal of Applied Polymer Science*, 65(3),469–479.https://doi.org/10.1002/(SICI)1097-4628(19970718)65: 3%3 C4 69::AID-APP6%3E3.0.CO;2-F

53. Moraes, E. B., Alvarez, M. E. T., Perioto, F. R., & Wolf-Maciel, M. R. (2009).Modeling and simulation for pervaporation process: An alternative for removing phenol from wastewater.*Separation Process Development Laboratory. School of Chemical Engineering. University of Campinas, Brazil*.https://doi.org/10.3303/CET0917271

54. Panda, S. R., & De, S. (2015). Preparation, characterization and antifouling properties of polyacrylonitrile/polyurethane blend membranes for water purification. *RSC Advances*, 5(30), 23599–23612. https://doi.org/10.1039/C5RA00736D

55. Hao, X., Pritzker, M., & Feng, X. (2009). Use of pervaporation for the separation of phenol from dilute aqueous solutions. *Journal of Membrane Science*, 335(1–2), 96–102. https://doi.org/10.1016/j.memsci.2009.02.036

56. Ding, C., Zhang, X., Li, C., Hao, X., Wang, Y., & Guan, G. (2016). ZIF-8 incorporated polyether block amide membrane for phenol permselective pervaporation with high efficiency. *Separation and Purification Technology*, 166, 252–261. https://doi.org/10.1016/j.seppur.2016.04.027

57. Wu, Y., Tian, G., Tan, H., & Fu, X. (2013). Pervaporation of phenol wastewater with PVDF–PU blend membrane. *Desalination and Water Treatment*, 51(25–27), 5311–5318. https://doi.org/10.1080/19443994.2013.768789

58. Xiao, M., Zhou, J., Zhang, Y., Hu, X., & Li, S. (2013). Pertraction performance of phenol through PDMS/PVDF composite membrane in the membrane aromatic recovery system (MARS). *Journal of Membrane Science*, 428, 172–180. https://doi.org/10.1016/j.memsci.2012.10.030

59. Kujawski, W., Warszawski, A., Ratajczak, W., Porębski, T., Capała, W., & Ostrowska, I. (2004). Application of pervaporation and adsorption to the phenol removal from wastewater. *Separation and Purification Technology*, 40(2), 123–132. https://doi.org/10.1016/j.seppur.2004.01.013

60. Mukherjee, R., & De, S. (2014). Adsorptive removal of phenolic compounds using cellulose acetate phthalate–alumina nanoparticle mixed matrix membrane. *Journal of Hazardous Materials*, 265, 8–19. https://doi.org/10.1016/j.jhazmat.2013.11.012

61. Pabby, A. K., & Sastre, A. M. (2013). State-of-the-art review on hollow fibre contactor technology and membrane-based extraction processes. *Journal of Membrane Science*, 430, 263–303. https://doi.org/10.1016/j.memsci.2012.11.060

62. Gabelman, A., &Hwang, S. T. (1999). Hollow fiber membrane contactors. *Journal of Membrane Science*, 159(1–2), 61–106. https://doi.org/10.1016/S0376-7388(99)00040-X

63. Pabby, A. K., & Sastre, A. M. (2013). State-of-the-art review on hollow fibre contactor technology and membrane-based extraction processes. *Journal of Membrane Science*, *430*, 263–303. https://doi.org/10.1016/j.memsci.2012.11.060

64. Praveen, P., & Loh, K. C. (2014). Solventless extraction/stripping of phenol using trioctylphosphine oxide impregnated hollow fiber membranes:Experimental & modeling analysis. *Chemical Engineering Journal*, *255*, 641–649. https://doi.org /10.1016/ j.cej.2014.06.097

65. Kujawski, W., Warszawski, A., Ratajczak, W., Porębski, T., Capała, W., & Ostrowska, I. (2004). Removal of phenol from wastewater by different separation techniques. *Desalination*, *163*(1–3), 287–296. https://doi.org/10.1016/S0011-9164(04)90202-0

66. Sawai, J., Higuchi, K., Minami, T., & Kikuchi, M. (2009). Permeation characteristics of 4-substituted phenols and anilines in aqueous solution during removal by a silicone rubber membrane. *Chemical Engineering Journal*, *152*(1), 133–138. https://doi.org/ 10.1016/j.cej.2009.04.003

67. Xiao, M., Zhou, J., Zhang, Y., Hu, X., & Li, S. (2013). Pertraction performance of phenol through PDMS/PVDF composite membrane in the membrane aromatic recovery system (MARS). *Journal of Membrane Science*, *428*, 172–180. https://doi. org/10.1016/j.memsci.2012.10.030

68. Koe, W. S., Lee, J. W., Chong, W. C., Pang, Y. L., & Sim, L. C. (2020). An overview of photocatalytic degradation: Photocatalysts, mechanisms, and development of photocatalytic membrane. *Environmental Science and Pollution Research*, *27*, 2522–2565. https://doi.org/10.1007/s11356-019-07193-5

69. Kim, S. H., Kwak, S. Y., Sohn, B. H., & Park, T. H. (2003). Design of TiO_2 nanoparticle self-assembled aromatic polyamide thin-film-composite (TFC) membrane as an approach to solve biofouling problem. *Journal of Membrane Science*, *211*(1), 157–165. https://doi.org/10.1016/S0376-7388(02)00418-0

70. Yang, S., Gu, J.S ., Yu, H.Y., Zhou, J., Li, S.F., Wu, X.M., & Wang, L. (2011). Polypropylene membrane surface modification by RAFT grafting polymerization and TiO_2 photocatalysts immobilization for phenol decomposition in a photocatalytic membrane reactor.*Separation and Purification Technology*, *83*,157–165. http://dx.doi. org/10.1016/ j.seppur.2011.09.030

71. Colón, G., Sanchez-Espana, J. M., Hidalgo, M. C., & Navío, J. A. (2006). Effect of TiO_2 acidic pre-treatment on the photocatalytic properties for phenol degradation. *Journal of Photochemistry and Photobiology A: Chemistry*, *179*(1–2), 20–27. https:// doi.org/10.1016/j.jphotochem.2005.07.007

72. Lupa, L., Cocheci, L., Trica, B., Coroaba, A., & Popa, A. (2020). Photodegradation of phenolic compounds from water in the presence of a Pd-containing exhausted adsorbent. *Applied Sciences*, *10*(23), 8440.https://doi.org/10.3390/app10238440

73. Rahmania, A., Rahimzadeha, H., & Beirami, S. (2019). *Desalination and Water Treatment*,144,224–32. https://doi.org/10.5004/dwt.2019.23664

74. Salim, N. E., Jaafar, J., Ismail, A. F., Othman, M. H. D., Rahman, M. A., Yusof, N., ...& Salleh, W. N. W. (2018). Preparation and characterization of hydrophilic surface modifier macromolecule modified poly (ether sulfone) photocatalytic membrane for phenol removal. *Chemical Engineering Journal*, *335*, 236–247. https://doi.org/ 10.1016/j.cej.2017.10.147

10 Removal of Heavy Metals from Petroleum Industry Wastewater/ Sludge Using Membrane Technology

Parashmoni Rajguru and Rajiv Goswami

10.1 INTRODUCTION

Heavy metals are metals or metalloids with a relatively higher density than water and pose potential toxic effects at different concentration levels. The elements with atomic weights between 63.5 and 200.6 gm/mole and densities >5 gm/cm^3 can be referred to as heavy metals [1]. All the heavy metals may not be toxic at trace levels, but excessive exposure to such metals may lead to toxicity. Examples of such heavy metals are copper (Cu), zinc (Zn), nickel (Ni), manganese (Mn), iron (Fe) and cobalt (Co), which are essential for the biochemical processes within the human body. Still, excessive exposure may cause toxicity [2,3]. On the other hand, some other heavy metals like arsenic (As), cadmium (Cd), lead (Pb), mercury (Hg) and chromium (Cr) are toxic even in trace amounts [4].

As reported in the literature, petroleum industry sludge/wastewater contains some toxic heavy metals like lead (Pb), cadmium (Cd), nickel (Ni), chromium (Cr), copper (Cu), vanadium (V), zinc (Zn), iron (Fe), arsenic (As), etc., along with different organic matters in high concentrations that are higher than the maximum permissible limits prescribed by the WHO [5–13]. These heavy metals can damage or change the functioning of different body organs, even in trace amounts. Table 10.1 shows the maximum permissible limits of different highly toxic heavy metals along with different concentrations obtained in petroleum industry wastewater and sludge samples. Tabulated data reveals that the concentrations may change depending on the operating process conducted by the refinery [8,14–18]. Heavy metal contamination may also occur due to fuel exploration [17,19]. Different approaches made by the researchers to remove these heavy metals from wastewater and sludge samples include chemical precipitation, coagulation, flocculation, floatation, ion exchange, adsorption and membrane separation. All of these methods have advantages along with some disadvantages [20–28].

DOI: 10.1201/9781003441359-10

TABLE 10.1
Maximum Limits of Heavy Metal Contamination and Reported Concentration in Different Petroleum Wastewater/Sludge Sample

Permissible Limits of Heavy Metals in Water System According to WHO

Pb (ppm)	Ni (ppm)	Cd (ppm)	Cu (ppm)	Cr (ppm)	V (ppm)	Zn (ppm)	As (ppm)	Fe (ppm)	References
0.01	0.02	0.003	1.00	0.05	–	3.00	0.01	0.10	[5]

Heavy metal concentration found in wastewater/sludge samples from different petroleum industries

Sample Type	Location	Various heavy metals reported with their concentration	Ref.
1. Raw Petroleum wastewater	Alfola petroleum area, west of Sudan	Pb-0.25 ppm; Ni-0.52 ppm; Cd-0.9 ppm; Cr-0.012 ppm	[10]
2. Processed wastewater from refinery	Nigeria	Pb-5.90 ± 6.04 ppb; Ni-86.95 ± 110.39 ppb; Cd-2.29 ± 2.37 ppb; V-0.13 ± 0.21 ppb	[14]
3. Secondary treated refinery wastewater	Kaduna Refining and Petrochemical Company (KRPC), Kaduna (Nigeria)	Pb-0.015 ppm; Cd-0.005 ppm; Cu-0.01 ppm; Cr-0.4 ppm; Zn-0.17 ppm; Fe-3.2 ppm; Pb-0.20 ± 0.115 ppm; Cd-0.042 ± 0.034 ppm; Cu-0.71 ± 0.421 ppm; Cr-0.60 ± 0.55 ppm; Zn-0.57 ± 0.439 ppm; Fe-3.13 ± 1.667 ppm	[15]
4. Petroleum refinery wastewater treatment pond sludge	Western Canada	Pb-64.7 ± 2.2 ppm; Ni-58.5 ± 3.5 ppm; Cd-2.0 ± 0.03 ppm; Cu-159 ± 6.7 ppm; Cr-70.0 ± 4.9 ppm; V-34.8 ± 1.3 ppm; Zn-2040 ± 36.5 ppm; As-68.7 ± 5.5 ppm	[16]
5. Petroleum refinery sludge samples	North Eastern India (IOCL Guwahati and Digboi)	Pb-2.68 ppm; Ni-34.93 ppm; Cd-0.06 ppm; Cu-12.92 ppm; Cr-45.39 ppm; Zn 34.70 ppm; As-2.24 ppm; Fe-150.47 ppm; Pb-4.02 ppm; Ni-7.11 ppm; Cd-0.16 ppm; Cu-2.39 ppm; Cr-8.07 ppm; Zn-131 ppm; As-1.27 ppm; Fe-302.97 ppm; Pb-2.62 ppm; Ni-38.72 ppm; Cd-0.05 ppm; Cu-12.89 ppm; Cr-8.38 ppm; Zn-68.14 ppm; As-2.22 ppm; Fe-101.13 ppm	[18]

10.2   CONVENTIONAL METHODS USED FOR THE SEPARATION OF HEAVY METALS FROM WATER SYSTEMS

In the chemical precipitation method, different chemical agents are used to make the heavy metals settle down by converting them from a dissolved state to a solid form [20]. The adsorption method is applied for the removal of heavy metals from wastewater by using different natural or synthetic adsorbents, including nanoparticles, polymeric materials, biomass, etc. [21,22]. Coagulating and flocculating agents are used in the coagulation-flocculation method to make agglomeration of the heavy metals, making them easier to separate from the water body [23]. The air flotation method is often used to remove heavy metals from wastewater. In this method, air is passed through the heavy metal-containing wastewater, forming bubbles to which the heavy metals adhere. Then these bubbles move towards the surface, where they can be easily separated [24,25]. Different ion exchange processes also have a high potential for the recovery of heavy metals from water systems [26]. All these mentioned approaches have the possibility of being sources of secondary pollution, thereby requiring a secondary treatment process. Among all these, membrane technology emerged as one of the most suitable options with high separation ability (in a solvent-free, greener way), good energy efficiency and low capital expenditure. Also, membrane technology can provide a one-step separation of heavy metals from wastewater and sludge, making it a promising application for wastewater treatment [5,27].

10.3   ROLE OF MEMBRANE TECHNOLOGY FOR SEPARATION OF HEAVY METALS FROM WATER SYSTEMS

Membrane technology for heavy metal removal was first introduced around the 1970s. Bhattacharyya et al. [28] investigated the removal of cadmium, selenium and arsenic over a pH range of 5–10, concentrations of metals ranging from 0.5 to 20 mM and transmembrane pressures of 2.8×10^5–5.6×10^5 N/M² using a charged non-cellulosic ultrafiltration membrane. They observed a high rejection of monovalent oxyanions as compared to the divalent following the Donnan mechanism. During the same time, another study was reported on treating wastewater with heavy metals using reverse osmosis technology to reuse the water after oxidation–reduction treatment [29]. In 1980, Strathmann et al. reported an experimental and theoretical approach for heavy metal removal where the removal efficiency of ultrafiltration was enhanced by using complexation, chelation and ion exchange properties [30]. Another study in 1981 by Bhattacharyya and Cheng reported the heavy metal [Zn(II), Cd(II), Cu(I) and Cu(II)] separation performance of ultrafiltration membranes when combined with a complexing/chelating agent, which differs depending on the size as well as charge of the metal complex [31]. From the reported work so far, membrane technology for heavy metal removal can be classified as ultrafiltration (UF) membrane, nanofiltration (NF) membrane, reverse osmosis (RO) membrane, forward osmosis (FO) membrane, liquid membrane (LM), electro-dialysis (ED) membrane, membrane distillation, etc. [1,2,32–34].

10.4 MEMBRANE-BASED MECHANISMS FOR HEAVY METAL REMOVAL

Heavy metal removal using membranes is mainly governed by size exclusion, charge effect and the adsorption mechanism and sometimes inclusion of two or more mechanisms may take place in membrane technology [31,35–38].

10.4.1 SIZE EXCLUSION MECHANISM

In membrane technology, size exclusion is one of the very important mechanisms for effective separation performance, where the pore diameter of a membrane must be optimised based on the particle size of the feed solution. Smaller pore size indicates higher selectivity of a membrane but decreases the flux through the membrane [39]. The pore diameter range of microfiltration membranes is 0.1–10 μm, whereas it is 0.01–1 μm for ultrafiltration membranes, and therefore, generally they are not very effective for heavy metal removal [40]. Therefore, sometimes different complexing or chelating agents are used to make the metal species larger, thereby enhancing the separation performance. The pore diameter of nanofiltration (NF) membranes ranges between 0.001 and 0.01 μm, which is in between ultrafiltration (UF) and reverse osmosis (RO) membranes and can remove heavy metals with varying efficiency. Due to their small pore diameters of 0.0001–0.001 μm, RO membranes are very effective for heavy metal removal [40]. A dense film over the membrane surface may also tighten the pore size, leading to high selectivity for heavy metal separation. Therefore, thin film composite membranes are very popular for heavy metal removal [7,40,41]. Illustrations of these membranes, including some of the reported research work, are discussed in this article.

10.4.2 ADSORPTION MECHANISM

The performance of an ultrafiltration membrane for heavy metal separation can be enhanced by using different adsorbent materials on its surface to increase the membrane's selectivity. Different heavy metals get adsorbed on the nanoporous adsorbent surface of the membrane [38]. This type of membrane is widely accepted because different types of bioadsorbents can be used that are non-hazardous, greener and can be reused easily [21,22,38]. The adsorbent surface can be modified by introducing functional groups such as carboxyl, lactone or phenyl groups. The adsorbent materials may be carbon-based, chitosan-based, bioadsorbent, magnetic materials, mineral or metal–organic framework, etc. [6]. Adsorptive membranes are of different types, such as polymeric, polymer–ceramic, electrospun nanofiber-based and nano-enhanced membranes [38]. Polymeric membrane offers good selectivity and reusability, where the inclusion of polymeric materials in the membrane system can be done easily but has low survival at high temperatures. Polymer–ceramic membranes can be designed easily with low expenditure, good chemical inertness and mechanical strength, but they provide less surface area and are not recyclable easily. Electrospun nanofiber membranes provide high flux but require additional modifications to obtain higher selectivity against smaller impurities [42]. Nano-enhanced membranes contain

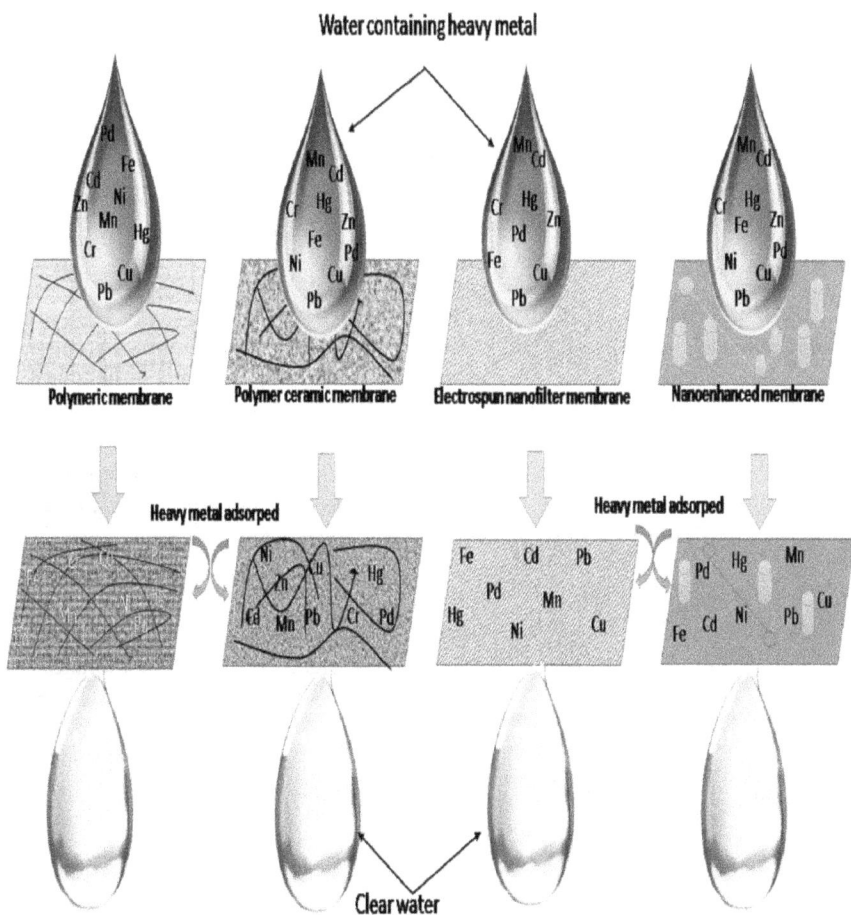

FIGURE 10.1 Pictorial representation of different membranes for heavy metal removal based on adsorption mechanism.

nanoparticles in the membrane system for better selectivity, have large surface areas providing different sites for good interaction and have high mechanical stability [43].

The adsorbent materials in the membrane system should be greener, non-hazardous, easily recyclable and have a high surface area that can chemically interact (better than physical interaction) with different pollutants. Srivastava et al. reported carbonaceous adsorbent material prepared from industrial waste slurry and performed adsorption performances against heavy metal ions [44]. Figure 10.1 represents the pictorial representation of heavy metal removal using different membranes.

10.4.3 CHARGE EFFECT MECHANISM

The surface charge of a membrane affects its selectivity for different heavy metal ions, which is known as the Donnan exclusion mechanism. Depending upon the pH, the membrane surface becomes either negatively charged, it will effectively reject anionic particles or positively charged and give better rejection of cationic particles. When the metal ion charge matches the charge on the membrane surface, then due to repulsion, these metal ions get rejected and are not permitted to pass through the membrane. That is, for the removal of cationic heavy metals, the membrane surface should be positively charged and vice versa [37–39].

10.4.4 SOLUTION-DIFFUSION MECHANISM

The solution-diffusion mechanism describes that permeants get dissolved in the membrane matrix and then diffuse through the membrane down the concentration gradient [45]. In the solution-diffusion model, pressure throughout the membrane is considered uniform and chemical potential across the membrane is considered a concentration gradient. Reverse osmosis (RO) membranes are mainly based on solution diffusion mechanisms (in the presence of hydraulic pressure). Fick's law is used to study the diffusion model following the concentration gradient (i.e., from a high to a low concentration region). Fick's law equation is given below.

$$J = - D(dC/dZ)$$

J is the flux per unit of time through the unit surface area of the membrane; D is the diffusion coefficient; C is the heavy metal concentration; Z is the dimension of the membrane.

In the forward osmosis (FO) process, water is extracted from a heavy metal-contaminated feed solution through a semi-permeable membrane. Here, diffusion of water molecules follows the normal osmotic rule (without any external hydraulic pressure) [35], i.e.,

$$\pi = \Sigma i M R T$$

where π is the osmotic pressure; i is the amount of molecules per mole of solute; M is the molarity of the solution; R is the gas constant and T is the temperature.

10.5 MEMBRANE TYPES FOR HEAVY METAL REMOVAL

10.5.1 COMPLEXATION ULTRAFILTRATION MEMBRANE

As explained, bare ultrafiltration membranes have very low efficiency for the separation of heavy metals, but this can be enhanced to a good extent when treated with chemical reagents. In this process, the heavy metals are allowed to form complexes with the chemical reagents used owing to different physical or chemical interactions, thereby increasing their diameter so that the size of the complex is greater than the pore diameter of the ultrafiltration membranes. This method of membrane design is known as complexation ultrafiltration membranes, and when complexation with heavy metals takes place using polymers, it is called polymer-enhanced ultrafiltration [46]. Cao et al. demonstrated heavy metal separation from wastewater using ultrafiltration membranes by enhancing their activity through polymer complexation. In the article, they extracted extracellular polymers from the sludges of the wastewater treatment plant itself and then allowed them to form a layer on the UF membrane through a filtration process, and they obtained very good heavy metal removal efficiency after this polymer incorporation [47]. The literature reveals that various microligands such as PAA (poly(acrylic) acids) [48–50], sodium salts of polyacrylic acids [48], Acrylic–maleic acid copolymer [51,52], PEI (polyvinylethyleneimine) [53–57], CMC (carboxy methyl cellulose) [58,59], PSS(poly (sodium 4-styrenesulfonate)) [60], PA-NH4(poly(ammonium acrylate)) [61] acts as complexing agent. These microligands, when introduced to different UF membranes such as polyacrylonitrile (PAN), polyether sulfone (PES), cellulose acetate (CA), polysulfone (PSf) and ceramic membranes, show effective removal of heavy metals. Petrov et al. reported a CMC-PAN UF membrane that showed effective removal of Cu, Pb, Ni and Cr up to 99% at pH ranges from 6.5 to 7 [62]. On the other hand, Mimoune et al. have reported another PVA (polyvinyl alcohol)-incorporated PES UF membrane that showed maximum removal efficiency of copper (Cu) up to 100% when pH was above 7 [63].

10.5.2 MICELLAR ENHANCED ULTRAFILTRATION MEMBRANE

It was observed that heavy metal removal through UF membranes is also effective up to an extent by using micelles-forming agents such as cationic and anionic surfactants [64–66]. Due to the electrical properties of surfactants, they can interact with the heavy metal ions to form large agglomerates (depending on pH) with a diameter greater than the pore diameter of the UF membrane. Huang et al. have reported using a PES UF membrane to treat wastewater containing heavy metals, where the efficiency of the membrane was enhanced using sodium dodecyl sulphate (SDS; an anionic surfactant) as a micelle-forming agent [66].

Different surfactants used in this type of membrane are SDS, cetyltrimethylammonium bromide (CTAB), cetylpyridinium chloride (CPC), etc. The reported literature showed good removal efficiency of this type of membrane for different heavy metal removal using very low amounts of cationic or anionic surfactants [67,68]. Secondary pollution is an important issue because of chemical surfactants, which can be overcome by using a membrane system [69,70]. In another study, 100% removal of Cu, Pb and Cd could be obtained using LAS and PES membrane [71]. Landaburu-Aguirre et al. revealed in their work that during the treatment of phosphorus-rich draining water, SDS was used for copper (Cu), cadmium (Cd), nickel (Ni) and zinc (Zn) removal without depending on pH change [72]. Different bioadsorbent scan be used because of their environment-friendly nature [72]. Cellulose acetate-EDTA membrane was used for Cu, Pb and Cd removal up to 97% efficiency [73].

10.5.3 NANOFILTRATION (NF) MEMBRANE

Although the size exclusion mechanism occurs in NF membranes, it is seen that the charge effect mechanism is the dominant one for this type of membrane. The nanofiltration (NF) membranes have pore sizes ranging from 0.001 to 0.01 micrometres [40], i.e., pore sizes between ultrafiltration and reverse osmosis membranes. Nanofiltration membranes were initially used to separate divalent salts and small organic residues in the 1980s [74]. It has sufficient pore size for removal of most of the heavy metals with a good flux value. Piperazine is one of the most commonly used monomers in NF membranes because it helps to provide a polyamide (PA) layer on the membrane surface with carboxylic and sulfonic groups (producing a negatively charged NF membrane). [75] On the other hand, PEI (polyethyleneimine), PAMAM (poly(amidoamine)) give rise to positively charged NF membranes for the removal of Ni(II), Pb(II), As(V), Cd(II), Cr(II) [76]. Nanofiltration membranes provide much better removal efficiency than ultrafiltration membranes but have lower flux values than ultrafiltration membranes. Therefore, UF membranes can be used for low concentrations of heavy metals, but NF membranes are preferable for high concentrations of heavy metals present in petroleum industry wastewater and sludge.

10.5.4 REVERSE OSMOSIS (RO) MEMBRANE

Due to the extremely small pore diameter, reverse osmosis membranes are highly effective for heavy metal removal, which is highly dependent on solution diffusion mechanisms. In reverse osmosis, water goes from a higher concentration to a lower concentration through a semipermeable membrane when operated with external pressure [77]. A reverse osmosis membrane for heavy metal separation was first designed by Ozaki et al. for copper, chromium and nickel removal using a low-pressure RO membrane (ULPROM/ES 20) and obtained good removal efficiency up to 98.75 % with an optimum pH range from 7 to 9 [78]. With low concentrations of heavy metals, i.e., lead, nickel, copper and chromium, very high removal rates could be obtained using a reverse osmosis membrane bioreactor [79]. In another study by Mohsen et al., EDTA was introduced to the reverse osmosis membrane system for the removal of copper and nickel to enhance the separation performance efficiency through the chelation of EDTA [80]. Feini et al. have reported that reverse osmosis membranes give very good heavy metal removal efficiency (i.e., high selectivity) but with a very low value of flux [81]. The reverse osmosis technique is useful for the removal of most of the heavy metals but requires capital investment and consumes high energy [77].

10.5.5 FORWARD OSMOSIS (FO) MEMBRANE

In the FO process, the feed solution from a highly saline draw solution is separated by a semipermeable membrane (a thin film composite membrane) creating a difference in osmotic pressure between the two, leading to the movement of water from the feed side to another side of the membrane. The process can be operated in two types: (a) active layer towards the feed solution and (b) active layer towards the highly saline draw solution. FO membrane for the removal of heavy metal was first used for the removal of arsenic and boron, and it was seen that metal rejection of FO membranes is better than RO membrane [82]. Recently, a FO nanocomposite membrane modified with sulfonated graphene oxide@metal–organic framework (SGO@UiO-66-TFN)

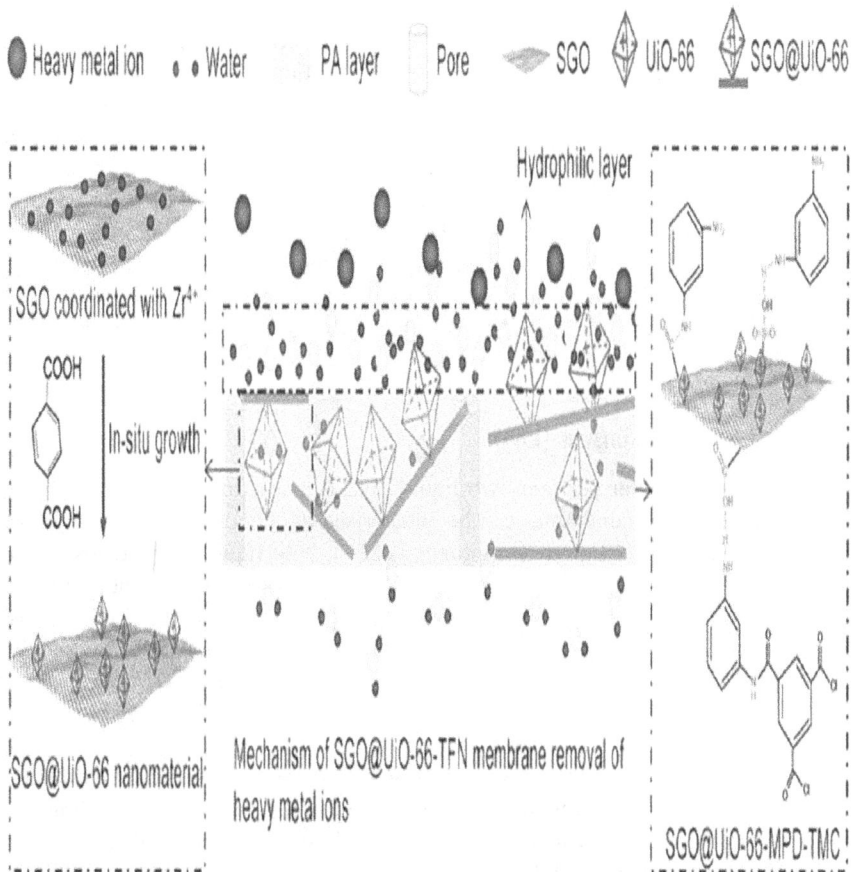

FIGURE 10.2 FO nanocomposite membrane modified with sulfonated graphene oxide@metal–organic framework (SGO@UiO-66-TFN) for removal of Cu(II) and Pb(II).

Source: [83].

was reported for Cu(II) and Pb(II) removal of more than 99.4% in two hours and more than 97.5% in ten hours with good water flux [83] (Figure 10.2). FO process shows very good heavy metal rejection without any hydraulic pressure, but the challenges are membrane selection, external and internal concentration polarisation and requirement of reconcentration of draw solution.

10.5.6 ELECTRODIALYSIS (ED) MEMBRANE

Electrodialysis membranes can be used for heavy metal removal and are electrically driven membranes where the cation exchange membrane is placed next to the anion exchange membranes and alternatively sequenced in parallel. When current is passed through the electrodes, positively charged ions move towards the cathode, which will pass through the negatively charged membrane and get retained in the positively charged membrane. Similarly, negatively charged ions get retained in the negatively charged membrane, facilitating separation of ions. Removal of Cd at different pH, temperature, concentration and with different types of electrolytes was studied using ED membrane [84]. Good removal of cadmium was obtained while maintaining excellent current efficiency and reasonable stack resistances over varied concentrations of $CdCl_2$ and $CdSO_4$. Recently, work was reported on electrodialysis configurations with an adsorptive membrane to remove Hg(II), Cu(II) and Fe(III) from the feed water sample and obtain very good removal (more than 99%) of the heavy metals [85]. It allows water purification without using chemicals over a wide pH range, but the economic viability is reduced due to the requirement of electric potential.

10.5.7 LIQUID MEMBRANE (LM)

It is a thin organic layer between two liquid phases that is immiscible to both feed and retentate. Liquid membranes can be categorised as supported liquid membranes (SLM), emulsion-based liquid membranes (ELM), bulk liquid membranes (BLM) and polymer inclusion liquid membranes (PILM) [86]. It is an efficient and highly selective technique, but durability is poor for this type of membrane [6]. BLM has a poor membrane area, and ELM is very unstable, so it is less efficient for the removal of heavy metals. In SLM, the liquid phases are held together by the capillary action of the micropores of the polymeric membrane and inorganic film. SLM contains a lower separation cost but is not long lasting. Some of the literature indicates the use of ionic liquid (IL) to improve the stability of the membrane system, where IL acts as a carrier and improves the conductivity and selectivity of the membrane for different ions, which is known as a supported ionic liquid membrane (SILM) [87–89]. Again, PILM has better durability than SILM as the liquid phase of the membrane is incorporated into the polymeric matrix, increasing its stability [90,91]. The flux of SILM and PILM membranes was studied, and it was found that after six days the flux is constant for PILM (which is lower than the SILM flux) but decreases in the case of SILM under the same conditions [90].

10.5.8 MEMBRANE DISTILLATION (MD)

Membrane distillation (MD) outperforms other pressure-driven membrane-based technologies such as ultrafiltration, nanofiltration and reverse osmosis because it is less expensive and has antifouling properties. Membrane distillation is mainly based on a hybrid thermally fuelled technique where hot and cold compartments are separated by a hydrophobic membrane of microporous structure, allowing only vapour to pass through it [92]. The pore size of the membrane used in MD ranges between 10 nm and 1 m; however, depending on the type of feed solution, the optimal pore size may differ [93].

Due to temperature differences on both sides, differences in vapour pressure occur between the two sides of the membrane, which act as the driving force for this type of membrane process [90]. The vapour is allowed to pass through the membrane and is then collected on the cold side upon condensation; the concentrated part of the feed solution is collected in the mould and allowed to cool [92]. Using the solid–liquid separation technique, crystallised parts are obtained, which are then concentrated in the membrane distillation unit, which is known as membrane distillation integrated with crystallisation. To reduce the heat loss from the membrane, an air gap is maintained between the membrane and the condensation surface, also known as the air gap membrane distillation (AGMD) process, which was reported to remove more than 96% of Hg(II), Pb(II) and As(III) from wastewater [94]. A direct-contact membrane distillation process (functionalised with alumina nanoparticles) was also reported for the removal of Cr(V), Cd(II) and Pb(II) from wastewater, where the presence of the alumina nanoparticles enhanced the membrane's performance with 99% salt rejection [95]. Recently, another approach to the solar-driven process of membrane distillation was reported, where a CB-PVA/PTFE nano-fibre membrane is used for more than 99% of salt rejection. The CB-PVA layer acts as a hydrophilic layer, providing the photothermal conversion efficiency and antifouling properties to the membrane, whereas the hydrophobic PTFE layer of the membrane allows selective permeation of vapour while inhibiting water and salt particles from passing through it [96]. The barrier to using MD as an effective tool for heavy metal removal is the price of commercial membranes; hence, further improvement of this technique is required.

10.6 CONCLUSION

Despite various membrane approaches, several issues and challenges have been identified by the researchers for its use in heavy metal removal. Various limitations such as low recyclability, high energy consumption and high process costs retard the growth of membrane technology up to a certain extent. As a result, very few membrane approaches are commercially viable.

The fouling strongly affects the durability of the membranes. Fouling of a membrane can be defined as the deposition of contaminants (present in the feed solution) on the membrane surface, thereby blocking the membrane pores (or decreasing the pore radius of the membrane), thereby inhibiting the flow of the permeate through the membrane. As a result, the maintenance cost and/or operational cost of the membranes became high as they required frequent cleansing. The fouling of a membrane can be decreased by simple cleansing procedures (such as backwashing), but this may be costly in many cases. Therefore, researchers have reported various methods, such as coating the membrane surface, incorporating nanoparticles into the membrane matrix, grafting onto the membrane surface, etc., which introduce antifouling behaviours to a membrane. Different pretreatment approaches can also be taken to maximise the lifespan of a membrane.

Different hybrid processes should be taken by combining various mechanisms in one membrane to achieve a highly efficient and durable membrane with highly advantageous physical, thermal and chemical properties. More attention should be given to biopolymers or the use of waste materials for membranes, avoiding the use of hazardous chemicals for the fabrication of membranes to minimise environmental damage. The limitations, like low chemical and thermal resistance and a short lifespan of the membrane owing to the use of biopolymers, can be overcome by incorporating the biopolymers with several materials having advantageous physical, chemical and thermal properties. The mentioned materials may include carbon-based or inorganic materials from the macroscale to the nanoscale.

A new horizon for membrane technology is possible now with various developments, including the incorporation of nanostructured materials with very high separation performance in combination with 3D printing-like production technology with very good computing possibilities. These approaches may lead to low-cost membrane modules with high flux and high removal efficiency for heavy metals, which ultimately increase the chance that such membrane technology will be adopted by various industries for the treatment of heavy metal-contaminated water.

REFERENCES

1. Sellaoui, L., Hessou, E.P., Badawi, M., Netto, M.S., Dotto, G.L., Silva, L.F.O., Tielens, F., Ifthikar, J., Bonilla-Petriciolet, A., & Chen, Z. (2021). Trapping of Ag^+, Cu^{2+}, and Co^{2+} by faujasite zeolite Y: New interpretations of the adsorption mechanism via DFT and statistical modeling investigation. *Chemical Engineering Journal, 420*, 127712. https://doi.org/10.1016/j.cej.2020.127712
2. Engwa, G.A., Ferdinand, P.U., Nwalo, F.N., & Unachukwu, M.N. (2019). Mechanism and health effects of heavy metal toxicity in humans. *Poisoning in the Modern World – New Tricks for an Old Dog, 10*, 70–90. https://doi.org/10.5772/intechopen.82511
3. Pandey, G., & Madhuri, S. (2014). Heavy metals causing toxicity in animals and fishes. *Research Journal of Animal, Veterinary and Fishery Sciences, 2*(2), 17–23. https://doi.org/10.1007/978-3-7643-8340-4_6
4. Rahman, Z., & Singh, V.P. (2019). The relative impact of toxic heavy metals (THMs) (arsenic (As), cadmium (Cd), chromium (Cr)(VI), mercury (Hg), and lead (Pb)) on the total environment: An overview. *Environmental Monitoring and Assessment, 191*, 1–21. https://doi.org/10.1007/s10661-019-7528-7
5. Kinuthia, G.K., Ngure, V., Beti, D., Lugalia, R., Wangila, A., & Kamau, L. (2020). Levels of heavy metals in wastewater and soil samples from open drainage channels in Nairobi, Kenya: Community health implication. *Scientific Reports, 10*(1), 8434. https://doi.org/10.1038/s41598-020-65359-5
6. Qasem, N.A., Mohammed, R.H., & Lawal, D.U. (2021). Removal of heavy metal ions from wastewater: A comprehensive and critical review. *NPJ Clean Water, 4*(1), 36. https://doi.org/10.1038/s41545-021-00127-0
7. Metwally, M.S., Al-Muzaini, S., Jacob, P.G., Bahloul, M., Urushigawa, Y., Sato, S., & Matsmura, A. (1997). Petroleum hydrocarbons and related heavy metals in the near-shore marine sediments of Kuwait. *Environment International, 23*(1), 115–121. https://doi.org/10.1016/S0160-4120(96)00082-7
8. Ismail, Z., & Beddri, A.M. (2009). Potential of water hyacinth as a removal agent for heavy metals from petroleum refinery effluents. *Water, Air, and Soil Pollution, 199*, 57–65. https://doi.org/10.1007/s11270-008-9859-9
9. Malamis, S., Katsou, E., Di Fabio, S., Frison, N., Cecchi, F., & Fatone, F. (2015). Treatment of petrochemical wastewater by employing membrane bioreactors: A case study of effluents discharged to a sensitive water recipient. *Desalination and Water Treatment, 53*(12), 3397–3406. https://doi.org/10.1080/19443994.2014.934112

10. Ezeldin, M., Nasir, S.A., Masaad, A.M., & Suleman, N.M. (2015). Determination of some heavy metals in raw petroleum wastewater samples before and after passing on Australis phragmites plant. *American Journal of Environmental Protection*, 4(6), 354–357. https://doi.org/10.11648/j.ajep.20150406.22

11. Varjani, S., Joshi, R., Srivastava, V.K., Ngo, H.H., & Guo, W. (2020). Treatment of wastewater from petroleum industry: Current practices and perspectives. *Environmental Science and Pollution Research*, 27, 27172–27180. https://doi.org/10.1007/s11356-019-04725-x

12. Peng, B., Yao, Z., Wang, X., Crombeen, M., Sweeney, D.G., & Tam, K.C. 2020. Cellulose-based materials in wastewater treatment of petroleum industry. *Green Energy & Environment*, 5(1), 37–49. https://doi.org/10.1016/j.gee.2019.09.003

13. de Jesús Treviño-Reséndez, J., Medel, A., & Meas, Y. (2021). Electrochemical technologies for treating petroleum industry wastewater. *Current Opinion in Electrochemistry*, 27, 100690. https://doi.org/10.1016/j.coelec.2021.100690

14. Marcus, A.C., & Ekpete, O.A. (2014). Impact of discharged process wastewater from an oil refinery on the physicochemical quality of a receiving waterbody in Rivers State, Nigeria. *Journal of Applied Chemistry*, 7, 01–08. https://doi.org/10.9790/5736-071210108

15. Mustapha, H. (2018). *Treatment of Petroleum Refinery Wastewater with Constructed Wetlands*. London: CRC Press. https://doi.org/10.1201/9780429450921

16. Hu, G., Li, J., & Hou, H. (2015). A combination of solvent extraction and freeze thaw for oil recovery from petroleum refinery wastewater treatment pond sludge. *Journal of Hazardous Materials*, 283, 832–840. https://doi.org/10.1016/j.jhazmat.2014.10.028

17. Sose, A.T., Kulkarni, S.J., & Sose, M.T. (2017). Oil industry –Analysis, effects and removal of heavy metals. *International Journal of Engineering Science Technologies*, 6, 254–257. https://doi.org/10.5281/zenodo.1012468

18. Roy, A., Sar, P., Sarkar, J., Dutta, A., Sarkar, P., Gupta, A., Mohapatra, B., Pal, S., & Kazy, S.K. (2018). Petroleum hydrocarbon rich oil refinery sludge of North-East India harbours anaerobic, fermentative, sulfate-reducing, syntrophic and methanogenic microbial populations. *BMC Microbiology*, 18(1), 1–22. https://doi.org/10.1186/s12866-018-1275-8

19. Ighalo, J. O., & Adeniyi, A. G. (2020). A comprehensive review of water quality monitoring and assessment in Nigeria. *Chemosphere*, 260, 127569. https://doi.org/10.1016/j.chemosphere.2020.127569

20. Wang, L.K., Vaccari, D.A., Li, Y., & Shammas, N.K. (2005). Chemical precipitation. *Physicochemical Treatment Processes*, 141–197. https://doi.org/10.1385/1-59259-820-x:141

21. De Gisi, S., Lofrano, G., Grassi, M., & Notarnicola, M. (2016). Characteristics and adsorption capacities of low-cost sorbents for wastewater treatment: A review. *Sustainable Materials and Technologies*, 9, 10–40. https://doi.org/10.1016/j.susmat.2016.06.002

22. Ibrahim, B.M. (2021). Heavy metal ions removal from wastewater using various low-cost agricultural wastes as adsorbents: A survey. *Zanco Journal of Pure and Applied Sciences*, 33(2), 76–91. https://doi.org/10.21271/ZJPAS.33.2.8

23. Teh, C.Y., Budiman, P.M., Shak, K.P.Y., & Wu, T.Y. (2016). Recent advancement of coagulation–flocculation and its application in wastewater treatment. *Industrial & Engineering Chemistry Research*, 55(16), 4363–4389. https://doi.org/10.1021/acs.iecr.5b04703

24. Kyzas, G.Z., & Matis, K.A. (2018). Flotation in water and wastewater treatment. *Processes*, *6*(8), 116. https://doi.org/10.3390/pr6080116

25. Pooja, G., Kumar, P.S., & Indraganti, S. (2022). Recent advancements in the removal/ recovery of toxic metals from aquatic system using flotation techniques. *Chemosphere*, *287*, 132231. https://doi.org/10.1016/j.chemosphere.2021.132231

26. Huang, X., Guida, S., Jefferson, B., & Soares, A. (2020). Economic evaluation of ion-exchange processes for nutrient removal and recovery from municipal wastewater. *NPJ Clean Water*, *3*(1), 7. https://doi.org/10.1038/s41545-020-0054-x

27. Ujang, Z., & Anderson, G.K. (1998). Performance of low pressure reverse osmosis membrane (LPROM) for separating mono-and divalent ions. *Water Science and Technology*, *38*(4–5), 521–528. https://doi.org/10.1016/S0273-1223(98)00553-8

28. Bhattacharyya, D., Moffitt, M., & Grieves, R.B. (1978). Charged membrane ultrafiltra-tion of toxic metal oxyanions and cations from single-and multisalt aqueous solutions. *Separation Science and Technology*, *13*(5), 449–463. https://doi.org/10.1080/014963 97808058294

29. Sato, T., Imaizumi, M., Kato, O., & Taniguchi, Y. (1977). RO applications in wastewater reclamation for re-use. *Desalination*, *23*(1–3), 65–76. https://doi.org/ 10.1016/S0011-9164(00)82509-6

30. Strathmann, H. (1980). Selective removal of heavy metal ions from aqueous solutions by diafiltration of macromolecular complexes. *Separation Science and Technology*, *15*(4), 1135–1152. https://doi.org/10.1080/01496398008076291

31. Bhattacharyya, D., & Cheng, C.S. (1981). Low pressure membrane separation process to remove heavy metal complexes. https://doi.org/10.13023/kwrri.rr.129

32. Sgarlata, C., Arena, G., Longo, E., Zhang, D., Yang, Y., & Bartsch, R.A. (2008). Heavy metal separation with polymer inclusion membranes. *Journal of Membrane Science*, *323*(2), 444–451. https://doi.org/10.3390/polym11111780

33. Asad, A., Sameoto, D., & Sadrzadeh, M. (2020). Overview of membrane tech-nology. In *Nanocomposite Membranes for Water and Gas Separation* (pp. 1–28). Oxford: Elsevier. https://doi.org/10.1016/B978-0-12-816710-6.00001-8

34. Attia, H., Alexander, S., Wright, C. J., & Hilal, N. (2017). Superhydrophobic electrospun membrane for heavy metals removal by air gap membrane distillation (AGMD). *Desalination*, *420*, 318–329. http://dx.doi.org/10.1016/j.desal.2017.07.022

35. Valladares Linares, R. (2014). Hybrid membrane system for desalination and wastewater treatment: Integrating forward osmosis and low pressure reverse osmosis. https://doi.org/10.4233/UUID:B9DC8FDE-B23D-4D14-9D09-8B2B7AA924F5

36. Hamid, M.F., Yusof, N., Ismail, N.M., & Azali, M.A. (2020). Role of membrane sur-face charge and complexation-ultrafiltration for heavy metals removal: A mini review. *Journal of Applied Membrane Science & Technology*, *24*(1). https://doi.org/10.11113/ amst.v24n1.170

37. Zhou, M.Y., Zhang, P., Fang, L.F., Zhu, B.K., Wang, J.L., Chen, J.H., & Abdallah, H. (2019). A positively charged tight UF membrane and its properties for removing trace metal cations via electrostatic repulsion mechanism. *Journal of Hazardous Materials*, *373*, 168–175. https://doi.org/10.1016/j.jhazmat.2019.03.088

38. Vo, T.S., Hossain, M.M., Jeong, H.M., & Kim, K. (2020). Heavy metal removal applications using adsorptive membranes. *Nano Convergence*, *7*, 1–26. https://doi.org/ 10.1186/s40580-020-00245-4

39. Hu, M.Z., Engtrakul, C., Bischoff, B.L., Lu, M., & Alemseghed, M. (2018). Surface-engineered inorganic nanoporous membranes for vapor and pervaporative separations of water–ethanol mixtures. *Membranes*, *8*(4), 95. https://doi.org/10.3390/membranes 8040095

40. OboteyEzugbe, E., & Rathilal, S. (2020). Membrane technologies in wastewater treatment: A review. *Membranes*, *10*(5), 89. https://doi.org/10.3390/membranes1 0050089

41. Davenport, D. M., Ritt, C. L., Verbeke, R., Dickmann, M., Egger, W., Vankelecom, I. F., & Elimelech, M. (2020). Thin film composite membrane compaction in high-pressure reverse osmosis. *Journal of Membrane Science*, *610*, 118268. https://doi.org/ 10.3390/membranes10050089

42. Selatile, M. K., Ray, S. S., Ojijo, V., & Sadiku, R. (2018). Recent developments in polymeric electrospun nanofibrous membranes for seawater desalination. *RSC Advances*, *8*(66), 37915–37938. https://doi.org/10.1039/C8RA07489E

43. Li, X., Sotto, A., Li, J., & Van der Bruggen, B. (2017). Progress and perspectives for synthesis of sustainable antifouling composite membranes containing in situ generated nanoparticles. *Journal of Membrane Science*, *524*, 502–528. https://doi.org/10.1016/ j.memsci.2016.11.040

44. Srivastava, S. K., Tyagi, R., & Pant, N. (1989). Adsorption of heavy metal ions on carbonaceous material developed from the waste slurry generated in local fertilizer plants. *Water Research*, *23*(9), 1161–1165. https://doi.org/10.1016/ 0043-1354(89)90160-7

45. Wijmans, J.G., & Baker, R.W. (1995). The solution-diffusion model: A review. *Journal of Membrane Science*, *107*(1–2), 1–21. https://doi.org/10.1016/0376-7388(95)00102-I

46. Jana, S., Saikia, A., Purkait, M.K., & Mohanty, K. (2011). Chitosan based ceramic ultrafiltration membrane: Preparation, characterization and application to remove Hg (II) and As (III) using polymer enhanced ultrafiltration. *Chemical Engineering Journal*, *170*(1), 209–219. https://doi.org/10.1016/j.cej.2011.03.056

47. Cao, D.Q., Wang, X., Wang, Q.H., Fang, X.M., Jin, J.Y., Hao, X.D., Iritani, E., & Katagiri, N. (2020). Removal of heavy metal ions by ultrafiltration with recovery of extracellular polymer substances from excess sludge. *Journal of Membrane Science*, *606*, 118103. https://doi.org/10.1016/j.memsci.2020.118103

48. Fu, F., & Wang, Q. (2011). Removal of heavy metal ions from wastewaters: A review. *Journal of Environmental Management*, *92*(3), 407–418. https://doi.org/10.1016/j.jenv man.2010.11.011

49. Canizares, P., Pérez, A., Camarillo, R., & Linares, J.J. (2004). A semi-continuous laboratory-scale polymer enhanced ultrafiltration process for the recovery of cadmium and lead from aqueous effluents. *Journal of Membrane Science*, *240*(1–2), 197–209. https://doi.org/10.1016/j.memsci.2004.04.021

50. Ounifi, I., Guesmi, Y., Ursino, C., Santoro, S., Mahfoudhi, S., Figoli, A., Ferjanie, E., & Hafiane, A. (2021). Antifouling membranes based on cellulose acetate (CA) blended with poly (acrylic acid) for heavy metal remediation. *Applied Sciences*, *11*(10), 4354. https://doi.org/10.3390/app11104354

51. Zhou, H., Qiu, Y.R. and Chen, Y.X., 2020. Recovery of Hg (II) from aqueous solution by complexation-ultrafiltration using rotating disk membrane and shear stability of PMA-Hg complex. *Journal of Central South University*, *27*(9), 2507–2514. https:// doi.org/10.1007/s11771-020-4471-2

52. Le, H.S., & Qiu, Y.R. (2020). Selective separation of Cd (II), Zn (II) and Pb (II) from Pb-Zn smelter wastewater via shear induced dissociation coupling with ultrafiltration. *Korean Journal of Chemical Engineering*, *37*, 784–791. https://doi.org/10.1007/s11 814-020-0509-2

53. Cañizares, P., Pérez, Á., & Camarillo, R. (2002). Recovery of heavy metals by means of ultrafiltration with water-soluble polymers: Calculation of design parameters. *Desalination*, *144*(1–3), 279–285. https://doi.org/10.1016/S0011-9164(02)00328-4

54. Molinari, R., Poerio, T., & Argurio, P. (2008). Selective separation of copper (II) and nickel (II) from aqueous media using the complexation–ultrafiltration process. *Chemosphere*, *70*(3), 341–348. https://doi.org/10.1016/j.chemosphere.2007.07.041

55. Aroua, M.K., Zuki, F.M., & Sulaiman, N.M. (2007). Removal of chromium ions from aqueous solutions by polymer-enhanced ultrafiltration. *Journal of Hazardous Materials*, *147*(3), 752–758. https://doi.org/10.1016/j.jhazmat.2007.01.120

56. Algieri, C., Chakraborty, S., & Candamano, S. (2021). A way to membrane-based environmental remediation for heavy metal removal. *Environments*, *8*(6), 52. https://doi.org/10.3390/environments8060052

57. Chou, Y.H., Choo, K.H., Chen, S.S., Yu, J.H., Peng, C.Y., & Li, C.W. (2018). Copper recovery via polyelectrolyte enhanced ultrafiltration followed by dithionite based chemical reduction: Effects of solution pH and polyelectrolyte type. *Separation and Purification Technology*, *198*, 113–120. https://doi.org/10.1016/j.seppur.2017.02.008

58. Barakat, M.A., & Schmidt, E. (2010). Polymer-enhanced ultrafiltration process for heavy metals removal from industrial wastewater. *Desalination*, *256*(1–3), 90–93. https://doi.org/10.1016/j.desal.2010.02.008

59. Han, B., Zhang, D., Shao, Z., Kong, L., & Lv, S. (2013). Preparation and characterization of cellulose acetate/carboxymethyl cellulose acetate blend ultrafiltration membranes. *Desalination*, *311*, 80–89. https://doi.org/10.1016/j.desal.2012.11.002

60. Chen, M., Shafer-Peltier, K., Randtke, S.J., & Peltier, E. (2018). Competitive association of cations with poly (sodium 4-styrenesulfonate)(PSS) and heavy metal removal from water by PSS-assisted ultrafiltration. *Chemical Engineering Journal*, *344*, 155–164. https://doi.org/10.1016/j.cej.2018.03.054

61. Ennigrou, D.J., Gzara, L., Romdhane, M.R.B., & Dhahbi, M. (2009). Cadmium removal from aqueous solutions by polyelectrolyte enhanced ultrafiltration. *Desalination*, *246*(1–3), 363–369. https://doi.org/10.1016/j.desal.2008.04.053

62. Petrov, S., & Nenov, V. (2004). Removal and recovery of copper from wastewater by a complexation–ultrafiltration process. *Desalination*, *162*, 201–209. https://doi.org/10.1016/S0011-9164(04)00043-8

63. Mimoune, S., & Amrani, F. (2007). Experimental study of metal ions removal from aqueous solutions by complexation–ultrafiltration. *Journal of Membrane Science*, *298*(1–2), 92–98. https://doi.org/10.1016/j.memsci.2007.04.003

64. Lin, W. (2020). *Micellar-and polymer-enhanced ultrafiltration for heavy metal and sulfate removal from aqueous solutions*. Doctoral dissertation, Memorial University of Newfoundland. https://doi.org/10.48336/g8dt-pj97

65. Jung, J., Yang, J.S., Kim, S.H., & Yang, J.W. (2008). Feasibility of micellar-enhanced ultrafiltration (MEUF) or the heavy metal removal in soil washing effluent. *Desalination*, *222*(1–3), 202–211. https://doi.org/10.1016/j.desal.2007.01.154

66. Huang, J., Yuan, F., Zeng, G., Li, X., Gu, Y., Shi, L., Liu, W., & Shi, Y. (2017). Influence of pH on heavy metal speciation and removal from wastewater using micellar-enhanced ultrafiltration. *Chemosphere*, *173*, 199–206. https://doi.org/10.1016/j.chemosphere.2016.12.137

67. Yaqub, M., & Lee, S. H. (2019). Heavy metals removal from aqueous solution through micellar enhanced ultrafiltration: A review. *Environmental Engineering Research*, *24*(3), 363–375. https://doi.org/https://doi.org/10.4491/eer.2018.249

68. Tortora, F., Innocenzi, V., Prisciandaro, M., Mazziotti di Celso, G., & Vegliò, F. (2016). Analysis of membrane performance in Ni and Co removal from liquid wastes by means of micellar-enhanced ultrafiltration. *Desalination and Water Treatment*, *57*(48–49), 22860–22867. https://doi.org/10.1080/19443994.2016.1180475

69. Baek, K., & Yang, T.W. (2004). Competitive bind of anionic metals with cetylpyridinium chloride micelle in micellar-enhanced ultrafiltration. *Desalination, 167,* 101–110. https://doi.org/10.1007/s11705-014-1407-0

70. Gwicana, S., Vorster, N., & Jacobs, E. (2006). The use of a cationic surfactant for micellar-enhanced ultrafiltration of platinum group metal anions. *Desalination, 199*(1–3), 504–506. https://doi.org/10.1016/j.desal.2006.03.192

71. Samper, E., Rodríguez, M., De la Rubia, M.A., & Prats, D. (2009). Removal of metal ions at low concentration by micellar-enhanced ultrafiltration (MEUF) using sodium dodecyl sulfate (SDS) and linear alkylbenzene sulfonate (LAS). *Separation and Purification Technology, 65*(3), 337–342. https://doi.org/10.1016/j.seppur.2008.11.013

72. Landaburu-Aguirre, J., Pongrácz, E., Sarpola, A., & Keiski, R.L. (2012). Simultaneous removal of heavy metals from phosphorous rich real wastewaters by micellar-enhanced ultrafiltration. *Separation and Purification Technology, 88,* 130–137. https://doi.org/10.1016/j.seppur.2011.12.025

73. Jung, J., Yang, J.S., Kim, S.H., & Yang, J.W. (2008). Feasibility of micellar-enhanced ultrafiltration (MEUF) or the heavy metal removal in soil washing effluent. *Desalination, 222*(1–3), 202–211. https://doi.org/10.1016/j.desal.2007.01.154

74. Lau, W.J., Ismail, A.F., Misdan, N., & Kassim, M.A. (2012). A recent progress in thin film composite membrane: A review. *Desalination, 287,* 190–199. https://doi.org/10.1016/j.desal.2011.04.004

75. Fang, W., Shi, L., & Wang, R. (2013). Interfacially polymerized composite nanofiltration hollow fiber membranes for low-pressure water softening. *Journal of Membrane Science, 430,* 129–139. https://doi.org/10.1016/j.memsci.2012.12.011

76. Gao, J., Sun, S.P., Zhu, W.P., & Chung, T.S. (2014). Chelating polymer modified P84 nanofiltration (NF) hollow fiber membranes for high efficient heavy metal removal. *Water Research, 63,* 252–261. https://doi.org/10.1016/j.watres.2014.06.006

77. Yadav, M., Gupta, R., Arora, G., Yadav, P., Srivastava, A., & Sharma, R.K. (2020). Current status of heavy metal contaminants and their removal/recovery techniques. In *Contaminants in Our Water: Identification and Remediation Methods* (pp. 41–64). American Chemical Society. https://doi.org/10.1021/bk-2020-1352.ch003

78. Ozaki, H., Sharma, K., & Saktaywin, W. (2002). Performance of an ultra-low-pressure reverse osmosis membrane (ULPROM) for separating heavy metal: Effects of interference parameters. *Desalination, 144*(1–3), 287–294. https://doi.org/10.1016/S0011-9164(02)00329-6

79. Dialynas, E., & Diamadopoulos, E. (2009). Integration of a membrane bioreactor coupled with reverse osmosis for advanced treatment of municipal wastewater. *Desalination, 238*(1–3), 302–311. https://doi.org/10.1016/j.desal.2008.01.046

80. Mohsen-Nia, M., Montazeri, P., & Modarress, H. (2007). Removal of Cu^{2+} and Ni^{2+} from wastewater with a chelating agent and reverse osmosis processes. *Desalination, 217*(1–3), 276–281. https://doi.org/10.1016/j.desal.2006.01.043

81. Feini, L.I.U., ZHANG, G., Qin, M., and ZHANG, H. (2008). Performance of nanofiltration and reverse osmosis membranes in metal effluent treatment. *Chinese Journal of Chemical Engineering, 16*(3), 441–445. https://doi.org/10.1016/S1004-9541(08)60102-0

82. Jin, X., She, Q., Ang, X., & Tang, C.Y. (2012). Removal of boron and arsenic by forward osmosis membrane: Influence of membrane orientation and organic fouling. *Journal of Membrane Science, 389,* 182–187. http://dx.doi.org/10.1016/j.memsci.2011.10.028

83. He, M., Wang, L., Zhang, Z., Zhang, Y., Zhu, J., Wang, X., Lv, Y., & Miao, R. (2020). Stable forward osmosis nanocomposite membrane doped with sulfonated graphene

oxide@ metal–organic frameworks for heavy metal removal. *ACS Applied Materials & Interfaces, 12*(51), 57102–57116. https://doi.org/10.1021/acsami.0c17405

84. Gering, K.L., & Scamehorn, J.F. (1988). Use of electrodialysis to remove heavy metals from water. *Separation Science and Technology, 23*(14–15), 2231–2267. https://doi. org/10.1080/01496398808058452

85. Uliana, A.A., Bui, N.T., Kamcev, J., Taylor, M.K., Urban, J.J., & Long, J.R. (2021). Ion-capture electrodialysis using multifunctional adsorptive membranes. *Science, 372*(6539), 296–299. https://doi.org/10.1126/science.abf5991

86. Hosseini, S.S., Bringas, E., Tan, N.R., Ortiz, I., Ghahramani, M., & Shahmirzadi, M.A.A. (2016). Recent progress in development of high performance polymeric membranes and materials for metal plating wastewater treatment: A review. *Journal of Water Process Engineering, 9*, 78–110. https://doi.org/10.1016/j.jwpe.2015.11.005

87. Malik, M.A., Hashim, M.A., & Nabi, F. (2011). Ionic liquids in supported liquid membrane technology. *Chemical Engineering Journal, 171*(1), 242–254. https://doi.org/10.1016/J.CEJ.2011.03.041

88. De Los Ríos, A.P., Hernández-Fernández, F.J., Lozano, L.J., Sánchez-Segado, S., Ginestá-Anzola, A., Godinez, C., Tomás-Alonso, F., & Quesada-Medina, J. (2013). On the selective separation of metal ions from hydrochloride aqueous solution by pertraction through supported ionic liquid membranes. *Journal of Membrane Science, 444*, 469–481. https://doi.org/10.1016/j.memsci.2013.05.006

89. Jean, E., Villemin, D., Hlaibi, M., & Lebrun, L. (2018). Heavy metal ions extraction using new supported liquid membranes containing ionic liquid as carrier. *Separation and Purification Technology, 201*, 1–9. https://doi.org/10.1016/j.seppur.2018.02.033

90. Zante, G., Boltoeva, M., Masmoudi, A., Barillon, R., & Trébouet, D. (2022). Supported ionic liquid and polymer inclusion membranes for metal separation. *Separation & Purification Reviews, 51*(1), 100–116. https://doi.org/10.1080/15422119.2020.1846564

91. Almeida, M.I.G., Cattrall, R.W., & Kolev, S.D. (2012). Recent trends in extraction and transport of metal ions using polymer inclusion membranes (PIMs). *Journal of Membrane Science, 415*, 9–23. https://doi.org/10.1016/j.memsci.2012.06.006

92. Zolotarev, P.P., Ugrozov, V.V., Volkina, I.B., & Nikulin, V.M. (1994). Treatment of wastewater for removing heavy metals by membrane distillation. *Journal of Hazardous Materials, 37*(1), 77–82. https://doi.org/10.1016/0304-3894(94)85035-6

93. Lou, X.Y., Xu, Z., Bai, A.P., Resina-Gallego, M., & Ji, Z.G. (2020). Separation and recycling of concentrated heavy metal wastewater by tube membrane distillation integrated with crystallization. *Membranes, 10*(1), 19. https://doi.org/10.3390/membranes10010019

94. Alkhudhiri, A., Hakami, M., Zacharof, M.P., Abu Homod, H., & Alsadun, A. (2020). Mercury, arsenic and lead removal by air gap membrane distillation: Experimental study. *Water, 12*(6), 1574. https://doi.org/10.3390/w12061574

95. Khraisheh, M., AlMomani, F., & Al-Ghouti, M. (2021). Electrospun Al_2O_3 hydrophobic functionalized membranes for heavy metal recovery using direct contact membrane distillation. *International Journal of Energy Research, 45*(6), 8151–8167. https://doi.org/10.1002/er.5710

96. Ju, J., Huang, Y., Liu, M., Xie, N., Shi, J., Fan, Y., Zhao, Y., & Kang, W. (2023). Construction of electrospinning Janus nanofiber membranes for efficient solar-driven membrane distillation. *Separation and Purification Technology, 305*, 122348. https://doi.org/10.1016/j.seppur.2022.122348

11 Enhanced Oil Recovery

Parashmoni Rajguru and Alimpia Borah

11.1 INTRODUCTION

Enhanced oil recovery (EOR) is of great importance to meet the high demand for oil as well as the economic growth of the oil industries [1–3]. Nowadays, hydrocarbons or hydrocarbon-based materials are mainly used as sources of energy. Many researchers have reported declines in the production of the reservoirs over time [3–5]. Oil recovery may include primary, secondary and tertiary phases. Primary oil recovery may occur under natural pressure without the injection of any fluids, including water or gas. When underground pressure is insufficient for the production of the residual oil, different secondary methods, such as fluid flooding, are carried out to move the oil forward towards the production site. Even after secondary treatment, a high amount of residual oil remains in the oil reservoirs. A large part of the reservoir remains unexplored due to technological limitations, and average recoveries of oil from the reservoirs globally range from 20% to 40% [5]. Therefore, tertiary treatment is very important to increase the productivity of the oil reservoirs, which can also be termed as EOR [3,6].

EOR increases the production of crude oil to a huge extent, which eventually lowers the price of crude oil. Using conventional extraction processes (primary and secondary techniques), only a very small fraction of the total oil in the reservoir is extracted (less than 30% in some cases). Production from the same reservoirs are increased up to 75% of the total oil in the reservoir using EOR techniques [7]. In this article, we have discussed different EOR approaches and smart water flooding as a cheap and environment-friendly approach for EOR. The article mainly focuses on the role of membrane technology in the production of smart water (ionically modified water) [8,9].

11.2 APPROACHES REPORTED FOR EOR

Researchers are always interested in finding novel, highly efficacious techniques for EOR. The various conventional and reported approaches include thermal [7] as well as non-thermal based approaches [10–12] which were addressed in the upcoming portion of this chapter.

11.2.1 Thermal Approaches for EOR

A thermal approach for EOR may include the injection of hot fluids like flooding with hot water or injecting steam [13]. Fire flooding [14] is also an acceptable approach for EOR. These approaches were used for reservoirs containing highly dense or viscous oil with a specific gravity of less than 20 API (American Petroleum Institute) [7]. Fire flooding or hot fluid flooding decreases the viscosity, interfacial tension and specific gravity of the heavy oil, which provides a driving force or pathway for the smooth flow of the heavy oil towards the surface or production well [15,16].

11.2.1.1 Cyclic Steam Stimulation

Cyclic steam stimulation [3,17,18] is a well-known thermal method for EOR. In this process, steam is injected into the well for a specific period of time. For equal distribution of the heat throughout the well, it is shut for some days. Then, the production from the well increased rapidly to a high extent.

11.2.1.2 Steam Flooding

It is also a thermal method for EOR. In this method, steam is continuously injected into the well, which reduces the viscosity of the oil present inside the well, thereby making easy recovery of the residual oil [3,19,20]. To recover heavy crude oil, steam-assisted gravity drainage using horizontal well technology can also be used, where steam is injected into the well and allowed to condense, forming hot water. It ultimately reduces the viscosity of the bitumen, and as a result, it flows to the bottom of the well due to gravity. The oil then moves towards the production well with the technology of a horizontal well.

11.2.1.3 In Situ Combustion

A portion of the residual oil inside the well is burned in the presence of oxygen or air to produce a very high amount of heat inside the well. It readily reduces the viscosity of the oil inside the well [21] facilitating easy recovery of oil. The types of in situ combustion may include reverse and forward combustion [22], highly pressurised air injection [23], etc. When combustion starts near the well of injection and the heat moves with the flow of air, it is called forward combustion. In cases of reverse combustion, the combustion starts near the production well, and the heat moves in the opposite direction to the airflow direction. In another method, air under high pressure is injected into the well, leading to the oxidation of the residual oil [24] at even lower temperatures inside the well. These approaches have some limitations, like the fact that they cost a lot since they use a lot of fuel and water with a high loss of thermal energy during EOR [7,25].

11.2.2 Non-Thermal Techniques for Enhanced Recovery of Oil

For crude oils with low to moderate viscosity, non-thermal techniques are generally applied for EOR [7]. Non-thermal methods are designed to decrease the tension at the interface enhancing the mobility of the crude oil. The non-thermal method of EOR is subclassified depending on the route of EOR, such as miscible flooding [26] and chemical flooding etc. [27–29].

11.2.2.1   Miscible Flooding

In this method, a fluid that is miscible with the residual oil of the well is used for flooding, and displacement of the mixture is done using piston-like technology [29,30]. In miscible flooding, miscible gases are injected into the production well or reservoirs. The residual oil and miscible gas' interfacial tension gets reduced, improving the mobility of the oil. The most common fluid, generally used for miscible flooding, is carbon dioxide, owing to its cost-effectiveness and the property of reducing the viscosity of oil [31]. The miscible slug process is used under the miscible flooding approach, where pentane or propane-like solvents are used in slug form [32]. Water or gases like nitrogen or methane are used to move the miscible slug forward. The technique is mostly used in sandstone and carbonate reservoirs. Continuous injection of gases like nitrogen, natural gas or fuel gas that has been enhanced with C2–C4 fractions is also used. These gases get condensed into the residual oil of the reservoir, creating a transition zone at a fairly high pressure. Multiple contacts of the injected gas with the reservoir oil lead to miscibility between the two. As a result, the volume of the oil phase increased and the viscosity of the same apparently decreased. As a result, it will move towards the production site with much less effort. Gas drive-through vaporisation can also be carried out, where C2–C6 fractions from the oil get vaporised and develop miscibility with the injecting gas. In order to vaporise the fractions, it must be noted that the injecting pressure is lower than the saturation pressure of the reservoir [3]. The application of this approach is restricted to reservoirs with high-pressure tolerances. The approach of CO_2 miscibility is also widely used for EOR. In addition to environmental concerns, CO_2 requires low pressure for its miscibility with crude oil. By using this method, the viscous instabilities are generally reduced. In another approach, nitrogen gas is also used in place of carbon dioxide; here, the minimal miscibility pressure required is high for nitrogen. Both carbon dioxide and nitrogen gas injection are generally used for medium-light and light residual crude oil in a reservoir with more than 30° API.

11.2.2.2   Chemical Flooding

Different chemical additives are added to the injecting fluid in order to enhance the movement of the injecting fluid or to decrease the interfacial tension at the interface of the injecting fluid and the reservoir oil. Chemical flooding may include polymer flooding [33], ionic or non-ionic surfactant flooding [34,35], flooding by a binary combination of surfactant and polymer [36], or flooding by a ternary combination of surfactant, polymer and alkali [37]. The fluid for chemical flooding may also include other auxiliary chemicals or their related combinations.

In chemical flooding, polymer flooding is done, where different polymers that are soluble in water are used to lower the permeability contrast and enhance the mobility ratio. Degradation of polymers or loss of polymers into the medium limits their use in EOR. Surfactant or alkaline flooding is also used to lower the interfacial tension between the residual oil and water. Loss of surfactants, their adsorption on the rock surface and their interaction with the composition of rock inside the well limit their applicability in EOR [38]. Micellar flooding is also one of the most successful chemical flooding approaches. The approach is mainly based on polymer slugs and microemulsion slugs, where brine is used to propel both. Both reservoir oil and water get miscible with microemulsions. The application of the approach is economically less viable due to costly chemicals and many reservoirs geology (rock surface) as well as the environment inside the well are not suitable for micro-flooding [39].

A low-cost and environmental-friendly approach of the microbial–EOR method can also be used, where metabolites of microorganisms decrease the viscosity of heavy crude oil inside the reservoir. Still, apparent procedural complexity and insufficient evidence from field tests and laboratory studies limit its applicability on a large scale [40].

11.3   MEMBRANES FOR EOR

By considering EOR from the perspectives of cost-effectiveness, environmental impact, ease of use and risk mitigation, water flooding is chosen above other chemical- or thermal-based techniques. The chemistry of injecting water plays a key role during water flooding for EOR. Therefore, the technology that can tackle the chemistry of injecting water is very much necessary [41]. For water flooding, salty water with varying hardness levels and low sulphate concentrations is generally required, which is very tough to achieve using conventional technologies. Some techniques reported where salty water sources, including seawater, were used to increase the salinity of the injecting water, but they may lead to uncontrollable hardness with unacceptable sulphate concentration [42,43]. Researchers have reported the use of KCl in injecting water to increase its hardness and make it efficient for EOR [44] but its use in large-scale, highly saline injection is not possible from a cost perspective.

To overcome all the mentioned issues, membrane technology stands best for controlling the hardness levels as well as the sulphate concentration of the injecting water. The type of membrane is selected depending on the level of hardness required. Owing to their large pore sizes (0.01–2 μm), ultrafiltration and microfiltration membranes [45,46] can only remove suspended particles and microbial contaminants [46]. Nanofiltration membranes with smaller pore sizes (0.2–2 nm) [47] can remove divalent ions from hard water to soften it. With pore sizes less than 0.001 μm, reverse osmosis (RO) membranes can show higher than 99% salt rejection [48]. Despite being known to have significant value-eroding qualities, seawater is frequently employed as an offshore injection fluid. Scale development due to chemical incompatibility between the injected seawater (with a high sulphate concentration) and the original formation water and reservoir rock containing barium and strontium is the most well-known issue [49].

Most of the oil reservoirs are fractured carbonate reservoirs (others are sandstone reservoirs), in which normal water flooding is ineffective for oil recovery due to the non-wettability of the rock surface [50,51]. Therefore, to make the rock surface wet, smart water is injected into the well [52]. Where smart water is nothing but ionically modified water, the concentration of monovalent and divalent ions is maintained at an optimum level (generally with reduced monovalent ion concentration and increased divalent ion concentration) depending on the rock type of the reservoir. Depending on the temperature and nature of the rock and fluid present in the carbonate and sandstone reservoirs, the composition of smart water may vary. Concentrations of divalent ions (which may be calcium, magnesium and sulphate) are high with low monovalent ion concentrations in the carbonate reservoir [53–56]. But in the case of sandstone reservoirs, in the low salinity brines, the presence of divalent ions is not preferred as it may inhibit the increase of pH, which is important to analyse the low salinity effect [56]. Total dissolved solids in smart water for carbonate and sandstone reservoirs should be 10,000–32,000 ppm and less than 5000 ppm, respectively. Upon injecting smart water on the oil-adhered rock surface, negatively charged, e.g., ions get adsorbed on the positively charged rock surface, which makes the positively charged (e.g., Ca^{2+}, Mg^{2+}) ions desorb the oil droplets from the rock surface by releasing the carboxylic group [57]. Figure 11.1 shows the pictorial representation

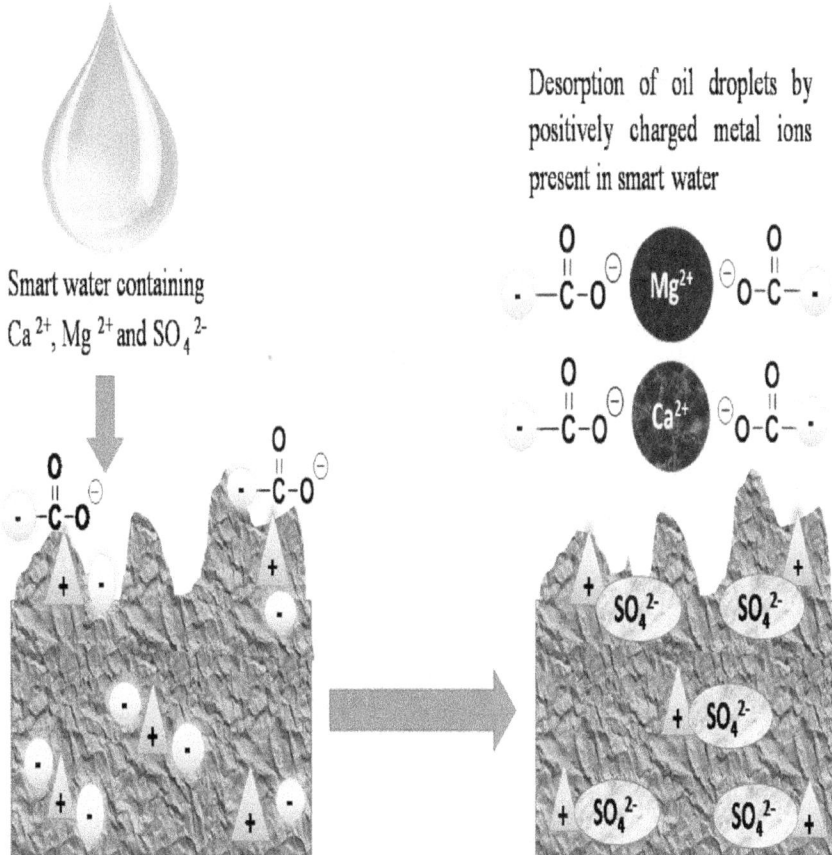

FIGURE 11.1 Pictorial representation of enhanced recovery of crude oil (containing carboxylic groups) from rock surfaces using ionically modified water (smart water).

of EOR of crude oil containing carboxylic group from rock surface using ionically modified water (smart water).

Sea water (SW) has the potential to be used as smart water after removing the monovalent ions and enriching it with divalent ions. Membranes can play a great role in maintaining ionic composition for the production of smart water. NF membrane is used to produce smart water as it provides high flux with low operating pressure to filter out the monovalent ions where the retentate is used as smart water [58]. Nanofiltration membranes have pore sizes ranging from 0.1 to 1 nm, and they are deliberately made charged (+Ve or −Ve), giving them a better rejection of . [58]. Pressure-retarded osmosis (PRO) membranes are also used to increase the amount of water available for flooding, making EOR more economical and leading to higher oil production [59]. Produced water (PW) is one of the major effluents of the petroleum industry, which is also a good candidate for EOR, which is beneficial from an environmental and economic point of view [60]. De-oiling of the produced water is done using floatation, hydrocyclones, dual media (by choosing a medium through which only water can pass, leaving behind the hydrocarbons) or nutshell filters and then the de-oiled produced water is allowed to pass through a nanofiltration membrane to separate calcium and barium ions. The retentate is combined with seawater, and permeate can also be added to seawater enriched with specific multivalent ions (phosphate, sulphate) to produce smart water. Thus, EOR has a great potential to increase the productivity of oil industries in a very green and cost-effective way. Figure 11.2 represents the EOR process by injecting smart water.

11.4 CONCLUSION

The future of oil recovery from non-renewable crude oil sources is represented by a paradigm shift towards membrane technology-based EOR. The large amount of research conducted over the past ten years and beyond is a ringing endorsement of the potential of membrane technology for EOR applications. Oil and gas industry-produced and processed water is one of the most difficult wastewaters to manage within acceptable pollution levels at reasonable costs due to its complexity. Undoubtedly, membrane technology is one of the leading candidates to address the underlying issue of produced wastewater treatment and its reuse as smart water.

The development of improved, solvent-resistant, high-flux membranes with acceptable selectivity and membrane modules at a cost that is affordable for technological adoption should be the main goal of advancement in membrane technology. Membrane materials and preparation procedures also play key roles in smart water production for the use of EOR.

The effective application of environmentally benign and low cost novel or hybrid materials with advanced physical, chemical and mechanical properties, integration of numerous cutting-edge and innovative methods with real-time monitoring and control, under various chemical and thermal scenarios, should be the primary goals of future membrane design strategies. The potential integration of several of the aforementioned approaches may enhance the economic viability and feasibility of the membrane based technology for smart water production with overall important

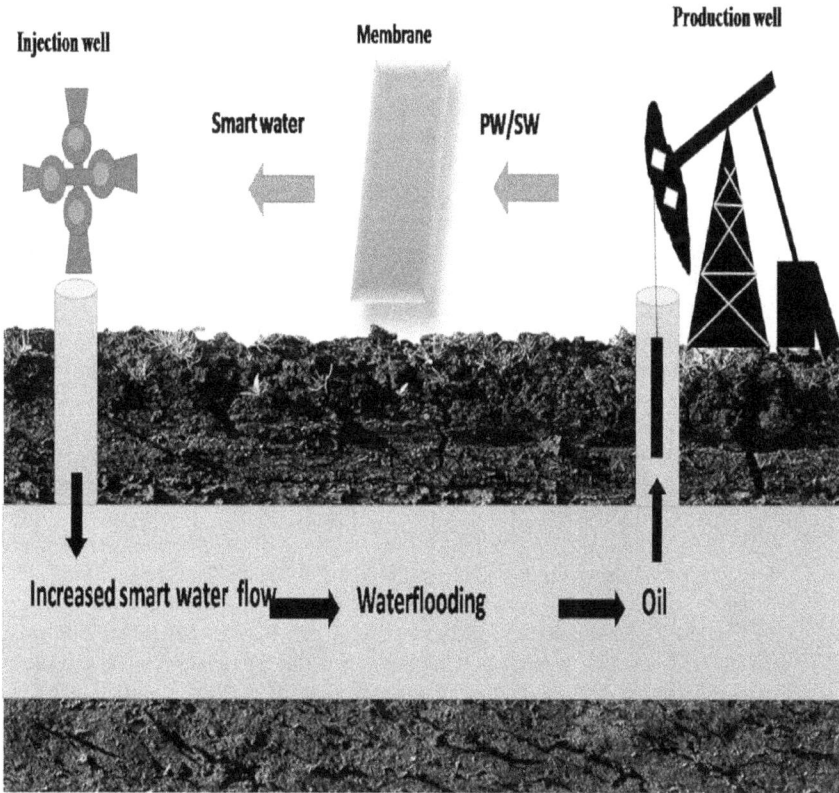

FIGURE 11.2 Representative diagram showing enhanced oil recovery (EOR) process by injecting smart water (smart water is obtained by using membrane technology).

benefits in terms of process efficiency, environmental impact, energy savings and lower costs, leading to broader industry adoption probability.

REFERENCES

1. Novak Mavar, K., Gaurina-Međimurec, N., & Hrnčević, L. (2021). Significance of enhanced oil recovery in carbon dioxide emission reduction. *Sustainability*, 13(4), 1800. https://doi.org/10.3390/su13041800
2. Guo, K., Li, H., & Yu, Z. (2016). In-situ heavy and extra-heavy oil recovery: A review. *Fuel*, 185, 886–902. https://doi.org/10.1016/j.fuel.2016.08.047
3. Thomas, S. (2008). Enhanced oil recovery: an overview. *Oil & Gas Science and Technology-Revue de l'IFP*, 63(1), 9–19. https://doi.org/10.2516/ogst:2007060
4. Xu, G., Yin, H., Yuan, H., & Xing, C. (2020). Decline curve analysis for multiple-fractured horizontal wells in tight oil reservoirs. *Advances in Geo-Energy Research*, 4(3), 296–304. https://doi.org/10.46690/ager.2020.03.07
5. Muggeridge, A., Cockin, A., Webb, K., Frampton, H., Collins, I., Moulds, T., & Salino, P. (2014). Recovery rates, enhanced oil recovery and technological limits.

Philosophical Transactions of the Royal Society A: Mathematical, Physical and Engineering Sciences, 372(2006), 20120320. https://doi.org/10.1098/rsta.2012.0320

6. Pogaku, R., Mohd Fuat, N. H., Sakar, S., Cha, Z. W., Musa, N., Awang Tajudin, D. N. A., & Morris, L. O. (2018). Polymer flooding and its combinations with other chemical injection methods in enhanced oil recovery. *Polymer Bulletin,* 75, 1753–1774. https://doi.org/10.1007/s00289-017-2106-z

7. Mokheimer, E., Hamdy, M., Abubakar, Z., Shakeel, M. R., Habib, M. A., & Mahmoud, M. (2019). A comprehensive review of thermal enhanced oil recovery: Techniques evaluation. *Journal of Energy Resources Technology,* 141(3). https://doi.org/10.1115/1.4041096

8. Austad, T. (2013). Water-based EOR in carbonates and sandstones: New chemical understanding of the EOR potential using "smart water". In *Enhanced Oil Recovery Field Case Studies* (pp. 301–335). Gulf Professional Publishing. https://doi.org/10.1016/B978-0-12-386545-8.00013-0

9. Nair, R. R. (2019). Smart Water for enhanced oil recovery from seawater and produced water by membranes. PhD Thesis, University of Stavanger. www.beredskapsradet.no/sites/default/files/inline-images/btmMpZGUnIYl3eg7RAqnayCB5kZWcRTc4WoMltM8uunOJxPDHy.pdf

10. Gunda, D., Ampomah, W., Grigg, R., & Balch, R. (2015, November). Reservoir fluid characterization for miscible enhanced oil recovery. In *Carbon Management Technology Conference.* OnePetro. https://doi.org/10.7122/440176-MS

11. Farajzadeh, R., Kahrobaei, S., Eftekhari, A. A., Mjeni, R. A., Boersma, D., & Bruining, J. (2021). Chemical enhanced oil recovery and the dilemma of more and cleaner energy. *Scientific Reports,* 11(1), 829. https://doi.org/10.1038/s41598-020-80369-z

12. Fani, M., Pourafshary, P., Mostaghimi, P., & Mosavat, N. (2022). Application of microfluidics in chemical enhanced oil recovery: A review. *Fuel,* 315, 123225. https://doi.org/10.1016/j.fuel.2022.123225

13. Dong, X., Liu, H., Chen, Z., Wu, K., Lu, N., & Zhang, Q. (2019). Enhanced oil recovery techniques for heavy oil and oilsands reservoirs after steam injection. *Applied Energy,* 239, 1190–1211. https://doi.org/10.1016/j.apenergy.2019.01.244

14. Hasan, M. M. (2021). Various techniques for enhanced oil recovery: A review. *Iraqi Journal of Oil & Gas Research,* 2(1). http://doi.org/10.55699/ijogr.2022.0201.1018

15. Liu, L. (2022). Explanation of heavy oil development technology. *Resources Data Journal,* 1, 17–23. https://doi.org/10.50908/rdj.1.0_17

16. Muzzafaruddin, K. (2019). Enhanced oil recovery. *International Journal of Petroleum and Petrochemical Engineering (IJPPE),* 5(4), 10–13. http://dx.doi.org/10.20431/2454-7980.0504002

17. Du, Y., Wang, Y., Jiang, P., Ge, J., & Zhang, G. (2013, July). Mechanism and feasibility study of nitrogen assisted cyclic steam stimulation for ultra-heavy oil reservoir. In *SPE Enhanced Oil Recovery Conference.* OnePetro. https://doi.org/10.2118/165212-MS

18. Wang, Y., Zhang, L., Deng, J., Wang, Y., Ren, S., & Hu, C. (2017). An innovative air assisted cyclic steam stimulation technique for enhanced heavy oil recovery. *Journal of Petroleum Science and Engineering,* 151, 254–263. https://doi.org/10.1016/j.petrol.2017.01.020

19. Alomair, O. A., & Alajmi, A. F. (2022). A novel experimental nanofluid-assisted steam flooding (NASF) approach for enhanced heavy oil recovery. *Fuel,* 313, 122691. https://doi.org/10.1016/j.fuel.2021.122691

20. Wang, Z., Li, S., & Li, Z. (2022). A novel strategy to reduce carbon emissions of heavy oil thermal recovery: Condensation heat transfer performance of flue gas-assisted steam flooding. *Applied Thermal Engineering, 205,* 118076. https://doi.org/10.1016/j.applthermaleng.2022.118076

21. Lu, T., Ban, X., Guo, E., Li, Q., Gu, Z., & Peng, D. (2022). Cyclic in-situ combustion process for improved heavy oil recovery after cyclic steam stimulation. *SPE Journal, 27*(03), 1447–1461. https://doi.org/10.2118/209207-PA

22. Storey, B. M., Worden, R. H., & McNamara, D. D. (2022). The geoscience of in-situ combustion and high-pressure air injection. *Geosciences, 12*(9), 340. https://doi.org/10.3390/geosciences12090340

23. Ushakova, A. S., Zatsepin, V., Khelkhal, M. A., Sitnov, S. A., & Vakhin, A. V. (2022). In situ combustion of heavy, medium, and light crude oils: Low-temperature oxidation in terms of a chain reaction approach. *Energy & Fuels, 36*(14), 7710–7721. https://doi.org/10.1021/acs.energyfuels.2c00965

24. Yuan, C., Pu, W. F., Ifticene, M. A., Zhao, S., & Varfolomeev, M. A. (2022). Crude oil oxidation in an air injection based enhanced oil recovery process: Chemical reaction mechanism and catalysis. *Energy & Fuels, 36*(10), 5209–5227. https://doi.org/10.1021/acs.energyfuels.2c01146

25. Alagorni, A. H., Yaacob, Z. B., & Nour, A. H. (2015). An overview of oil production stages: Enhanced oil recovery techniques and nitrogen injection. *International Journal of Environmental Science and Development, 6*(9), 693. https://doi.org/10.7763/IJESD.2015.V6.682

26. Li, L., Zhou, X., Su, Y., Xiao, P., Chen, Z., & Zheng, J. (2022). Influence of Heterogeneity and Fracture Conductivity on Supercritical CO_2 Miscible Flooding Enhancing Oil Recovery and Gas Channeling in Tight Oil Reservoirs. *Energy & Fuels, 36*(15), 8199–8209. https://doi.org/10.1021/acs.energyfuels.2c01587

27. Nagatsu, Y., Abe, K., Konmoto, K., & Omori, K. (2020). Chemical flooding for enhanced heavy oil recovery via chemical-reaction-producing viscoelastic material. *Energy & Fuels, 34*(9), 10655–10665. https://doi.org/10.1021/acs.energyfuels.0c01298

28. Jalali, A., MohsenatabarFirozjaii, A., & Shadizadeh, S. R. (2019). Experimental investigation on new derived natural surfactant: Wettability alteration, IFT reduction, and core flooding in oil wet carbonate reservoir. *Energy Sources, Part A: Recovery, Utilization, and Environmental Effects,* 1–11. https://doi.org/10.1080/15567036.2019.1670285

29. Almobarak, M., Wu, Z., Zhou, D., Fan, K., Liu, Y., & Xie, Q. (2021). A review of chemical-assisted minimum miscibility pressure reduction in CO_2 injection for enhanced oil recovery. *Petroleum, 7*(3), 245–253. https://doi.org/10.1016/j.petlm.2021.01.001

30. Zhao, Y., Zhang, Y., Lei, X., Zhang, Y., & Song, Y. (2020). CO_2 flooding enhanced oil recovery evaluated using magnetic resonance imaging technique. *Energy, 203,* 117878. https://doi.org/10.1016/j.energy.2020.117878

31. VK, R., Chintala, V., & Kumar, S. (2021). Recent developments, challenges and opportunities for harnessing solar renewable energy for thermal Enhanced Oil Recovery (EOR). *Energy Sources, Part A: Recovery, Utilization, and Environmental Effects, 43*(22), 2878–2895. https://doi.org/10.1080/15567036.2019.1639850

32. OJHA, D. (2023). Enhanced Oil Recovery. http://dx.doi.org/10.5281/zenodo.6816528

33. Sarmah, S., Gogoi, S. B., Jagatheesan, K., & Hazarika, K. (2022). Formulation of a combined low saline water and polymer flooding for enhanced oil recovery. *International Journal of Ambient Energy, 43*(1), 1089–1097. https://doi.org/10.1080/01430750.2019.1683068

34. Kumari, R., Kakati, A., Nagarajan, R., & Sangwai, J. S. (2019). Synergistic effect of mixed anionic and cationic surfactant systems on the interfacial tension of crude oil-water and enhanced oil recovery. *Journal of Dispersion Science and Technology*, *40*(7), 969–981. https://doi.org/10.1080/01932691.2018.1489280

35. Wang, Z., Lin, M., Jin, S., Yang, Z., Dong, Z., & Zhang, J. (2019). Combined flooding systems with polymer microspheres and nonionic surfactant for enhanced water sweep and oil displacement efficiency in heterogeneous reservoirs. *Journal of Dispersion Science and Technology*. https://doi.org/10.1080/01932 691.2019.1570850

36. Xu, Y., Wang, T., Zhang, L., Tang, Y., Huang, W., & Jia, H. (2023). Investigation on the effects of cationic surface active ionic liquid/anionic surfactant mixtures on the interfacial tension of water/crude oil system and their application in enhancing crude oil recovery. *Journal of Dispersion Science and Technology*, *44*(1), 214–224. https://doi.org/10.1080/01932691.2021.1942034

37. Sarmah, S., & Gogoi, S. B. (2022). Design and application of an alkaline–surfactant–polymer (ASP) slug for enhanced oil recovery: A case study for a depleted oil field reservoir. *Petroleum Science and Technology*, *40*(18), 2213–2237. https://doi.org/10.1080/10916466.2022.2038623

38. Liu, D., Zhang, X., Tian, F., Liu, X., Yuan, J., & Huang, B. (2022). Review on nanoparticle-surfactant nanofluids: Formula fabrication and applications in enhanced oil recovery. *Journal of Dispersion Science and Technology*, *43*(5), 745–759. https://doi.org/10.1080/01932691.2020.1844745

39. Massarweh, O., & Abushaikha, A. S. (2020). The use of surfactants in enhanced oil recovery: A review of recent advances. *Energy Reports*, *6*, 3150–3178. https://doi.org/10.1016/j.egyr.2020.11.009

40. Zhang, J., Gao, H., & Xue, Q. (2020). Potential applications of microbial enhanced oil recovery to heavy oil. *Critical Reviews in Biotechnology*, *40*(4), 459–474. https://doi.org/10.1080/07388551.2020.1739618

41. Olayiwola, S. O., & Dejam, M. (2019). A comprehensive review on interaction of nanoparticles with low salinity water and surfactant for enhanced oil recovery in sandstone and carbonate reservoirs. *Fuel*, *241*, 1045–1057. https://doi.org/10.1016/j.fuel.2018.12.122

42. Zhang, P., Tweheyo, M. T., & Austad, T. (2007). Wettability alteration and improved oil recovery by spontaneous imbibition of seawater into chalk: Impact of the potential determining ions Ca^{2+}, Mg^{2+}, and SO_4^{2-}. *Colloids and Surfaces A: Physicochemical and Engineering Aspects*, *301*(1–3), 199–208. https://doi.org/10.1016/j.colsu rfa.2006.12.058

43. Nair, R. R., Protasova, E., Bilstad, T., & Strand, S. (2016, September). Reuse of produced water by membranes for enhanced oil recovery. In *SPE Annual Technical Conference and Exhibition*. OnePetro. https://doi.org/10.2118/181588-MS

44. Salunkhe, B., Schuman, T., Al Brahim, A., & Bai, B. (2021). Ultra-high temperature resistant preformed particle gels for enhanced oil recovery. *Chemical Engineering Journal*, *426*, 130712. https://doi.org/10.1016/j.cej.2021.130712

45. Hua, F. L., Tsang, Y. F., Wang, Y. J., Chan, S. Y., Chua, H., & Sin, S. N. (2007). Performance study of ceramic microfiltration membrane for oily wastewater treatment. *Chemical Engineering Journal*, *128*(2–3), 169–175. https://doi.org/10.1016/j.cej.2006.10.017

46. Arévalo, J., Ruiz, L. M., Parada-Albarracín, J. A., González-Pérez, D. M., Pérez, J., Moreno, B., & Gómez, M. A. (2012). Wastewater reuse after treatment by MBR.

Microfiltration or ultrafiltration? *Desalination*, *299*, 22–27. https://doi.org/10.1016/j.desal.2012.05.008

47. He, M., Li, W. D., Chen, J. C., Zhang, Z. G., Wang, X. F., & Yang, G. H. (2022). Immobilization of silver nanoparticles on cellulose nanofibrils incorporated into nanofiltration membrane for enhanced desalination performance. *NPJ Clean Water*, *5*(1), 64. https://doi.org/10.1038/s41545-022-00217-7

48. Maddah, H. A., Alzhrani, A. S., Bassyouni, M., Abdel-Aziz, M. H., Zoromba, M., & Almalki, A. M. (2018). Evaluation of various membrane filtration modules for the treatment of seawater. *Applied Water Science*, *8*, 1-13. https://doi.org/10.1007/s13201-018-0793-8

49. Pedenaud, P., Hurtevent, C., & Baraka-Lokmane, S. (2012, May). Industrial experience in sea water desulfation. In *SPE International Conference on Oilfield Scale*. OnePetro. https://doi.org/10.2118/155123-MS

50. Liu, D., Zhong, X., Shi, X., Qi, Y., & Qi, Y. (2015). Enhanced oil recovery from fractured carbonate reservoir using membrane technology. *Journal of Petroleum Science and Engineering*, *135*, 10–15. https://doi.org/10.1016/j.petrol.2015.08.008

51. Xu, Z. X., Li, S. Y., Li, B. F., Chen, D. Q., Liu, Z. Y., & Li, Z. M. (2020). A review of development methods and EOR technologies for carbonate reservoirs. *Petroleum Science*, *17*, 990–1013. https://doi.org/10.1007/s12182-020-00467-5

52. Hao, J., Mohammadkhani, S., Shahverdi, H., Esfahany, M. N., & Shapiro, A. (2019). Mechanisms of smart waterflooding in carbonate oil reservoirs: A review. *Journal of Petroleum Science and Engineering*, *179*, 276–291. https://doi.org/10.1016/j.petrol.2019.04.049

53. Rezaeian, M. S., Mousavi, S. M., Saljoughi, E., & Amiri, H. A. A. (2020). Evaluation of thin film composite membrane in production of ionically modified water applied for enhanced oil recovery. *Desalination*, *474*, 114194. https://doi.org/10.1016/j.desal.2019.114194

54. Gopani, P. H., Singh, N., Sarma, H. K., Mattey, P., & Srivastava, V. R. (2021). Role of monovalent and divalent ions in low-salinity water flood in carbonate reservoirs: An integrated analysis through zeta potentiometric and simulation studies. *Energies*, *14*(3), 729. https://doi.org/10.3390/en14030729

55. Al Murayri, M. T., Sulaiman, D. S., Al-Kharji, A., Al Kabani, M., Sorbie, K. S., Ness, G., … & Salehi, M. (2021, December). Scale mitigation for field implementation of alkaline-surfactant-polymer ASP flooding in a heterogeneous high temperature carbonate reservoir with high divalent cation concentration in formation water. In *Abu Dhabi International Petroleum Exhibition & Conference*. OnePetro. https://doi.org/10.2118/207573-MS

56. Nair, R. R., Protasova, E., Strand, S., & Bilstad, T. (2018). Membrane performance analysis for smart water production for enhanced oil recovery in carbonate and sandstone reservoirs. *Energy & Fuels*, *32*(4), 4988–4995. https://doi.org/10.1021/acs.energyfuels.8b00447

57. Awolayo, A., Sarma, H., & AlSumaiti, A. (2016). An experimental investigation into the impact of sulfate ions in smart water to improve oil recovery in carbonate reservoirs. *Transport in Porous Media*, *111*, 649–668. https://doi.org/10.1007/s11242-015-0616-4

58. Nair, R. R., Protasova, E., & Bilstad, T. (2016). Smart water for enhanced oil recovery by nano-filtration. *Journal of Petroleum & Environmental Biotechnology*, *7*, 1–8. https://doi.org/10.4172/2157-7463.1000273

59. Janson, A., Dardor, D., Al Maas, M., Minier-Matar, J., Abdel-Wahab, A., & Adham, S. (2020). Pressure-retarded osmosis for enhanced oil recovery. *Desalination*, *491*, 114568. https://doi.org/10.1016/j.desal.2020.114568

60. Mansour, M. S., Abdel-Shafy, H. I., & El Azab, W. I. (2020). Innovative reuse of drinking water sludge for the treatment of petroleum produced water to enhance oil recovery. *Egyptian Journal of Petroleum*, *29*(2), 163–169. https://doi.org/10.1016/j.ejpe.2020.02.002

12 Petroleum Industry Sulphur Emission and Removal Using Membrane Technology

Sudeepta Baruah, Parashmoni Rajguru and Krishna Kamal Hazarika

12.1 INTRODUCTION

Sulphur is one of the most available elements on earth and is used by various industries as a raw material [1]. Sulphur may exist in different forms, such as elemental sulphur, hydrogen sulphide and organic sulphides, such as sulphate, sulphide, thiophene, thiol, sulphone, sulphonate, sulfoxide, etc. All crude oil comprises sulphur compounds; however, the exact level may vary depending on the source of the crude oil. Products that are derived from petroleum and also from crude oil, sulphur is found in a number of different forms, including high-molecular-mass heterocyclic compounds, mercaptans, hydrogen sulphide, sulphides, thiophene derivatives, disulfides, etc. In both inorganic and organic forms, sulphur is present in petroleum coke, ranging from 1 to 10% by weight. Approximately 0.1% of petroleum feedstock was reported to contain elemental sulphur [2]. Since most of the energy used by refineries gets from a portion of the hydrocarbons they consume, some of the feed sulphur is released from the refinery in the flue gases from boilers, furnaces and other plants that combust refinery fuels (Figure 12.1).

Sulphur emissions during petroleum production operations have adverse impacts on the environment. Sulphur dioxide contamination of the atmosphere can no longer be supported; it is now universally acknowledged. So much focus has been given to eliminating sulphur from the fuels we generally burn. Therefore, removing or reducing the emissions of sulphur from petroleum industry operations and processing is a great concern for environmental as well as economic benefits. Crude oil contains sulphur contents ranging from 0.05 to 6 wt.%, and natural gas contains approximately 0–0.2% of sulphur. Sulphur compounds tend to damage various catalysts used in processing crude oil and cause corrosion issues in pipelines, pumping equipment and refinery equipment, leaving them undesirable in the refining process. Shi and Wu have reported that in crude oil, the presence of elemental sulphur generally ranges from 0.1 to 6 wt.% and in some crude oil, it may increase up to 14% [3]. The combustion

DOI: 10.1201/9781003441359-12

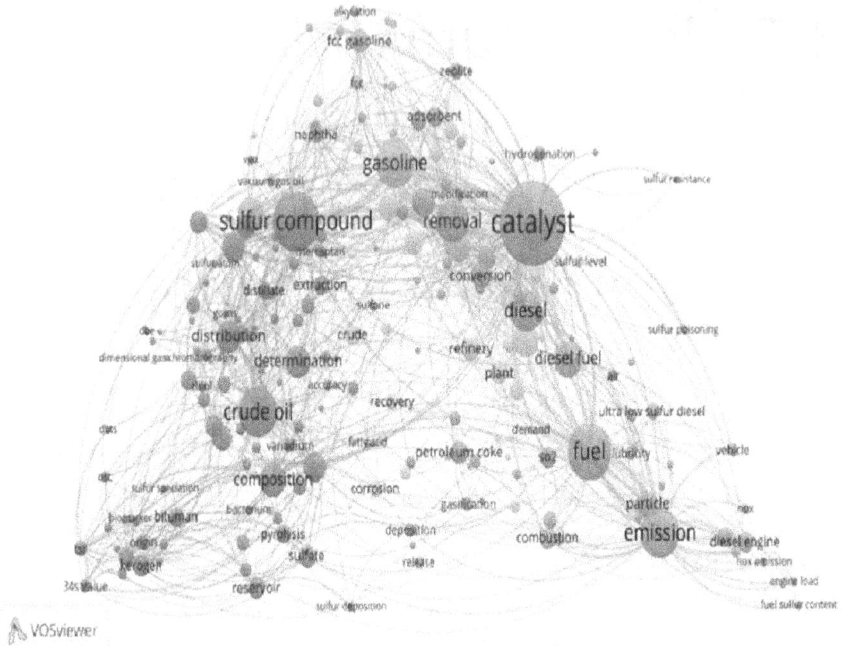

FIGURE 12.1 Bibliometric networks associated to petroleum sulphur compounds [3].

of sulphur-containing fuel oil leads to the production of acid gases, which are highly harmful to the environment as well as to mankind.

Environmental restrictions that place strict limits on the amount of sulphur allowed in transportation fuels are currently the main driving force behind the decrease of sulphur in fuels. To ensure that energy is supplied at a competitive price, new and efficient ways must be developed to remove the sulphur from lower-quality feedstocks. This chapter discusses the various desulfurization approaches currently being studied globally to remove sulphur compounds from liquid fuels, as well as the viability of using membrane methods to do so.

Thiophenes and sulphides are major sulphur compounds present in some of the petroleum products. Sulphoxides or sulphones are obtained by oxidizing sulphur compounds present in petroleum, making them easier to separate from hydrocarbons. Thus, thiophenes and sulphides are separated from petroleum hydrocarbons using a selective oxidation process [4]. Owing to the high electron density of sulphur atoms, they can easily form ligands with metal atoms, and depending on this principle, ligand exchange chromatography can also be used to separate sulphur compounds from petroleum hydrocarbons [5]. Sulphur compounds in gasoline are recognized as a sound matter for the petroleum industry as an environmental concern. Thus, the presence of sulphur in petroleum products leads to the corrosion of equipment, catalyst poisoning and air pollution. In many countries, strict limitations have been applied to the sulphur content of different types of fuels, taking it as a very serious issue [6]. As a result, there is a need for research into ideas and long-lasting solutions.

12.2   HEALTH AND ENVIRONMENT HAZARD FROM SULPHUR EMISSION

Sulphur has numerous detrimental effects on the environment and human health. Sulphur emissions degrade the quality of the surrounding air. Sulphur dioxide damages our ecology by being absorbed by plants and soils. Sulphur dioxide has a negative impact on the growth of plants and can harm crops. Greater concentrations of sulphur dioxide may hinder photosynthesis, although minimal concentrations of SO_2 around 0.3 ppm have little impact on it [7]. Sulphur dioxide damages plants severely, and the effects may be long term. Dead tissue on the leaf margins that is readily visible is a sign of acute damage. Areas of the leaf blade that are brownish-red or faded white are indications of chronic damage [8]. The rate of sulphur dioxide absorption by the leaves depends on the duration of exposure, concentration and internal leaf parameters linked to the amount of the activity of the plant and controlled by the time of the year, humidity, temperature, leaf turgor, etc. [9]. Sulphur dioxide concentrations would have to be 5 ppm or more for the majority of the individuals to react; however, some sensitive people may have modest symptoms at concentrations of 1–2 ppm. At sulphur dioxide concentrations of 1500 $\mu g/m^3$ over a 24-h period, mortality may increase. When exposed to 500–715 $\mu g/m^3$ of sulphur dioxide over the course of a day, there may be an increase in death rates (Figure 12.2). Adverse health effects have been reported when 24-h average levels of sulphur dioxide surpass 300 $\mu g/m^3$, or average yearly concentrations of sulphur dioxide reach 115 $\mu g/m^3$ [7].

Smog aerosols are generated as a direct consequence of sulphur dioxide. Sulphur dioxide is easily dissolved in water and is present in the atmosphere, resulting in sulphurous acid (H_2SO_3). According to the USEPA (United States Environmental Protection Agency), 1998, sulphate aerosol, also known as acid aerosol, is produced from around 30% of the sulphur dioxide in the atmosphere. Human health, ecosystems, agriculture and regional and global climates are all impacted by sulphur aerosols. Another sulphur oxide, sulphur trioxide (SO_3), is

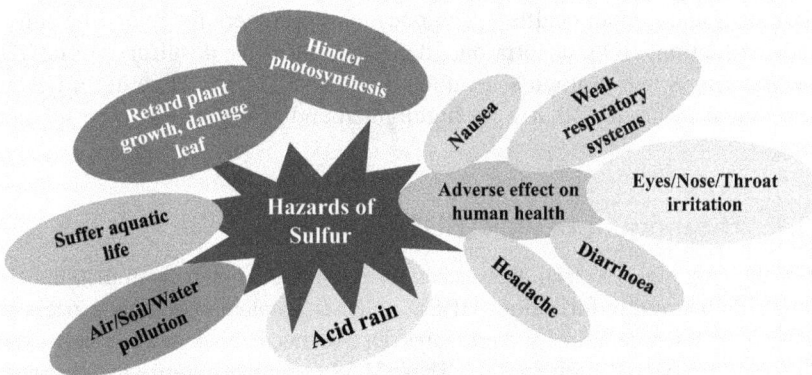

FIGURE 12.2   Hazardous effect of sulphur on environment.

either directly released into the environment or generated from sulphur dioxide and quickly transformed into sulphuric acid (H_2SO_4). The primary component of acid rain is sulphuric acid, which is created when sulphur dioxide reacts with water and oxygen. Building materials like marble, limestone and dolomite are chemically eroded by acids. Chemical deterioration of historical landmarks and works of art is particularly a matter of concern. Paper and leather can also be damaged by sulphurous acid and sulphuric acids, which are created when moisture reacts with sulphur dioxide and sulphur trioxide. Deforestation results from acid rain. The flora, including forests and agroforestry systems, is adversely impacted by sulphur oxide emissions. Waterways that are getting more acidic threaten sensitive ecosystems and aquatic life. Fish and other aquatic species may harm humans and animals due to the lower pH of acid rain and greater aluminium concentrations in surface water. Most fish eggs will not mature at pH below 5; lower pH levels can kill fish. Acid rain has the potential to gravely harm the soil. The nutritional value of soil is degraded by the action of acid rain. The important microorganisms present in soil cannot survive at lower pH as it leads to the dysfunction of their enzymes. The eyes, nose and throat may get irritated after exposure to sulphur dioxide. Sulphur side effects might cause headaches, vertigo, nausea and diarrhoea. Sulphur dioxide produces sulphur particles, which if repeatedly breathed in, can lead to bronchitis and asthma. Minor SO_x exposure can have negative effects on the respiratory system, especially in kids and people with weak respiratory systems. When SO_x reacts with atmospheric components, the minute tiny particles created in the atmosphere can profoundly enter the lungs and induce malignant diseases, including lung cancer and other dangerous respiratory ailments. Life can be threatened while exposing to higherSO_2 concentration even at a very short time of duration. The fact that sulphur dioxide is believed to be more hazardous than carbon dioxide is another significant effect worth highlighting.

12.3 APPROACHES FOR REMOVAL OF SULPHUR

It is crucial to get rid of the sulphur and lessen its toxicity since it harms the environment and human health. Techniques by researchers for removing sulphur include extraction, HDS, adsorption, alkylation, oxidative desulfurization (ODS) and employing some biological treatments are also known. Here are a few sulphur removal techniques that were highlighted and discussed in some scholarly articles.

12.3.1 Hydrodesulfurization (HDS)

One of the ongoing industrial approaches for removing sulphur from distilled fuels is known as hydrodesulfurization (HDS), which is a notable catalytic process that operates at high temperatures and pressures. Due to this, HDS is an extremely expensive method for deep desulfurization. These HDS techniques are frequently used in the petroleum refining sector. In the HDS process, according to Demirbas's literature [10], the different sulphur compounds present are catalytically treated with hydrogen to produce hydrogen sulphide. A portion of the hydrogen sulphide is converted by air into sulphur dioxide; the following overall reaction depicts the production of sulphur:

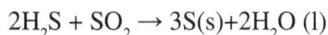

$$2H_2S + SO_2 \rightarrow 3S(s) + 2H_2O \ (l)$$

Throughout the HDS process, organic sulphur compounds are broken down into H_2S using hydrogen or a hydrogen donor. The catalytic interaction of hydrogen with sulphur compounds in the charge stock generates hydrogen sulphide, which is easily detachable from the oil and is widely used to desulfurize petroleum and petroleum fractions. Co–Mo or Ni–Mo sulphides are usually used catalysts on alumina support in the HDS technique [10]. Researchers have attempted to merge the benefits of microwave heating with HDS and have found significant results. HDS techniques are frequently employed in the industry to enhance fuel quality. In addition, HDS is ineffective at eliminating heterocyclic compounds of sulphur like dibenzothiophene (DBT) and its derivatives such as 4,6-dimethyldibenzothiophene. Among sulphur compounds, dimethyldibenzothiophene (DMDBT) is the least reactive. DMDBT significantly limits the deep desulfurization of gasoline, having sulphur concentrations ranging from 500 to 10 ppm.

12.3.2 Oxidative Desulfurization (ODS)

In order to remove stubborn sulphur compounds from petroleum distillate fractions, ODS, a very effective and practical alternative to HDS, has been found to attract interest rapidly [11]. In ODS, compounds containing sulphur are converted into sulfones through an oxidation reaction involving an oxidant such as hydrogen peroxide (H_2O_2), sulphuric acid (H_2SO_4), etc. Because of the greater polarity of the sulfone molecule, it is then readily removed from the fuel.

Using a catalyst that is supported by manganese dioxide and magnetically reduced graphene oxide nanocomposite ($MnO_2/MrGO$) in the existence of an $H_2O_2/HCOOH$ oxidation system, Ahmad et al. examined the ODS, i.e., ODS, of modelled and actual samples of oil. Dibenzothiophene was found to be most effectively removed from modelled oil samples under 40 °C temperature, 1 hour of reaction time, 80 mg of catalyst per 0.01 L and H_2O_2 (2 mL)/HCOOH. $MnO_2/MrGO$ displayed intense activity of desulfurization to 80% under these conditions [11].

12.3.3 Extraction

The research community has put emphasis on extraction, a promising desulfurization approach [12]. No exceptionally high temperatures or pressures are necessarily required in the extraction process, which is advantageous and efficient. One of the acknowledged green approaches is using ionic liquid for the extraction method [13]. Ionic liquids are thought to be efficient in terms of technology and the environment. This is owed to the fact that it is done with minimal operating conditions and without employing any components. Ahmed et al. [12] performed extraction desulfurization (EDS) with ionic liquid, which is regarded as one of the best solvents, in one of his work. Tetrabutylphosphonium methanesulfonate ionic liquid was used in the oxidation-extraction desulfurization (OEDS) technique by Ahmad and his group to deep desulfurize a commercial diesel fuel under ideal experimental conditions. Utilizing water and n-hexane as the primary and secondary regeneration solvents, the ionic liquid was successfully restored [12].

Sulphur compounds are also extracted using some polar organic solvents. With the help of aprotic solvents like dimethylformamide, acetonitrile, dimethylacetamide, dimethyl sulfoxide, N-methylpyrrolidone, thiophenes and also their derivatized products are obtained from petroleum-based feedstock. These solvents have a poor level of selectivity, a good degree of dissolving of aromatic sulphur compounds and a capacity to combine with them to form stable π-complexes [14]. N-methylpyrrolidone is the most popular extractant due to its greater capacity than most of the other solvents. Hydrocarbon loss during the extraction process is a significant issue that is solved by adjusting the extraction temperature and adding co-solvent. Although the solvent specificity and raffinate production is enhanced by adding H_2O or a non-polarized solvent, this frequently results in less sulphur being removed from petroleum feed.

Besides all these techniques, alkali treatment is a desirable desulfurization method due to its accessibility and low chemical cost. Polycyclic sulphur compounds and heavy mercaptans, known to be weakly soluble in H_2O as well as weak acids, are more difficult to remove from petroleum products with sodium hydroxide than light mercaptans and hydrogen sulphide.

12.3.4 ADSORPTIVE DESULFURIZATION AND BIO-DESULFURIZATION (BDS)

The additional desulfurization processes that have the ability to produce super-clean fuels include adsorptive desulfurization and bio-desulfurization (BDS). Adsorption desulfurization is a quicker and simpler method of eliminating sulphur from diesel [15].

The adsorbents employed in the adsorption process scoop up the sulphur with preference. Removing selectively the low-concentrated compounds found in liquids using the adsorption method is effective. An adsorbent (solid in nature) that preferentially adsorbs the compounds containing sulphur is brought into contact with the fuel. The large surface area for adsorption is provided by the active adsorbent being put on a porous, non-reactive substrate. When sulphur molecules bind to the adsorbent on the substrate and stay there independently, adsorption takes place. One of the top adsorbents employed by numerous researchers, even by our ancestors, is activated carbon. It was reported that the concentration of sulphur in diesel fuel was decreased by more than 50% when it was desulfurized using an adsorption technique involving activated carbon [15].

Biological desulfurization, also known as biodesulfurization, is a well-known laboratory process for removing sulphur from coals using bacterial cultures [16]. Bio-desulfurization (BDS) has gained much attention recently owing to its environmentally friendly handling of fossil fuels. However, using the BDS procedure is severely hampered by how slowly the removal process proceeds.

The following research articles that use these techniques are acknowledged:

Iruretagoyena and Montesano [17] presented their work on utilizing nanostructured adsorbents for selective sulphur removal from liquid fuels. A broad survey of particular adsorption processes for the elimination of sulphur compounds is provided in this chapter. The application of nanostructured materials as sulphur adsorbents is highlighted. Reactive adsorption desulfurization and adsorption desulfurization are two primary categories of selective adsorptions for the elimination of sulphur compounds. A review of the three primary sulphur adsorption mechanisms given were bulk inclusion in reactive adsorption desulfurization, direct interactions between sulphur and adsorption sites and π-complexation. In addition, promising classes of nanostructured adsorbents such as MOFs, zeolites, carbon nanostructured adsorbents and mesoporous silica are used for selective adsorptions to remove sulphur [17].

Celis-Garc and associates [18], whose research goal was to combine sulphate reduction and sulphide oxidation in one reactor to convert soluble sulphur (sulphate) into insoluble sulphur (elemental sulphur). A 2.3 L down-flow fluidized bed reactor was employed to achieve this. Sulphate and oxygen (air) acted as electron acceptors, and lactate functioned as electron donors. Through a linked anaerobic/aerobic process in one reactor with lactate, sulphate and oxygen (air) as substrates, elemental sulphur could be produced. The key to the procedure was the formation of a biofilm with sulphate-reducing and sulphide-oxidizing activity. About 50% of the sulphate was converted to elemental sulphur at low airflow rates [18].

In order to desulfurize gasoline, jet fuel and diesel fuel at 25 °C, Xiaoliang et al. [19] created an adsorbent. The findings show that adsorbents based on transition metals created in this study are efficient at removing sulphur compounds from fuels, including refractory sulphur compounds. Selective adsorption for removing sulphur (SARS) is a selective adsorption technique that removes sulphur less than 1 wt.% of fuel mass while leaving the remaining 99 wt.% of sulphur-free fuel mass unchanged. In order to address the demands for fuel cell applications and ultra-clean transportation fuels, this study investigates a room-temperature adsorption technique that may efficiently remove the sulphur content in diesel fuel, gasoline and jet fuel at minimal funding and operational cost. Finding a selective adsorbent for the sulphur compounds while preventing the adsorption of the co-existing olefins and aromatic hydrocarbons is a significant issue for isolating the sulphur compounds from the fuels [19].

The evaluation of the biodesulfurization process was examined by Nazari et al. biodesulfurization (BDS) is a biological process that has been proposed for the desulfurization of sulphur ring components. It is a non-destructive technique to remove sulphur from petroleum hydrocarbons in mild conditions and may be used in combination with HDS. Biodesulfurization necessitates a new challenge to increase desulfurization activity through genetic engineering techniques and bioreactor development in order to transform from a science fiction methodology to a practical industrial method for reducing sulphur from fossil fuels [20].

12.3.5 FLUID CATALYTIC CRACKING

The process by which the high boiling-point, high-molecular weight hydrocarbon fractions of petroleum (crude oils) are transformed into lighter and more valuable products like olefinic gases, gasoline and other petroleum products is known as fluid

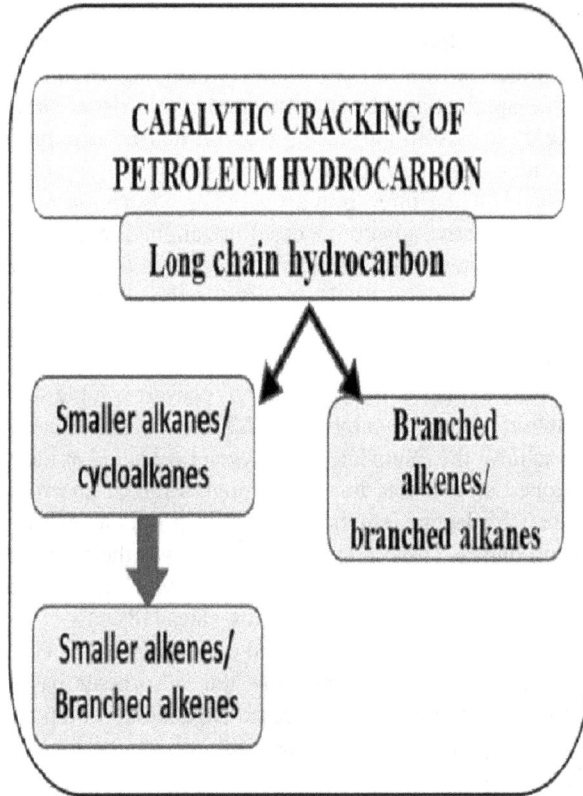

FIGURE 12.3   Catalytic cracking in petroleum industry.

catalytic cracking (FCC). The cyclic process of FCC, which entails reaction, product separation and regeneration of catalytic cracking creates products with higher economic value than those produced by thermal cracking [21]. Wen et al. describe how sulphur is removed in situ condition from naphtha during the fluid catalytic cracking process used to reform gasoline. A confined fluidized bed reactor was utilized to perform gasoline desulfurization using commercial fluid catalytic cracking naphtha. According to the results, the sulphur content of gasoline was reduced due to proper breaking ability and better hydrogen transfer activity (Figure 12.3). Sulphur was reduced by 67.37% using the Technical Pilot Scale Riser method [22].

12.3.6   SULPHUR EXTRACTION THROUGH SUPER CRITICAL FLUID

Any substance that exhibits liquid and in place of 'and' 'as well as' will be more proper for this sentence gaseous properties above its critical pressure and critical temperature is considered as supercritical fluid. It has the ability to dissolve materials like a liquid and can pass through solids like a gas. One of the most widely used supercritical fluids that can replace organic solvents in a variety of industrial processes is water.

Water has a substantially lower dielectric constant under supercritical circumstances; because of this, subcritical and supercritical water exhibits remarkable solubility for organic molecules that include significant non-polar units and functions as an organic solvent. At the solvent's critical pressure and temperature, supercritical fluid treatment is often done in a batch reactor system that is stirred mechanically or in a rocking batch reactor. With higher pressure, more soluble material is generated. The treatment of supercritical fluids is influenced by several factors, including temperature, pressure and extraction duration, chemical composition of the extracted material and type of solvent. The core aspect influencing sulphur elimination is reaction temperature. The sulphur from crude oil releases into the liquid and gas effluents as the temperature rises. While sulphur removal in liquid effluents is unaffected by reaction temperature in gas effluents, it is reliant on reaction temperature in gas effluents. When under supercritical conditions, the pressure increased while the sulphur removal in liquid effluents increased [23]. One of the promising non-conventional technologies for desulfurization of crude oil is the use of supercritical water. Supercritical water desulfurization has its potential as a method for removing sulphur from fuel sources like heavy oil and bitumen. The solvent capacity of supercritical water is influenced by the water density, and as the water density rises, so does the tendency for sulphur removal in liquid effluent.

Additionally, there are several studies where the following procedures are used for sulphur removal: the wet removal process, a theoretical model, employing ionic liquid, electrochemical engineering and electrochemistry.

Hales computed the sulphur scavenging rates using the general wet removal process, which split into several pathways. These include combined scavenging and chemical reactions, direct sulphate scavenging and direct sulphur dioxide scavenging. In most contexts, the chemical conversion step is anticipated to be the rate-limiting step for sulphur scavenging via the pathway of SO_2 absorption followed by aqueous-phase conversion. Given the previously mentioned features of sulphur deposition by wet processes, it would seem crucial to have the ability to estimate sulphur compound wet deposition rates based on source and rainfall data. However, more preferably, these calculations are important to foretell the long-term impact of greater industrialization and energy production. This skill is useful for forecasting the regional distribution of sulphate concentrations in both air and water. A comprehensive model with this capability might be used to forecast the combined impacts of transport, deposition and conversion on the future of atmospheric sulphur compounds [24].

Sarkar et al. attempted with a view to pick up a clear grip on the process of SO_2 absorption in water. It is necessary to create a generalized theoretical model applying the residence time distribution accession for forecasting the removal ability in a co-current gas–liquid scrubber. The suggested model takes into account the distribution of SO_2 concentrations in both the liquid and gas phases as a consequence of droplet diameter, liquid pH, inlet gas ppm, gas and liquid flow rates and altitude of the scrubber. Excellent agreement between experimental findings and the proposed theoretical model demonstrates that this method might attain removal efficiencies of approximately 75–99% [25].

Paucar et al. reviewed the techniques of denitrogenation and desulfurization of fuels and petroleum using ionic liquids. Additionally, they have investigated the efficacy and selectivity of ionic liquids as well as their environmental, biodegrad-ation, bioaccumulation and corrosivity-related limits [26].

The development of SO_2 de-pollution procedures can benefit from the use of electrochemical methods, which do not demand the constant use of chemical reagents. An electrochemical approach that couples the electrolysis phenomenon upstream to the sulphuric acid plant in order to remove sulphur dioxide from fuel gases. Taieb et al. modified the hybrid cell Westinghouse, where an electrolysis cell produces hydrogen at the cathode and H_2SO_4 from SO_2 and H_2O at the anode at the same time, and water is broken down to H_2 and O_2 by employing the electrode potential [27].

In their work, Ntagia et al. and his group first introduce the field of environmental electrochemistry and the fundamentals of electrochemical engineering. The electrochemical characteristics of the sulphur components that could be employed in electrochemical therapy are then discussed. Furthermore, abiotic and biotic electrochemistry principles are briefly discussed for the treatment of wastewaters containing sulphur [28].

Esmaeil and Ahmad reviewed the difficulties in removing elemental sulphur from residues and during the leaching of copper and zinc sulphides. The research that has been done to lessen the passivating effect of molten sulphur during the leaching of copper sulphide and zinc sulphide, as well as to eliminate the generated elemental sulphur from the leach residues, has been summarized in this article [29].

Most of the aforementioned techniques use high temperatures and high pressures and, most importantly, consume a lot of time. So, there is a need to have sustainable solutions to ensure the sustainability of the environment, as high temperatures and pressure are not ideal for a growing sustainable environment. Therefore, using a simpler yet effective technology to remove sulphur is a matter of concern in the ongoing phase of research. Membrane separation is a process with lots of potential benefits compared to the mostly used conventional approaches.

12.4   ROLE OF MEMBRANE TECHNOLOGY IN SULPHUR REMOVAL

Membrane technology offers a huge prospective for eliminating sulphur compounds as it is a non-solvent based, greener approach with one step of separation. So, membrane-based separation is a convenient strategy for desulfurizing fuels in the petrochemical industry. Low energy usage, flexibility, material selectivity, environmental friendliness and reduced space needs are some of the benefits of employing membrane technology. Researchers have given emphasis on studying membranes, as they are more efficient from a technological perspective (Figure 12.4).

Being non-solvent-based greener way with one-step separation, membrane technology has great potential for the removal of sulphur compounds. Mainly, the pervaporation technique is used for the desulfurization of gasoline (for the removal of organic sulphur), whereby partial vaporizing of the liquid is allowed to pass through a non-porous dense membrane, which leads to selective permeation of the components of the liquid. Owing to their higher affinity and higher diffusion rate, sulphur compounds permeate through the dense membrane preferentially. The organic sulphur compounds absorb into the membrane and then diffuse through the membrane

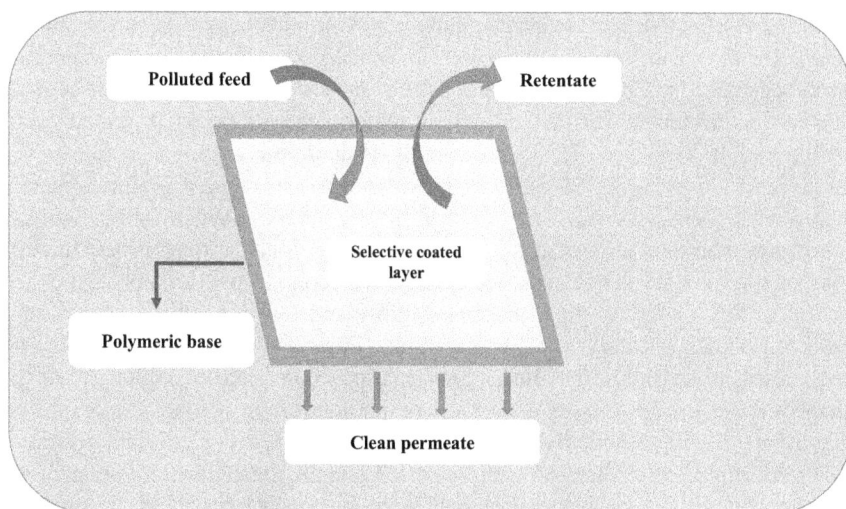

FIGURE 12.4   Schematic representation of membrane separation.

and desorbed on another side of the membrane, then evaporate and get condensed at the cold permeate side. To maintain a low absolute pressure, a vacuum force is applied at the permeate side, and the feed solution must come in contact with the dense membrane for a continuous separation process [30].

Due to its superior mechanical and thermal characteristics, polyimides are suitable materials for membranes. Additionally, polyimides have good chemical resistance to organic liquids, excellent radiation resistance and no flammability, which makes them desirable for industrial separation applications as hollow fibres or flat membranes. Different industrial applications, including the treatment of wastewater, biogas and natural gas are made using polyimides. In educational as well as industrial exploration for the sulphur removal process, membrane-based separation has been drawing much more attention.

A polyethylene glycol(PEG)/polyethersulfone(PES) composite membrane was reported by Kong et al. for pervaporative desulfurization of FCC gasoline by the use of PEG as an active layer on the support of PES in the sight of maleic anhydride as a cross linker, and it was observed that with increasing the content of the cross linker, permeation through the membrane decreases as low cross-linked membranes get swelled by gasoline, at cross-linking agent amounts of 17% and 8%, the membrane reached a steady state at 6 h and 3 h, respectively, where it was observed that flux gradually increases upon decreasing the sulphur enrichment factor [31].

For desulfurization through a membrane-based separation process called pervaporation, some new membranes consisting of polyimides and POSS (polyhedral oligomeric silsesquioxane) compounds were studied. Through a reactive organic substituent, POSS (polyhedral oligomeric silsesquioxane) compounds can operate as a cross-linking agent between polymer chains, improving the membrane characteristics. Compared to the HDS process, the pervaporation setup operates without a catalyst at low temperatures and pressures. Pervaporation is a more reliable, cost-effective and environmentally responsible technique that uses minimal energy. The flux and the selectivity are the two basic aspects of the pervaporation membrane-based separation technique. Konietzny et al. aim to fabricate a nano-scale compound embedded in a polymeric membrane with better performance for desulfurization. They used POSS because of its enormous significance across several research domains. POSS units are made of a Si_8O_{12} cubic cage-like structure, whereas POSS reagents are made up of a durable silicon–oxygen framework. The POSS framework has been described as a zero-dimensional or sphere-like structure. It has an approximate diameter of 15 Å. POSS compounds have a number of exceptional qualities, including great thermal stability and low density. The insertion of POSS compounds with organic side groups into conventional polymers and thermoplastic resins is feasible. Cross-linking makes it possible to create a 3D network that enhances the chemical stability and excellent thermal stability of the membrane. The monomers (2,3,5,6-tetramethyl-1,4-phenylene diamine) and (3,5-diaminobenzoic acid) were employed as diamino components, while the dianhydride component was (4,4'-(hexafluoroisopropylidene) diphthalic anhydride). 49:1 copolyimide composite mentioned was (4,4'-(hexafluoroisopropylidene) diphthalic anhydride)-(2,3,5,6-tetramethyl-1,4-phenylene diamine)/(4,4'-(hexafluoroisopropylidene) diphthalic anhydride)-(3,5-diaminobenzoic acid). Tetrahydrofuran (THF) was applied to dissolve the copolyimide for 3–5 h at room temperature. Filtered inseparable particles were eliminated before adding the solution to the glycidyl–POSS and mixing for an additional 2 h. The copolyimide employed in this research contained 1 wt.% and 10 wt.% of glycidyl–POSS depending on the solid polymer. After that, the solution was casted on a metal plate. THF (tetrahydrofuran) was then evaporated, and distilled water was used to remove the membrane from the metal plate. The membrane is then dried for 24 h at 150 °C and 80 mbar in a vacuum oven. The typical membrane thickness ranges from 28 to 32 μm. They observed finally that the hybrid membranes' permeability improved at higher temperatures while their overall selectivity remained mostly unaffected [32].

Researchers have reported the pervaporation process of mixtures containing n-heptane and thiophene. The chemical imidization method was used for the cyclization of poly(amic acid), gaining through the solubilization of fluorine-containing aromatic dianhydride, 2,2'-bis (3,4-dicarboxyphenyl) hexafluoropropane dianhydride, with aromatic diamine. Having the good solubility property of polyimide in strong bipolar solvents, a polyimide asymmetric membrane of n-heptane/thiophene mixtures was prepared. The permeation flux and sulphur enrichment factor of the polyimide membranes are in the range of 0.56–1.68 kg/m² h and 3.12–2.24, respectively, and were studied for n-heptane/thiophene mixtures at 40–77 °C [33]. Pervaporation desulfurization of refinery naphtha was done by White and his team using multi-block copolymers of polyurea/urethane on polytetrafluoroethylene substrate of pore size 0.2 μm. A polysiloxane membrane with enough flux and selectivity to distinguish between a sulphur-enriched permeate fraction and a sulphur-deficient retentate fraction when treated with a pervaporation set-up [34]. The earliest membrane was obtained using a solution of toluene diisocyanate-terminated polyethylene adipate in 4-dioxane added to another solution containing 4-4'-methylene dianiline dissolved in 4-dioxane, and then the mixture solution was casted on a polytetrafluoroethylene substrate. The synthesis of the second membrane was the same, just by taking N,N-dimethylformamide in place of 4-dioxane. An enrichment factor (ratio of sulphur content in feed to sulphur content in permeate) of 7.53 was obtained for the feed sample containing 1065 µg/g of thiophenic compounds for the first membrane. 9.58 enrichment factor was obtained for the feed sample containing 419 µg/g of thiophenic compounds for the second copolymer. The pervaporation efficiency and swelling of the PEG (polyethylene glycol) membrane was investigated by Linn and his team. N-methyl pyrrolidone (NMP) was used by them to dissolve the PEG (polyethylene glycol) polymer, resulting in a homogeneous solution containing 12 wt.% polymer at 25 °C. After filtering and degassing, the solution was casted onto a polyethersulfone ultrafiltration (UF) membrane support. Then, a particular membrane was dried in a vacuum drying oven for 24 h after the evaporation of the solvent. To prepare a cross-linking membrane, at room temperature, the PEG polymer is combined with maleic anhydride as a cross linker and trimethylamine as a catalyst to generate a homogenous mixture containing 12 wt.% polymer, casted on an ultrafiltration polyethersulfone membrane. Before and after cross linking with maleic anhydride, they observed that upon addition of maleic anhydride, flux decreased to 26.36 kg/m^2 h with enrichment factor 1, whereas without the cross linker, maleic anhydride flux was 63 kg/ (m^2 h) with enrichment factor 3.05 [35]. Likewise, various researchers have reported modified materials for membrane pervaporation separation of sulphur compounds to get high selectivity as well as flux, which include blending, cross linking, grafting, copolymerization and inclusion of adsorbents such as nanoparticles, zeolites, etc. According to published research, Xu et al. designed a tetraethyl orthosilicate (TEOS) cross-linked polydimethylsiloxane (PDMS)/ceramic membrane for the pervaporation approach of sulphur removal. They casted the membrane on a ZrO_2/Al_2O_3 ceramic membrane support after boiling the support for 30 minutes, washed with deionized water, and also cross linking it with 20% TEOS. The cross-linked PDMS was then coated by the dipping method on the ceramic support, which then dried for 24 h at 25 °C. The cross-linking agent is favoured for selectivity and to control the behaviour of swelling of the membrane. After doping the cross linker to the membrane, the sulphur enrichment factor improved while the flux rate dropped. In a scientific brief manner, a reticular spatial structure was created by the addition of cross-linking chemicals to the PDMS polymeric solution, which was advantageous for the ability of the membrane to resist its swelling in gasoline. As the interchain free volume decreased, the macromolecules and chain segments' mobility also decreased, which is what caused the permeation flux to drop more than before. The reason for the dropping off more slowly of the permeation flow of thiophene species than that of the hydrocarbon species was that they had a greater affinity for the membrane. Consequently, the cross-linking agent's inclusion enhanced the sulphur enrichment factor throughout the pervaporation process. The overall flux risen as the feed temperature and sulphur concentration was risen, whereas the sulphur enrichment factor declined, according to experimental data. The total flux and sulphur enrichment factor were in favour due to a comparatively low permeate pressure. In the meantime, as the feed flow rate increased, the total flux and sulphur enrichment factor slightly climbed [36]. Li et al. used polydimethylsiloxane (PDMS) as a base polymer because of its greater ability to separate organic compounds. The π-complexation between inorganic metal ions and the thiophene aromatic ring causes deep desulfurization. In order to enhance the separation property of the PDMS membrane, Ni^{2+} was chosen as the inorganic component. The stability and longevity of nickel is significantly greater. The weaker interactions formed by Ni^{2+} are strong enough to retain the adsorbate and exhibit better adsorption–desorption properties. So, performance will be slightly improved by adding $Ni^{2+}Y$ zeolite to the PDMS membrane. PDMS-$Ni^{2+}Y$ was used as the active layer, and polysulfone ultrafiltration membrane was used as the support layer to produce good efficiency and lessen swelling. The inorganic filler material of the active PDMS layer was $Ni^{2+}Y$ zeolite owing to its low binding energies, excellent stability, relatively inexpensive and availability. Good desulfurization performance was noted with the prepared membrane, where the sulphur feed concentration was 500 mg/L at 30 °C [37].

12.5 CONCLUSION

The suitability of any technology for a given application depends on various elements, including the operating environment, cost, weight, space restrictions, feed volume, outlet specifications, technological maturity, etc. [38]. Utilizing membrane technology has benefits such as being easy to operate, having minimal environmental impact, not producing secondary pollutants, etc. The process performance relies on the membrane material selectivity, their working efficiency, temperature, pressure and energy requirement, material-to-feed ratio, etc. In some cases, cross linking, coating and plasticizing the polymeric membrane would help to have better removing performance [38]. Flat sheet membranes are favoured for maximum laboratory-scale applications, while hollow fibre membranes are much more productive, especially for industry-level applications. Over the past few years, numerous studies on the fabrication, characterization and applications of hollow fibre membranes have been performed; the most frequently used membrane materials are polysulfone, polyvinylidene fluoride, polytetrafluoroethylene, polyethersulfone and polyacrylonitrile. Hollow fibre membrane separation is probably cost-effective and efficient for large-scale removal applications. Industries prefer hollow fibre membrane modules because of their superior separation capabilities, ease of handling during module fabrication, bigger surface area per unit volume and self-supporting capacity [39]. A flat sheet membrane module needs more space and porous supports to be used in various applications. A hollow fibre membrane module has a higher degree of flexibility than a flat sheet membrane module. Though membranes (Flat sheet membrane and hollow fibre membrane) have a limited lifespan, they can be reused through the backwashing process.

This chapter presented aspects regarding the removal of sulphur by membrane technology, which provides the best results among the other separation or removal methods owing to its low cost, high selectivity, flexibility, greener one-step separation and many more. Desulfurizing sulphur from petroleum fuels is a major concern that needs to be addressed as it leads to environmental pollution and is hazardous to nature. Membrane-based technologies or techniques are practical choices among the many alternatives for eliminating sulphur traces that cause an offensive odour in the air. Their benefits include easy scaling and installation, selectivity, direct flows of odorous chemicals, complete automation with measurable operating parameters (pH, temperature, ionic strength) and low running expenses.

REFERENCES

1. Ceccotti, S. P., Morris, R. J., & Messick, D. L. (1998). A global overview of the sulphur situation: Industry's background, market trends, and commercial aspects of sulphur fertilizers. *Sulphur in Agroecosystems*, 175–202. https://doi.org/10.1007/978-94-011-5100-9_6
2. Katasonovaa, O. N., Savoninaa, E. Yu., & Maryutinaa, T. A. (2021). Extraction methods for removing sulfur and its compounds from crude oil and petroleum products. *Russian Journal of Applied Chemistry*, *94*(4), 411–436. https://doi.org/10.1134/S107042722 1040017
3. Shi, Q., & Wu, J. (2021). Review on sulfur compounds in petroleum and its products: State-of-the-art and perspectives. *Energy & Fuels*, *35*(18), 14445–14461. https://doi.org/10.1021/acs.energyfuels.1c02229
4. Payzant, J. D., Montgomery, D. S., & Strausz, O. P. (1986). Sulfides in petroleum. *Organic Geochemistry*, *9*(6), 357–369. https://doi.org/10.1016/0146-6380(86)90117-8
5. Nishioka, M., Campbell, R. M., Lee, M. L., & Castle, R. N. (1986). Isolation of sulphur heterocycles from petroleum and coal-derived materials by ligand exchange chromatography. *Fuel*, *65*(2), 270–273. https://doi.org/10.1016/0016-2361(86)90019-0
6. Mortaheb, H. R., Ghaemmaghami, F., & Mokhtarani, B. (2012). A review on removal of sulfur components from gasoline by pervaporation. *Chemical Engineering Research and Design*, *90*(3), 409–432. https://doi.org/10.1016/j.cherd.2011.07.019
7. Tewari, A., & Shukla, N. P. (1991). Air pollution-adverse effects of sulfur dioxide. *Reviews on Environmental Health*, *9*(1), 39–46. https://doi.org/10.1515/REVEH.1991.9.1.39
8. Painter D. E. (1974). *Air Pollution Technology* (pp. 46–48). Reston, VA: Reston Publishing Company and Prentice-Hall Company. https://archive.org/details/airpollutiontech0000pain
9. McCabe, L. C. (1952). *Air Pollution: Proceedings* (pp. 41–90). New York: McGraw-Hill. https://nepis.epa.gov/Exe/ZyPURL.cgi?Dockey=9101TEC6.TXT
10. Demirbas, A. (2016). Sulfur removal from crude oil using supercritical water. *Petroleum Science and Technology*, *34*(7), 622–626. https://doi.org/10.1080/10916 466.2016.1154871
11. Ahmad, W., Ur Rahman, A., Ahmad, I., Yaseen, M., Mohamed Jan, B., Stylianakis, M. M., & Ikram, R. (2021). Oxidative desulfurization of petroleum distillate fractions using manganese dioxide supported on magnetic reduced graphene oxide as catalyst. *Nanomaterials*, *11*(1), 203. https://doi.org/10.3390/nano11010203
12. Ahmed, O. U., Mjalli, F. S., Al-Wahaibi, T., Al-Wahaibi, Y., & AlNashef, I. M. (2016). Efficient non-catalytic oxidative and extractive desulfurization of liquid fuels using ionic liquids. *RSC Advances*, *6*(105), 103606–103617. https://doi.org/10.1039/C6R A22032K
13. Mohumed, H., Rahman, S., Imtiaz, S. A., & Zhang, Y. (2020). Oxidative-extractive desulfurization of model fuels using a pyridinium ionic liquid. *ACS Omega*, *5*(14), 8023–8031. https://doi.org/10.1021/acsomega.0c00096
14. Bedda, K., Hamada, B., Semikin, K. V., & Kuzichkin, N. V. (2019). Desulfurization of light cycle oil by extraction with polar organic solvents. *Petroleum & Coal*, *61*(6), 1352–1360. www.researchgate.net/publication/337032946

15. Al Zubaidy, I. A., Tarsh, F. B., Darwish, N. N., Majeed, B. S. S. A., Al Sharafi, A., & Chacra, L. A. (2013). Adsorption process of sulfur removal from diesel oil using sorbent materials. *Journal of Clean Energy Technologies*, *1*(1), 66–68. https://doi.org/10.7763/JOCET.2013.V1.16

16. Demirbas, A., & Balat, M. (2004). Coal desulfurization via different methods. *Energy Sources*, *26*(6), 541–550. https://doi.org/10.1080/00908310490429669

17. Iruretagoyena, D., & Montesano, R. (2018). Selective sulfur removal from liquid fuels using nanostructured adsorbents. *Nanotechnology in Oil and Gas Industries: Principles and Applications*, 133–150. Cham: Springer. https://doi.org/10.1007/978-3-319-60630-9_5

18. Celis-García, L. B., González-Blanco, G., & Meraz, M. (2008). Removal of sulfur inorganic compounds by a biofilm of sulfate reducing and sulfide oxidizing bacteria in a down-flow fluidized bed reactor. *Journal of Chemical Technology & Biotechnology: International Research in Process, Environmental & Clean Technology*, *83*(3), 260–268. https://doi.org/10.1002/jctb.1802

19. Ma, X., Sun, L., & Song, C. (2002). A new approach to deep desulfurization of gasoline, diesel fuel and jet fuel by selective adsorption for ultra-clean fuels and for fuel cell applications. *Catalysis Today*, *77*(1–2), 107–116. https://doi.org/10.1016/S0920-5861(02)00237-7

20. Nazari, F., Kefayati, M. E., & Raheb, J. (2017). The study of biological technologies for the removal of sulfur compounds. *Journal of Sciences, Islamic Republic of Iran*, *28*(3), 205–219. http://jsciences.ut.ac.ir

21. Pinheiro, C. I., Fernandes, J. L., Domingues, L., Chambel, A. J., Graça, I., Oliveira, N. M., & Ribeiro, F. R. (2012). Fluid catalytic cracking (FCC) process modeling, simulation, and control. *Industrial & Engineering Chemistry Research*, *51*(1), 1–29. https://doi.org/10.1021/ie200743c

22. Wen, Y., Wang, G., Xu, C., & Gao, J. (2012). Study on in situ sulfur removal from gasoline in fluid catalytic cracking process. *Energy & Fuels*, *26*(6), 3201–3211. https://doi.org/10.1021/ef300499j

23. Park, J. H., Joung, Y. O., & Park, S. D. (2007). Sulfur removal from coal with supercritical fluid treatment. *Korean Journal of Chemical Engineering*, *24*, 314–318. https://doi.org/10.1007/s11814-007-5043-y

24. Hales, J. M. (1978). Wet removal of sulfur compounds from the atmosphere. *Atmospheric Environment*, *12*, 389–399. https://doi.org/10.1016/B978-0-08-022932-4.50044-6

25. Sarkar, S., Meikap, B. C., & Chatterjee, S. G. (2007). Modeling of removal of sulfur dioxide from flue gases in a horizontal cocurrent gas–liquid scrubber. *Chemical Engineering Journal*, *131*(1–3), 263–271. https://doi.org/10.1016/j.cej.2006.12.013

26. Paucar, N. E., Kiggins, P., Blad, B., De Jesus, K., Afrin, F., Pashikanti, S., & Sharma, K. (2021). Ionic liquids for the removal of sulfur and nitrogen compounds in fuels: A review. *Environmental Chemistry Letters*, *19*, 1205–1228. https://doi.org/10.1007/s10311-020-01135-1

27. Taieb, D., & Brahim, A. B. (2013). Electrochemical method for sulphur dioxide removal from flue gases: Application on sulphuric acid plant in Tunisia. *Comptes Rendus Chimie*, *16*(1), 39–50. https://doi.org/10.1016/j.crci.2012.08.009

28. Ntagia, E., Prévoteau, A., & Rabaey, K. (2020). Electrochemical removal of sulfur pollution. In *Environmental Technologies to Treat Sulfur Pollution: Principles and Engineering* (2nd ed., pp. 247–276). London: IWA Publishing. http://iwaponline.com/ebooks/book/chapter-pdf/772258/9781789060966_0247.pdf

29. Jorjani, E., & Ghahreman, A. (2017). Challenges with elemental sulfur removal during the leaching of copper and zinc sulfides, and from the residues; A review. *Hydrometallurgy, 171*, 333–343. https://doi.org/10.1016/j.hydromet.2017.06.011
30. Fihri, A., Mahfouz, R., Shahrani, A., Taie, I., & Alabedi, G. (2016). Pervaporative desulfurization of gasoline: A review. *Chemical Engineering and Processing-Process Intensification, 107*, 94–105. https://doi.org/10.1016/j.cep.2016.06.006
31. Kong, Y., Lin, L., Zhang, Y., Lu, F., Xie, K., Liu, R., Guo, L., Shao, S., Yang, J., & Shi, D. (2008). Studies on polyethylene glycol/polyethersulfone composite membranes for FCC gasoline desulphurization by pervaporation. *European Polymer Journal, 44*(10), 3335–3343. https://doi.org/10.1016/j.eurpolymj.2008.07.034
32. Konietzny, R., Koschine, T., Raetzke, K., & Staudt, C. (2014). POSS-hybrid membranes for the removal of sulfur aromatics by pervaporation. *Separation and Purification Technology, 123*, 175–182. https://doi.org/10.1016/j.seppur.2013.12.024
33. Wang, L. H., Zhao, Z. P., Li, J. D., & Chen, C. X. (2006). Synthesis and characterization of fluorinated polyimides for pervaporation of n-heptane/thiophene mixtures. *European Polymer Journal, 42*, 1266–1272. https://doi.org/10.1016/j.eurpolymj.2005.12.013
34. White, L. S., Wormsbecher, R. F., & Lesemann, M. (2002). Membrane separation for sulfur reduction, Patent US 0211706 A1; 2004. https://patents.google.com/patent/US20040211706
35. Lin, L., Kong, Y., Wang, G., Qu, H., Yang, J., & Shi, D. (2006). Selection and crosslinking modification of membrane material for FCC gasoline desulfurization. *Journal of Membrane Science, 285*(1–2), 144–151. https://doi.org/10.1016/j.memsci.2006.08.016
36. Xu, R., Liu, G., Dong, X., & Jin, W. (2010). Pervaporation separation of n-octane/thiophene mixtures using polydimethylsiloxane/ceramic composite membranes. *Desalination, 258*(1–3), 106–111. https://doi.org/10.1016/j.desal.2010.03.035
37. Li, B., Xu, D., Jiang, Z., Zhang, X., Liu, W., & Dong, X. (2008). Pervaporation performance of PDMS-Ni^{2+} Y zeolite hybrid membranes in the desulfurization of gasoline. *Journal of Membrane Science, 322*(2), 293–301. https://doi.org/10.1016/j.memsci.2008.06.015
38. Ma, Y., Guo, H., Selyanchyn, R., Wang, B., Deng, L., Dai, Z., & Jiang, X. (2021). Hydrogen sulfide removal from natural gas using membrane technology: A review. *Journal of Materials Chemistry A, 9*(36), 20211–20240. https://doi.org/10.1039/D1TA04693D
39. Imtiaz, A., Othman, M. H. D., Jilani, A., Khan, I. U., Kamaludin, R., Iqbal, J., & Al-Sehemi, A. G. (2022). Challenges, opportunities and future directions of membrane technology for natural gas purification: A critical review. *Membranes, 12*(7), 646. https://doi.org/10.3390/membranes12070646

13 Application of Membrane Technology in Oil and Gas Fields

*Shrisha S. Raj, Nazia Shaik and
Sundergopal Sridhar*

13.1 INTRODUCTION

A nation's economic growth and employment mostly depend on petrochemical industries in developing countries. Petrochemical industries are derived from petroleum and natural gas, which can be used in different areas [1]. During this process, the generation of oil wastewater generated large amounts of contaminants in the ecosystem [2]. These contaminants have shown high toxicity, and hence, efficient methods must be investigated in order to prevent water resources from getting polluted [3]. The generation of oil wastewater from several industries, such as dairy, metal processing, poultry, edible oil refineries, tannery industries, restaurants and petrochemical industries [4]. Before the wastewater is discharged into the environment, it is mandatory to treat the wastewater [5]. The quantity of wastewater required to be treated before it is recycled or unleashed into the ecosystem has risen in sophistication as a consequence of current drilling improvements like hydraulic fracturing, sand tar and enhanced oil recovery (EOR).

Petrochemical wastewater contains several amounts of organic and inorganic components. Thus, a suitable treatment is needed for the disposal of wastewater for reuse using different methods [6]. Different methods of conventional processes, such as flocculation, anaerobic treatment, the activated sludge process, adsorption, coagulation and chemical precipitation, are used to treat wastewater inthe petroleum industry [7]. During the coagulation and adsorption process, it has produced a lot of sludge along with less rejection of the pollutants present in the wastewater. This process is widely used in the treatment of wastewater and drinking water to remove organic pollutants. Only a few studies have been focused on the treatment of oily wastewater [8]. The major drawback was that the operation cost was high, and it produced secondary pollutants as a product [9]. Another activated sludge process has been widely used for oil wastewater; in this technique, suspended bacterial biomass is taken to be responsible for the elimination of pollutants. However, using this technique, only a few pollutants, such as phosphorus, nitrogen and organic carbon substances, can be removed, but it is environmentally friendly and costeffective.

DOI: 10.1201/9781003441359-13

Refinery wastewater and [the] treatment [of produced water (PW)] by oil companies have been conventionally done by chemical and physical processes. During gas and oil production, a potential impact on species, such as population and ecosystem, may influence ecological parameters like biomass, biodiversity and productivity.

Moreover, many developments are still needed for current methods, which are facing problems of low efficiency and high cost. Besides, these traditional methods for the treatment of produced water have been discussed in the section below. Figure 13.1 shows the generation of produced water from oil to water using the oil drilling process. This PW treatment is discussed clearly in the section below.

Produced water (PW) is pungent water wedged below the underground arrangements and brought to the floor/surface during oil and gas progression and creation. Normally, PW carries a lot of impurities, as shown in Figure 13.2, such

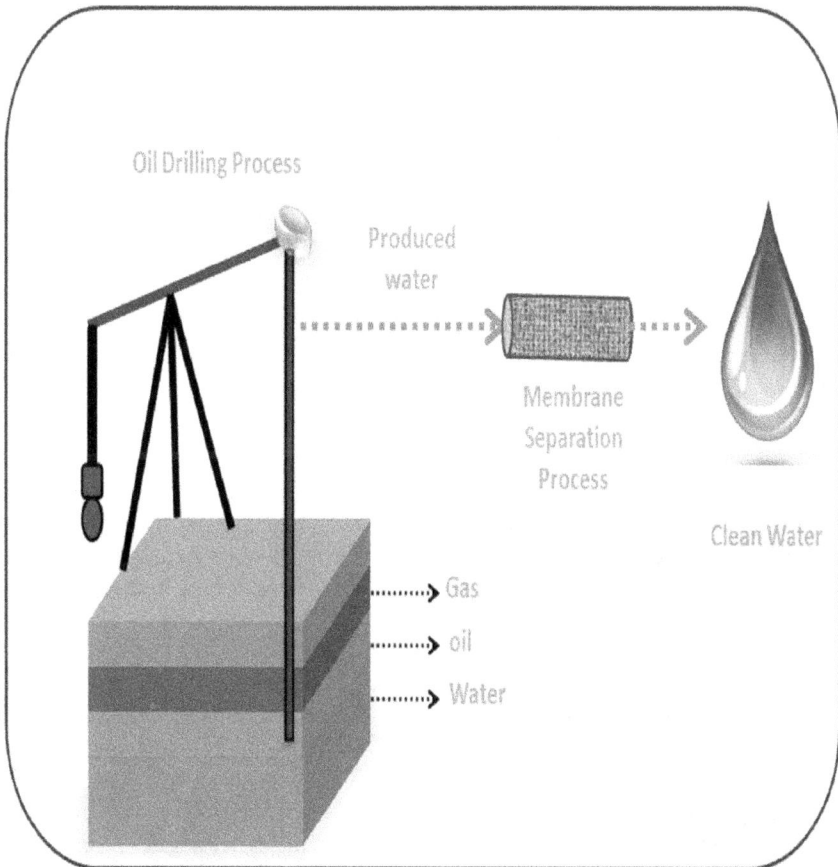

FIGURE 13.1 Schematic representation for production of produced water from gas and oil well.

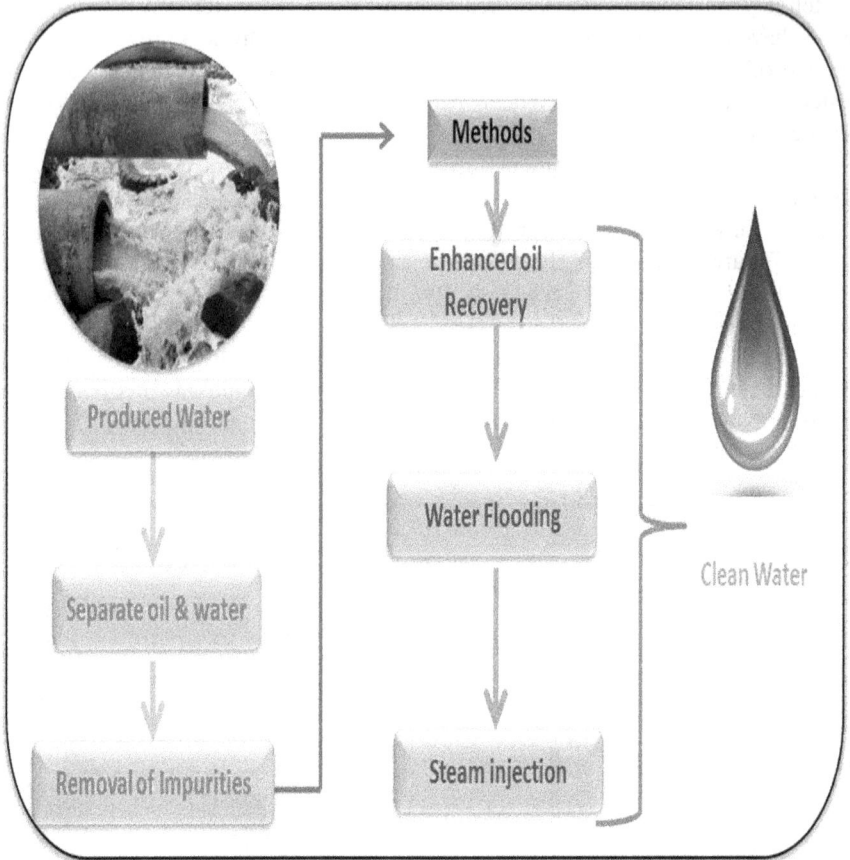

FIGURE 13.2   Treatment of produced water.

as organic and inorganic matter, which refers to oil and grease, chemical oxygen demand (COD), total suspended solids (TSS), biochemical oxygen demand (BOD), total dissolved solids (TDS) and gases [10].With an estimated global volume-to-product ratio of 3:1, PW is the biggest waste stream in the oil and gas sector. Before it can be used for any purpose, PW must first be cleaned of its many impurities and it is often created in vast quantities as a byproduct and produced in large amounts. The main problem in PW treatment is the high salt concentration, which is indicated by the TDS and the total organic carbon (TOC). The majority of PW possess more salt than seawater, with TDS concentrations between 1000 and 400,000 ppm [11].

In addition, steadily inflexible ecological recommendations and economical boundaries require the utilisation of similarly developed remedy strategies. Moreover, an increasing number of stringent environmental rules and financial constraints necessitate the use of superior treatment methods, which have been used and discussed in the below section.

13.2   ENHANCED OIL RECOVERY

The EOR method, which is a division of the improved oil recovery (IOR) technique, can restore a huge amount of oil from the subsurface. The anticipated recovery and the alternative methods will be pretentious by manytechnological and economical factors. Specific membranes have been used for increased oil recovery with commercial success, while others have received significant research attention. Most of the methods for recovery have already been evaluated, such as miscible carbon dioxide (CO_2) for light oil reservoirs and steam injection-based techniques for tar sands and heavier oils (if the asset is appropriate for such utilises).

Duan et al. explained that a large proportion of the recovery techniques have already been valued, such as steam injection-based methods in tar sands and heavier oils (if the reservoir provides approving conditions for such applications) and miscible CO_2 for light oil reservoirs [12–13]. Yongrui et al. used an integrated process like Fenton's oxidation and biological process to treat polymer flooding. However, using this hazardous oxidative method can produce a byproduct;for this, an additional safety separation is required [14].

Past research has identified many ways to recover oil using the EOR process, which is costly and difficult and has been worked out for only some processes. On the other hand, the EOR process plays an important role in energy mounting and definite supply areas for oil production [15]. Therefore, several researchers recommended that there is a requirement to develop a new technology for recuperating oil. In conclusion, there is an immediate requirement to discover a suitable technique for complicated and exclusively viscous streams. However, this technique is fundamentally not valid for feed that has TDS concentrations above 20 g/L.

13.2.1   WATER FLOODING

Among the various IOR techniques that have been adopted successfully over the years was water flooding. Pressure is regulated and modulated in this technique through primary depletion mode. Water that is made available through reservoir connate more often does not differ with varying compositions of sorts.Still, it significantly differs in composition from the water that is available for the injection. It has been established through numerous parametric laboratory investigations of crude oil recovery that water flood recoveries for the connate and injected brine of the same or different compositions greatly change depending on the brine composition. The difference between injected and connate brines has so far not been incorporated in various laboratory tests that have been designed for the prediction of water flood performance [16].

Gladkov et al. have established through numerous parametric laboratory investigations of crude oil recovery that water flood recoveries for the connate and injected brine of the same or different compositions greatly change depending on the brine composition [17]. Zhao et al. used a model called semi-supervised learning (SSL) to categorise the leftover oil from the water-flooding of oil fields using 2D experiments on a fixed glass micro-model [18]. Roueche and Karacan used predictive and probabilistic methods for the recovery of oil from the identified residual oil zone. Basic techniques are not taken into consideration due to the geological characteristics of the reservoir, such as interlayers and the contact related to the planar structure [19]. In total, there are a variety of ways to improve oil output, such as water injection, water shut-off, subdividing the injection-production section to improve the impact of areal scrubbing and cyclic water injection [20]. We must still work out a way to suggest an approach for stabilising the hydrocarbon recovery rate [21].

The most beneficial use of PW in the oil and gas industry has been maintainingEOR and reservoir pressure since past flooding. Depending on environmental regulatory needs, the surplus PW can either be discharged to the ocean or reinjected into disposal wells. Generally, a relatively small fraction of PW is treated to be reused or recycled worldwide [22].

13.2.2 Steam Injection

Ever since 1950, one of the most widely used thermal recovery processes has been steam injection. The recovery process depends mainly on many factors, as the effectiveness of any recovery process depends on various factors, viz., viscosity reduction, wettability alteration, gas expansion, etc. This process has been successfully used in both light and heavy reservoirs. Through various chemical additions in integrated form, the effectiveness of the process can be increased. In addition to chemical additions, and hydraulic fracturing, the application of horizontal drilling techniques havebeen successfully used. By using modern technology, the effectiveness of the recovery factor was increased from 15% to 40%. SI is an elegant technology because it offers a speedy payout at a high success rate due to enhancing field development experiences. On the other hand, compared to other stream floods, in terms of the final recovery factors, SI is still poor or uncompetitive [23].

In petrochemical industries, so many synthetic compounds, such as propylene and ethylene, are the most significantly used in the synthesis of polyolefins like acrylonitrile, styrene, ethylene dichloride, polypropylene, etc. Hence, in recent years, the separation of the oil from the oil wastewater has been a major factor, so an advanced separation technique has been used. It has become usual to treat produced wastewater from refineries and other sources through membrane technology (MT). Researchers have concentrated on enhancing the competence and long-term viability of the membrane for the period of wastewater treatment operations. The refinery and produced wastewater treatment were accomplished by most important primary and secondary treatment from the past works, followed by concentrating on several pressure-driven membrane technologies like membrane distillation (MD), microfiltration (MF), ultrafiltration (UF), nanofiltration (NF) and reverse osmosis (RO) [24,25].

MT is the usual term for several separation techniques. Over the past decades, it has become the most important separation technology. The most important application of MT is that it produces stable water without any impurities with a simple, well-arranged and easy process, and it uses low energy as compared to other technologies. MT has become economically feasible in most countries and also has a wide range of applications like seawater desalination, food processing, biotechnology and petroleum [26,27]. Generally, membranes are selective barriers or thin layers that can work by applying potential gradients such as pressure, concentration, temperature or electrical difference to separate undesirable materials. They are classified depending on their pore size, morphology and materials [28]. The membrane's performance is determined by flux, selectivity and permeability.

Depending on the particle size, pressure-driven membrane separation processes such as MF, UF, NF and RO can remove the fine suspended solids in oil refinery wastewater [29]. The materials used for membranes, such as polyvinylidene fluoride (PVDF) and polyacrylonitrile (PAN), are hydrophobic, as shown in Table 13.1. The membrane is hydrophobic, which makes it highly liable to organic fouling by oil and grease, with the specification of <1 mg/L oil and grease. RO membranes with extremely small pores of less than 0.001 μm, can separate salts and metals from oil effluent to produce clean water. Compared to MF or UF membranes, RO membranes are particularly susceptible to damage from oil effluent [30].

TABLE 13.1
A Literature Review on Refinery and Produced Wastewater Treatment Using Membrane Treatment

Membrane Category	Materials and Properties of the Membrane	Parameter Studies	Water Source	References
Microfiltration	Polyacrylonitrile and ceramic	Oil–water separation	Synthetic produced water	[31]
	hydrophobic PVDF	TOC rejection	Real and synthetic produced water	[32]
Ultrafiltration	Hydrophilic cellulose	Fouling resistance and TOC removal	Synthetic produced water	[33,34]
	PVP hydrophilised polyethersulfone	TSS, COD, TOC, O&G removal	Real produced water	[35]
Nanofiltration and forward osmosis	Commercial MF, UF, NF and RO	Cost estimation and effect of several membrane combinations	Oil and gas well-produced water	[36]
	Commercial RO and NF	A comparison study of RO and NF membranes and on fouling	Produced water	[37,38]
	RO membrane	A pretreatment electrocoagulation step and after that, the solution passes to the RO membrane	Real produced water	[39]

13.2.3 WATER QUALITY TREATMENT SPECIFICATIONS AND COMPATIBILITY

Some of the water parameters, such as TDS, BOD, suspended solids (SS), TOC, TSS, total metals, oil and grease (O&G), as well as some of the general physical parameters, such as taste, odour, hardness, turbidity and pH, include acids and alkalis [40]. The pollutants present in the wastewater are becoming more complex and attention also needs to be paid to the removal of contaminants present in the effluent [41]. Some of the water quality parameters, like biological, physical and chemical parameters, are shown in Table 13.2 [42].

13.2.3.1 Recommendations for Reusing Effluent Water

Reusing water is a major development for future purposes and is an alternative amenity worldwide [43]. Reusing of recycled water from oil wastewater with new rapid development technologies plays a key role [44]. Reuse of water from some industries, such as paper and pulp, textiles, oil and gas industries and food industries A requirement for water systems allows using recycled water at a specific point in the manufacturing process of the industries [45].

This recycled water can be used mainly in the below-mentioned areas such as:

- Cooling systems
- Cleaning of water tanks
- Boiler feed tanks
- Other uses, such as fire-fighting [46]

TABLE 13.2
Water Quality Parameters

S.No	Water Quality Parameters		
	Biological	**Physical**	**Chemical**
1.	Bacteria	Turbidity	pH
2.	Algae	Temperature	Acidity
3.	Viruses	Colour	Alkalinity
4.	Protozoa	Taste and odour	Chloride
5.	–	Solids	Chlorine residual
6.	–	Conductivity	Sulphate
7.	–	–	Nitrogen
8.	–	–	Fluoride
9.	–	–	Iron and manganese
10.	–	–	Copper and zinc
11.	–	–	Hardness
12.	–	–	Dissolved oxygen
13.	–	–	BOD
14.	–	–	COD
15.	–	–	Inorganic matters
16.	–	–	Organic constituents
17.	–	–	Radioactive substances

The recycled effluent water is to be used in the recirculation of the cooling system. Initially, the effluent consists of a relatively high level of BOD, COD, SS and ammonia. Hence, it is difficult to control the level in the below section, so the methods have been discussed [47].

13.2.3.2 Experimental Studies of Oil Wastewater at Different pH

Some of the experimental studies were carried out with the synthetic oil–water separation with 1000 ppm oil concentration present in the feed. Studies were carried out at different pH such as 6,7,8 and 9 with UF 10 kDa membrane. Figure 13.3 shows a graph between pH and flux. The recovery of water at different pH levels with 1000 ppm oil concentration, which was treated with an UF 10 kDa membrane, is shown in Table 13.3.

13.3 WELL STIMULATION

Acidizing is frequently used for activating efficiently. Compared to a vertical well, this method stands out the mostwhen employed in a horizontal well. With the aid of a longer borehole length, an expanded side in the reservoir characteristics close to the wellbore, and possibly an additional mechanical capacity through which to identify the fluids in the wellbore, the fluid circulation in a horizontal well is utilised. To accurately plan acidizing treatment for horizontal wells, a comprehensive fluid-establishing model that includes reservoir acidizing simulators is required.

This chapter describes a horizontal well with fluid model placement. This model predicts where infused fluids will be set by monitoring the outer layers of different fluids in the wellbore. It can trace a variety of interfaces for a variety of injection stages in a horizontal well. The model allows movement of the tubing tail during injection when using coiled tubing. It also resists concurrent injections from a tubing string and the annulus. The model's fluid allocation output may be used as input information for sandstone acidizing design in a reservoir acidizing model [48].

13.4 REMOVAL OF ORGANIC MATTERS, SUSPENDED SOLIDS ANDTDS

Pollutants such as inorganic salts and organic matter that are in TDS are an important concern for people and aquatic systems. The organic and inorganic salt materials present in TDS pose a risk to marine ecosystems and the water supply for people. Some inorganic salts include magnesium, sodium, potassium, sulphates, chlorides and bicarbonates [49]. For natural aquatic life, the TDS should be at a certain limit. Still, in the case of human activities such as mining, agriculture and urbanisation, the TDS should be exactly significant [50]. However, the oil and gas industries will discharge a high amount of TDS, which faces challenges in our daily lives [51]. Hence, resource recovery should be identified and some of the methods are also discussed in this chapter of the book. The biggest by-product generated by oil and gas production operations is wastewater, which is commonly referred to as produced water [52]. Produced water is the largest by-product generated by the petroleum industry

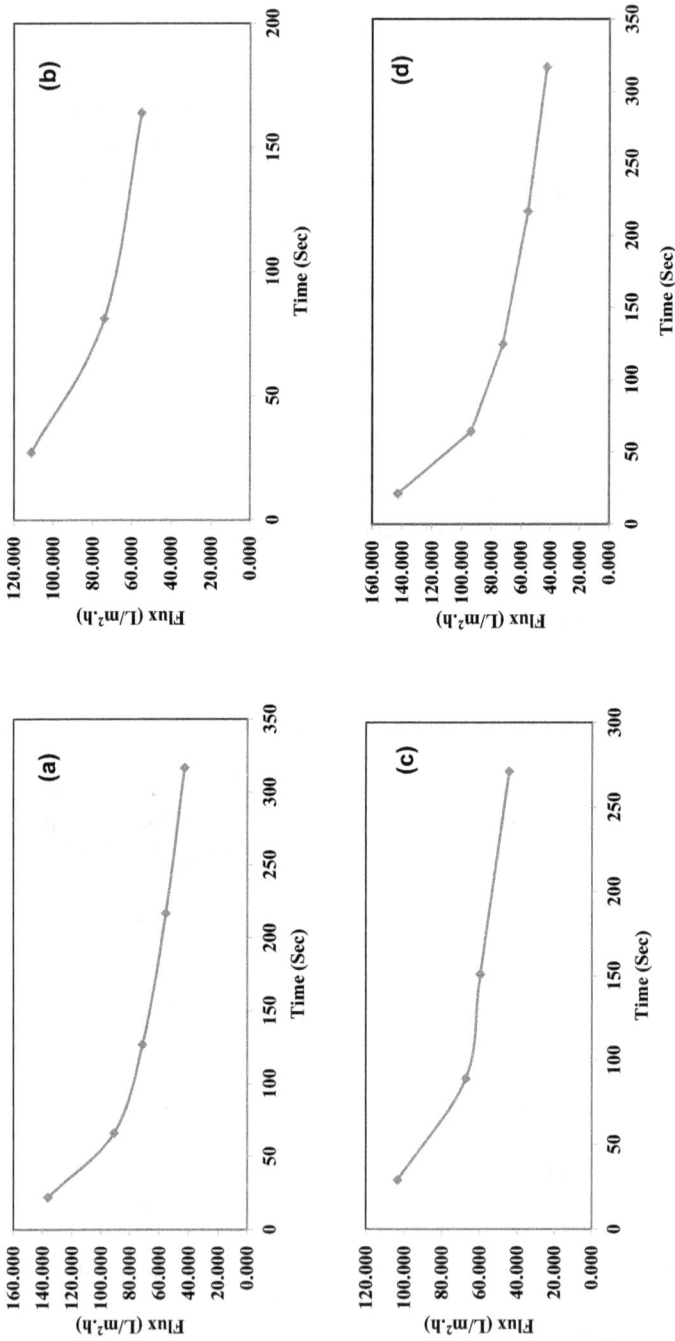

FIGURE 13.3 Graph between pH and flux with UF 10 kDa membrane of (a) pH-6; (b) pH-7; (c) pH-8 and (d) pH-9 of oil 1000 ppm concentration present in oil and water mixture.

TABLE 13.3
Water Recovery From Oil and Water Mixture Using UF 10 kDa Membrane at Different pH

Feed Volume	pH	Different pH	% Recovery
		Permeate Volume Collected (L)	
5	6	4.5	90
3.5	7	3	85.71
5	8	4	80
5	9	4.4	90

during oil and gas extraction [52]. Produced water is a complex matrix that frequently contains TDS. The composition of PW differs from the geographic location of the field and the type of extraction. Generally, it has ten times higher salinities than seawater's (35,000 mg/L), with TDS concentrations between 10,000 and 400,000 mg/L. Higher amounts of trace metals, naturally occurring radioactive material (NORM) and dissolved organic matter (DOM) are caused by interactions between minerals and water that occur naturally and intentionally. These interactions also result from chemical additions made during manufacturing (DOM).

13.5 RECOVERY OF LITHIUM (LI)

Lithium (Li) demand is rising rapidly worldwide due to the deployment of electric vehicles. Due to the expanding global use of electric cars, demand for lithium (Li) is predicted to rise significantly over the coming years, along with costs [53]. Li-recovery technologies, including those of adsorbents, membrane-based treatments and electrolysis-based systems, can be used to sustainably recover Li from gas and oil-produced water. Li is a precious metal that is widely recognised for its potential use in thermonuclear fusion, its current usage in Li-ion batteries and a variety of other applications. Li is used to make glass, medicinal items, lubricants and CO_2 adsorbents for use in aeroplanes and submarines. Since there is now an unstable balance between supply and demand for this alkali metal, the value of Li has started climbing significantly in recent years as a result of high demand and poor supply [54]. Li recovery can be sustainably recovered from the gas- and oil-produced water. Treatment of wastewater can be influenced by technologies such as adsorption, membranes and electrodialysis, which are also assessed in the discussion below. These technologies are adequate for the recovery of Li from wastewater [55].

13.5.1 ADSORPTION BASED

Among the widely used adsorbents are lithium-ion sieves (LIS). These adsorbents were developed by Volkhin in 1971. The development of this lithium in sieves is a two-stage process. In the primary stage, ions are first added by redox or any of the ion exchange processes. In the second stage, the ions are removed from the main structure through any viable process. The result is ion-specific vacancies for the target ions that are primarily affected due to the ion screening phenomenon in the memory effect. Three primary families are aluminium hydroxide ion sieves, lithium manganese oxide ion sieves and lithium titanic oxide ion sieves. We all know that lithium can easily diffuse or can be absorbed readily into the non-stoichiometric crystalline networks of manganese (Mn), titanium or aluminium hydroxide. Thus, we can have viable chemical processes for the production of an intermediate form of mixed lithium, Mn and titanium oxide (LMTO) [56].

Seip et al. prepared two Mn-based Li selective adsorbents through a co-precipitation technique to extract lithium from flow-back and PW released from hydraulically fractured gas and oil wells [57]. Due to their crystalline or layered characteristics, hydroxide sorbents and metal oxides are selective for Li because they let Li into ion exchange sites while sterically excluding bigger ions. These substances let out hydrogen ions while discharging Li ions in neutral and high pH solutions, and they free hydrogen ions while releasing Li ions in acidic solutions [58].

13.5.2 Membrane Based

The most vital type of membrane-based separation technology is to recover Li from an aqueous environment. Among several technologies used for Li recovery, membrane-based technology, driven by pressure, temperature gradient and electrical field, has recently gained popularity because of its low impact and safety for the environment. The different membrane-based methods used in recent years for Li recovery are the NF membrane process, the ion-sieve membrane process, the supported liquid membrane (SLM) process, membrane distillation crystallisation, etc. NF membrane system is based on the mechanisms of steric hindrance and Donnan exclusion. NF membranes such as Desal-5 DL, Desal DK, DK-1812 and NF90 are used for Li recovery. Wen et al. [59] recovered LiCl from dilute saline by utilising a Desal-5 DL NF membrane for the first time. The separation factor of Li^+ over Mg^{2+} could reach 3.5. Later, Li et al. [60] concentrated on increasing filtration stability by evaluating the effects of pH in addition to salinity on the parting of Mg^{2+}/Li^+. Due to critical concentration polarisation and high viscosity, there was a significant increase in feed salinity and a decrease in flux.

13.5.3 Electrolysis Based

For the production of organic acids, the desalting of seawater and the treatment of industrial pollutants, electrodialysis (ED) is largely preferred. The main mechanism of ED is the electrical potential difference as a driving force for moving ions. The main advantage of ED is that it is a green process and highly selective towards monovalent ions. It is an electrochemical membrane-based separation process in which cation and anion exchange membranes are positioned alternatively. As the electricity is applied, the cations and anions move towards the relevant ion exchange membrane to the correcting electrode. The traditional ion exchange membranes cannot divide the ions with the same charges. The progress of the monovalent ion exchange membrane, which separates monovalent ions, is very significant for the use of ED to recover Li from seawater or salt lake brine. Normally, ED membranes do not distinguish between unlike ions, even though a few differences in transport rates all the way through the membrane can be observed [61].

Flux was affected by pH, but separating Li^+ and Mg^{2+} under lower pH is always good due to increased dielectric exclusion of multivalent ions [62]. In a double-stage NF device, the pH of the solution that was acquired was crucial. After extraction by a two-stage NF system at a lower pH, the Mg^{2+}/Li^+ ratio decreased from an initial value of 13 to 0.1. To attain an immense and accurate Li^+ isolation, the metal–organic framework (MOF)-based NF membrane was created. The greatest separation selectivity for $S_{Li, Mg}$ was obtained in 1815, and the flux was kept at a high level of approximately 6.7 mol m^{-2} h^{-1} [63]. The polyamide skin layers of widely available NF membranes made of polyamide are negatively charged. Therefore, the creation of positively charged NF membranes and their benefits are greatly utilised for Li regeneration. The NF membrane is the only commercialised membrane technology used for Li recovery in large-scale applications.Still, fouling of the membrane is a disadvantage that results in reduced permeability and selectivity of the membrane.

SLMtechnology is also one of the membrane-based technologies used for Li recovery due to its high selectivity, low energy consumption and solvent usage. Sharma et al. [64] explained how the extractants tri-n-butyl phosphate and di-2-ethyl hexyl phosphoric acid were utilised for the selective extraction of Li^+ from synthetically prepared seawater by using a hollow fibre SLM module. The synergistic effect was shown between tri-n-butyl phosphate and di-2-ethyl hexyl phosphoric acid in equilibrium studies. The equilibrium constant values for Li^+, K^+ and Na^+ ions were found to be 95.4 × 10^{-5} m^3/k mol, 3.69 × 10^{-5} m^3/k mol and 4.6 × 10^{-5} m^3/k mol, respectively. The low concentrations of Li^+, K^+ and Na^+ ions in the feed phase were used in hollow fibre-SLMexperiments and showed high flux towards Li^+ ions. Later, at higher concentrations of K^+ and Na^+ in the feed segment, the flux of Li^+ ions decreased. Nonetheless, the low stability of SLM limits their commercial application.

Nie et al. [65] explored the possibility of selective electrodialysis (SED) for Li extraction from the artificially synthesised brine solution. The synthesised brine solution had a Mg^{2+}/Li^+ mass ratio of around 150. Li^+ recovery of 95 % is gained, and the Mg^{2+}/Li^+ mass ratio is reduced to 8 after processing through SED. Compared to the NF membrane system, the SED process has shown scientific superiority for the fractionation of Mg^{2+}/Li^+ with a high mass ratio. The SED-based technology process is a precious method that is relatively environmentally friendly and economical. In the future, research should be carried out on the design and development of SED systems and the fabrication of ion exchange membranes with low resistance and high selectivity to understand an effective Li recovery from Salt Lake brine.

13.5.4   Gas Permeation for CO_2 Removal From Natural Gas

The composition of natural gas changes from one location to another, and its quality depends on the concentration of pollutants. It mostly consists of methane (CH_4), higher hydrocarbons, CO_2, nitrogen and small amounts of helium, hydrogen sulphide, argon, oxygen and water vapour. Removing$CO_2$and hydrogen sulphide (H_2S) is necessary before feeding the natural gas to a pipeline. These pollutants are corrosive, and hydrogen sulphide is noxious as well. Sour gas can be treated for the removal of CO_2, and H_2S by adsorption, absorption and membrane. Membrane processes have been confirmed to be technically and inexpensively better than competing technologies in several industrial applications. This procedure employs a cross-flow mode of operation wherein a portion of the feed gas is collected as permeate, and the rest is collected as a reject stream.

The membrane used for the separation should contain both characteristics for separation characteristics and resistance to hydrocarbons present in the natural gas. Blending with another polymer can enhance the properties of the membrane; this also increases the separation factor. It is desirable to have a PEBAX 2533 gas permeation membrane with high selectivity, strong resistance and high gas permeation rates for separated gases. A skid-mounted pilot plant was designed by EIL for the separation of CO_2 and H_2S using membranes and was commissioned at ONGC, Hazira and test trials were carried out, as shown in Figure 13.4(a) and (b). This involves the usage of an activated carbon filter and a micron filter for the separation of dust particles [66]. The samples were analysed for the percentage of CH_4, CO_2 and H_2S, and the results are shown in Table 13.4.

The entire experiment was carried out at room temperature (30 °C). The feed and permeate lines were expatriated by vacuum pump (Hind High Vacuum Co. Model ED-18) Figure 13.5(a). With a mass flow controller (MFC), the feed gas was put progressively into the upper chamber, and the outflow valve was maintained closed until the dial gauge showed the pressure that was needed. The infused gas was drawn into SS 316 gas sample containers with a 100 ml capacity for further investigation using nitrogen or hydrogen as the carrier gas (Figure 13.5(b) and (c)). After attaining a stable state, samples were taken. Toensure stability, experiments were carried out repeatedly. A needle valve was employed to regulate and maintain the carrier gas flow rate, and a soap bubble meter connected to the permeate line's end was used for measuring it. The permeate flow was collected after the membrane and feed gas were in equilibrium for three hours. A gas chromatograph was used to determine the feed and permeate stream composition throughout an average of 6–8 h of sample collection. At higher pressures, the permeate flow was large enough for direct measurement in a soap bubble meter, and hence, carrier gas was not utilised for these experiments. The method of calculation of the permeability coefficient from experimental data is given below [66,67].

Calculation procedure

For pure gases
Permeate Flux, F = (permeate flow rate)/membrane area   [Units: cc/cm² s]
Permeance, $P = F/\Delta p$ [Units: cc /cm² s cm Hg]
Δp = partial pressure difference [cm Hg]
Selectivity, $\alpha = P_1/P_2$
P_1 = permeance of propylene
P_2 = permeance of propane.

Membrane Characteristics

P. CO_2	7.11	Permeability	Barrers
P. CH_4	0.40	Permeability	Barrers
tm	1.00	Thickness	Microns
Q. CO_2	1.945E-02	Permeance	M3/Sq M-Hr. Atm
Q. CH_4	1.094E-03	Permeance	M3/Sq M-Hr. Atm
S	17.78	Selectivity	

	Stage 1				Stage 2				Final
	F1	F2	P1	R1	F3	F4	P2	R2	R3
Flow, M³/hr	100.00	70.00	7.00	63.00	30.00	37.00	3.70	33.30	96.30
CO_2 %V	0.050	0.050	0.331	0.019	0.050	0.103	0.560	0.053	0.030
CH_4 %V	0.950	0.950	0.669	0.981	0.950	0.897	0.440	0.947	0.970
P, Atm	60.00	60.00	1.00	59.00	60.00	58.00	57.00	1.00	57.00
Stage Cut			0.1000				0.1000		
Pr. Drop			1.00				1.00		
P High, Atm			60.00				58.00		
P Low, Atm			1.00				1.00		
A, M²			7.4370E+01				2.7666E+01		

Total Area (M²)	102
CH_4 Loss, %	1.72
CO_2 Rejection, %	41.40

Constants						
	K1	0.70743	1403.38	K1	-1.0536	1729.02
	K2	-1.2536	920.973	K2	-0.8016	569.752
	K3	0.44879	-0.7888	K3	0.05018	1.04825

Simulation Stage 1 Stage2

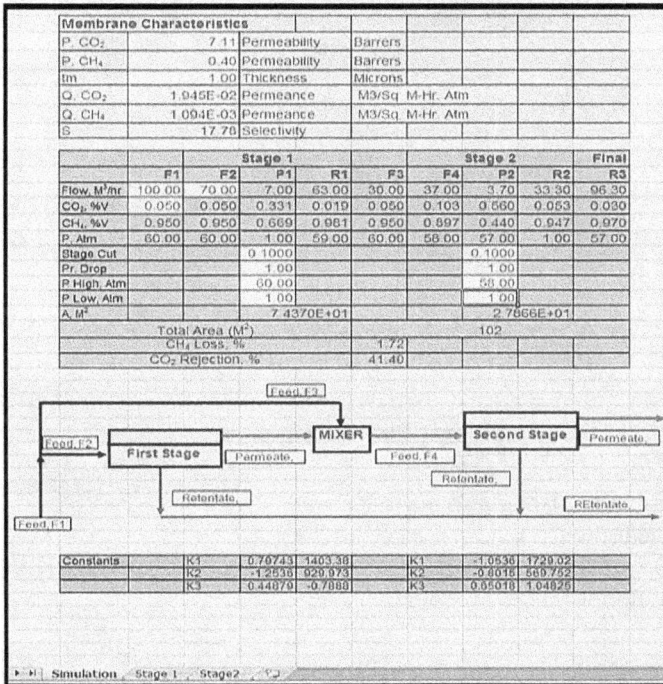

FIGURE 13.4 (a) Skid-mounted pilot plant for separation of carbon dioxide and H_2S at ONGC, Hazira and(b) simulation results for separation of carbon dioxide from natural gas (CH_4) at a feed flow of 100 NM³/h.

TABLE 13.4
The Samples Were Analysed for % CH_4, CO_2 and H_2S

Quality Parameters	Feed	Permeate	Reject
CH_4 (conc. %)	81.7	75.4	84.93
CO_2 (conc. %)	6.0	10.81	3.3
H_2S ppm	187	295	62

FIGURE 13.5 (a) Biogas plant; (b) results of biogas purification by polysulfone and (c) matrimid hollow fibremembrane modules.

13.5.4.1   Case Study on Biogas Purification

Poultry Litter Feed: 350 kg/day
Feed Preparation: 10% solids in water
Biogas Production: 5 L/min
Biogas Composition: 45% CH_4 + 49% CO_2 + 6% CO_2 + 1000 ppm H_2S
Approx Calorific Value: 1.6 kWh

Propylene is one of the products of fluid catalytic cracking (FCC) of hydrocarbon feedstock, which also contains paraffin propane. The C_3 splitter is the tallest distillation column in any refinery with 95 stages, and along with the C_2 splitter for ethane/ethylene separation, the total energy consumption of these two columns in all refineries put together in India could account for the power consumption of an entire

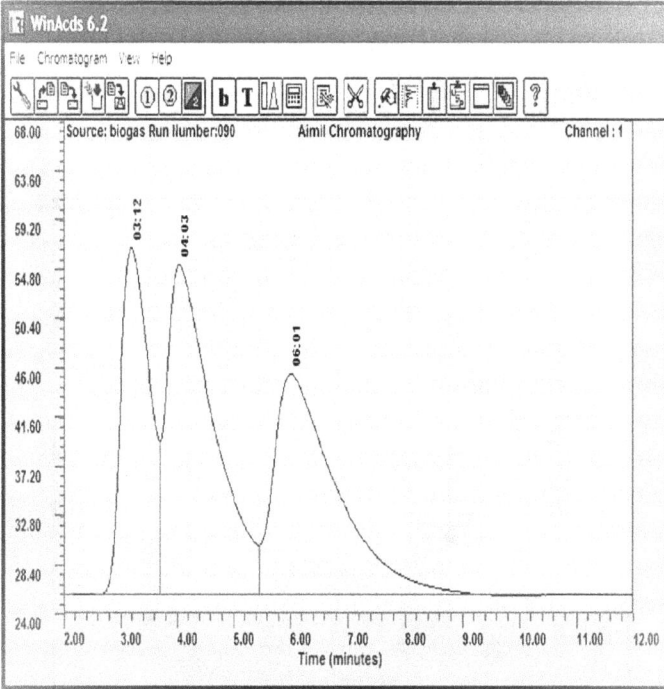

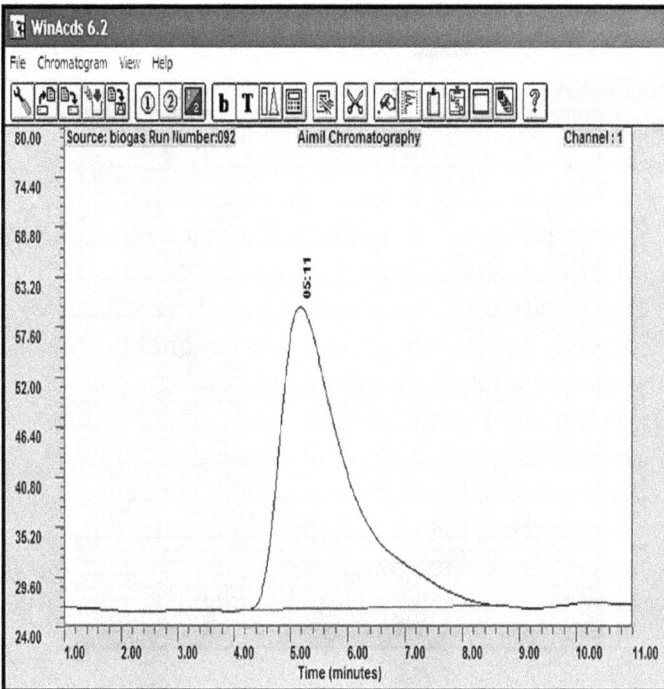

FIGURE 13.5   (Continued)

country such as Singapore. This is due to the same physical properties and molecular sizes of propane and propylene (Figure 13.6(a) and (b)). With the advent of membrane separation processes, many industrial operations have been modified to enrich gases with respect to the desired components.

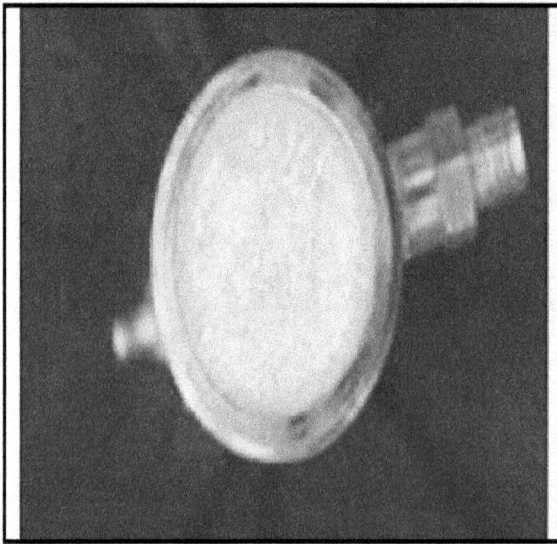

FIGURE 13.6   (a) and (b) matrimid hollow fibremodule.

TABLE 13.5
Results of Biogas Purification by Polysulfone and Matrimid Hollow FibreMembrane Modules

Sample Name	GC file no.	% CH_4	% CO_2	% N_2
Biogas feed	Biogas.090	40.89	32.832	26.27
Permeate (polysulfone)	Biogas.091	33.54	20.692	45.74
Permeate (Matrimid)	Biogas.092	1	99	1

Polyether-*block*-amide (Pebax) membrane was investigated for propylene recovery and further modified with the incorporation of silver metal ions to improve its selectivity with respect to propylene, since the silver would be present in an ionic state as this polymer matrix solubilises the salts well. Silver appeared to facilitate the selective transport of the olefin through reversible p-bond complications, as seen in Table 13.5. The $AgBF_4$ membrane was further scaled up into a spiral-wound module.

The next examples are provided and represent the efficacy of the separation features of the blended polyamide, wherein the data was collected from a pilot plant commissioned at ONGC, Hazira. The test rig, which consisted of a single gas separation membrane module, was 8 inches in diameter, 39.3 in long, and had an active membrane area of 30 m^2. Blended polyamide membranes with different ratios were conducted in a lab scale with an active membrane area of 42.43 cm^2. All runs were conducted at temperatures above 25 °C.

The polymer solution is prepared by dissolving 5-wt.%Pebax 2533 in a solvent mixture of water and ethanol with a ratio of around 70:30, continuously stirring at a higher temperature. The casting of the polymer solution and subsequent evaporation of the solvent mixture obtained the membrane. The effective thickness of the top layer is 5 mm. In this process, the feed natural gas, at a pipeline pressure of 60 kg/cm^2enters the required membrane separator unit at a feed flow rate of 52.9 nm^3/h. To ensure the partial pressure difference, i.e., the driving force across the synthesised membrane, is always smaller than the feed pressure, The permeate stream flow rate is 18.9 nm^3/h. The reject stream is sent to the adsorption process to reduce the H_2S concentration to the permissible limit. The permeate stream is enriched in CO_2 and hydrogen sulphide.

13.5.4.2 Olefin–Paraffin Separation

13.5.4.2.1 Synthesis of Silver-Incorporated Pebax (Ag–Pebax) Membrane
Using solution casting and solvent evaporation procedures, the hybrid membranes of silver–Pebax were synthesised on PSSU's ultra-porous support. Hybrid membranes of silver–Pebax were synthesised on the PSSU ultra-porous support. Around 5.4 g of silver tetrafluoroborate and 15 g of Pebax polymer were dissolved in 80 mL of isobutanol and cast on PSSU support to the required thickness using a doctor's blade. The solvent was evaporated in an oven at 110 °C for 3–5 min to get the Ag–Pebax membrane, with the thickness of the top particular layer ranging between 15 and 20 mm. Similarly, silver perchlorate ($AgClO_4$) was also used to prepare mixed matrix

FIGURE 13.7   Silver-incorporated Pebax membranes.

membranes of Pebax. The extent of silver loading varied from 5% to 33% of the polymer weight. The silver-incorporated Pebax membrane is shown in Figure 13.7.

13.5.4.2.2   Carbon-Loaded Membranes

Another category of mixed matrix membranes was synthesised by loading Pebax with 100% and 50% activated carbon of extremely fine particle size, about 10 mm. The reason for the loading of 50% is that the polymer–carbon weight ratio was 1:0.5. The method for membrane synthesis was the same as in the case of Ag–Pebax. Figure 13.8(a)depicts a photograph of the gas separation manifold, and Figure 13.8(b) and Table 13.6 show the spirally wound silver-loaded Pebax membranes for olefin separation.

13.6   CONCLUSION

In the oil and gas industry, one of the major issues affecting the ecosystem is the discharge of the produced water directly from the petrochemical industry. Hence, wastewater treatment should be carried out so that it can be recycled and should bring it to an acceptable range. Some of the treatment methods are discussed in the book chapter. Some of the methods are discussed, such as water flooding, steam injection and EOR, but the recovery of water and metals is not within an acceptable range. Hence, the main challenge is solving the inherent problem of produced wastewater treatment using advanced separation processes like membrane technology. If the concentration of the impurities is high in the produced water, further research should be done in future developments. A pre-treatment step should be required before being

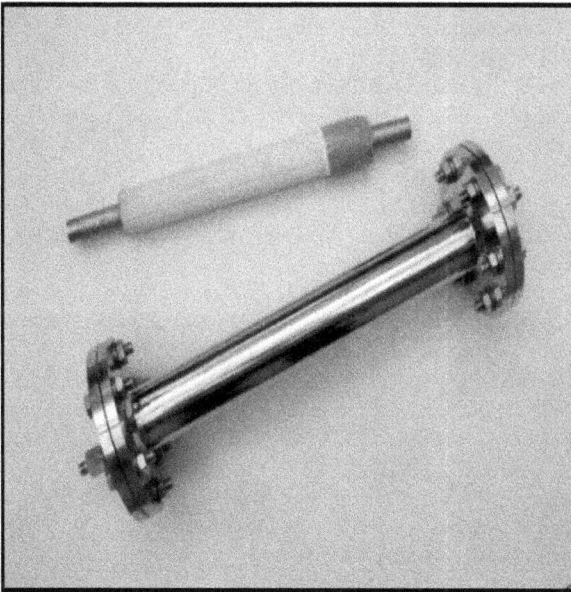

FIGURE 13.8   (a) Photograph of the gas separation manifold and (b) spirally wound silver-loaded Pebax membranes for olefin separation.

TABLE 13.6
Results for Pebax Spiral Wound Modules Loaded with AgBF$_4$ (36% Loading)

S.No.	Pressure(Kg/cm^2)	Permeance (K) (GPU) Propane	Selectivity (K_{C3H6}/K_{C3H8}) Propylene	
1.	2	0.32×10^{-2}	4.99×10^{-2}	15.4
2.	5	0.61×10^{-2}	15.6×10^{-2}	25.5

fed to the membrane separation process. More attention is being given to the reuse of water worldwide, which is also an essential step in preventing environmental damage.

REFERENCES

1. Wang, F. H., Hao, H. T., Sun, R. F., Li, S. Y., Han, R. M., Papelis, C., & Zhang, Y. (2014). Bench-scale and pilot-scale evaluation of coagulation pre-treatment for wastewater reused by reverse osmosis in a petrochemical circulating cooling water system. *Desalination, 335*(1), 64–69. https://doi.org/10.1016/j.desal.2013.12.013
2. Ali, A., Quist-Jensen, C. A., Drioli, E., & Macedonio, F. (2018). valuation of integrated microfiltration and membrane distillation/crystallization processes for produced water treatment. *Desalination, 434*(SI), 161–168. https://doi.org/10.1016/j.desal.2017.11.035
3. Piekutin, J., & Skoczko, I. (2016). Removal of petroleum compounds from aqueous solutions in the aeration and reverse osmosis system. *Desalination and Water Treatment, 57*(26), 12135–12140. https://doi.org/10.1080/19443994.2015.1048732
4. Sanghamitra, P., Mazumder, D., & Mukherjee, S. (2021). Treatment of wastewater containing oil and grease by biological method-a review. *Journal of Environmental Science and Health, Part A,56*(4), 394–412. https://doi.org/10.1080/10934 529.2021.1884468
5. Akinapally, S., Dheeravath, B., Panga, K. K., Vurimindi, H., & Sanaga, S. (2021). Treatment of pesticide intermediate industrial wastewater using hybrid methodologies. *Applied Water Science,11*, 1–7. https://doi.org/10.1007/s13201-021-01387-4
6. Wei, X., Zhang, S., Han, Y., & Wolfe, F. A. (2019). Treatment of petrochemical wastewater and produced water from oil and gas. *Water Environment Research,91*(10), 1025–1033. https://doi.org/10.1002/wer.1172
7. Varjani, S., Joshi, R., Srivastava, V. K., Ngo, H. H., & Guo, W. (2020). Treatment of wastewater from petroleum industry: Current practices and perspectives. *Environmental Science and Pollution Research,27*, 27172–27180. https://doi.org/10.1007/s11356-019-04725-x
8. Jung,C., Phal, N., Oh, J., Chu, K. H., Jang, M., &Yoon, Y. (2015). Removal of humic and tannic acids by adsorption–coagulation combined systems with activated biochar. *Journal of Hazardous Materials,300*, 808–814. https://doi.org/10.1016/j.jhaz mat.2015.08.025
9. Zhang, M. H., Dong, H., Zhao, L., Wang, D. X., & Meng, D. (2019). A review on Fenton process for organic wastewater treatment based on optimization perspective. *Science of the Total Environment, 670*, 110–121. https://doi.org/10.1016/j.scitot env.2019.03.180

10. Zhang,T., Gregory, K., Hammack, R. W., & Vidic, R.D. (2014).Co-precipitation of radium with barium and strontium sulfate and its impact on the fate of radium during treatment of produced water from unconventional gas extraction. *Environmental Science &Technology,48*,4596–4603. https://doi.org/10.1021/es405168b

11. Ahmad, N.A., Goh, P.S., Yogarathinam, L.T., Zulhairun, A.K.,& Ismail, A.F. (2020).Current advances in membrane technologies for produced water desalination. *Desalination, 493*, 114643.https://doi.org/10.1016/j.desal.2020.114643

12. Duan, M., Ma,Y., Fang, S., Shi, P., Zhang, J., & Jing, B. (2014). Treatment of wastewater produced from polymer flooding using polyoxyalkylated polyethyleneimine. *Separation and Purification Technology,133*, 160–167. https://doi.org/10.1016/j.seppur.2014.06.058

13. Ricceri, F., Farinelli, G., Giagnorio, M., Zamboi, A., & Tiraferri, A. (2022). Optimization of physico-chemical and membrane filtration processes to remove high molecular weight polymers from produced water in enhanced oil recovery operations. *Journal of Environmental Management,302*, 114015.https://doi.org/10.1016/j.jenvman.2021.114015

14. Yongrui, P., Zheng, Z., Bao, M., Li, Y., Zhou, Y., & Sang, G. (2015). Treatment of partially hydrolyzed polyacrylamide wastewater by combined Fenton oxidation and anaerobic biological processes. *Chemical Engineering Journal,273*, 1–6. https://doi.org/10.1016/j. cej.2015.01.034

15. Morrow, N., & Buckley, J. (2011). Improved oil recovery by low-salinity waterflooding. *Journal of Petroleum Technology,63*(05), 106–112. https://doi.org/10.2118/129421-JPT

16. Tang, G. Q., & Morrow, N. R. (1997). Salinity, temperature, oil composition, and oil recovery by waterflooding. *SPE Reservoir Engineering,12*(04), 269–276. https://doi.org/10.2118/36680-PA

17. Yongrui, P., Zheng, Z., Bao, M., Li, Y., Zhou, Y., & Sang, G. (2015). Treatment of partially hydrolyzed polyacrylamide wastewater by combined Fenton oxidation and anaerobic biological processes. *Chemical Engineering Journal,273*, 1–6. https://doi.org/10.1016/j. cej.2015.01.034

18. Gladkov, A., Sakhibgareev, R., Salimov, D., Skiba, A., & Drozdov, O. (2017,October). Application of CRM for production and remaining oil reserves reservoir allocation in mature west Siberian waterflood field. In *SPE Russian Petroleum Technology Conference*. OnePetro. https://doi.org/10.2118/187841-MS

19. Zhao, Y., Jiang, H., Li, J., Wang, C., Gao, Y., Yu, F., & Su, H. (2017,November). Study on the classification and formation mechanism of microscopic remaining oil in high water cut stage based on machine learning. In *Abu Dhabi International Petroleum Exhibition & Conference*. OnePetro. https://doi.org/10.2118/188228-MS

20. Roueché, J. N., & Karacan, C. Ö. (2018, April). Zone identification and oil saturation prediction in a waterflooded field: Residual oil zone, East Seminole Field, Texas, USA, Permian Basin. In *SPE Improved Oil Recovery Conference*. OnePetro. https://doi.org/10.2118/190170-MS

21. Lu, X. G., &Xu, J. (2017, October). Waterflooding optimization: A pragmatic and cost-effective approach to improve oil recovery from mature fields. In *SPE/IATMI Asia Pacific Oil & Gas Conference and Exhibition*. OnePetro. https://doi.org/10.2118/186431-MS

22. Adham, S., Hussain, A., Minier-Matar, J., Janson, A., &Sharma, R. (2018). Membrane applications and opportunities for water management in the oil & gas industry. *Desalination, 440*, 2–17. https://doi.org/10.1016/j.desal.2018.01.030

23. Dong, X., Liu, H., Chen, Z., Wu, K., Lu, N., & Zhang, Q. (2019). Enhanced oil recovery techniques for heavy oil and oilsands reservoirs after steam injection. *Applied Energy, 239*, 1190–1211. https://doi.org/10.1016/j.apenergy.2019.01.244

24. Nair, R. R., Protasova, E., Strand, S., & Bilstad, T. (2018). Membrane performance analysis for smart water production for enhanced oil recovery in carbonate and sandstone reservoirs. *Energy &Fuels,32*(4), 4988–4995. https://doi.org/10.1021/acs.energyfuels.8b00447

25. Kantzas, A., Chatzis, I., & Dullien, F. A. L. (1988,April). Enhanced oil recovery by inert gas injection. In *SPE Enhanced Oil Recovery Symposium*, OnePetro. https://doi.org/10.2118/17379-MS

26. Asad, A., Sameoto, D., & Sadrzadeh, M. (2020). Overview of membrane technology. In *Nanocomposite Membranes for Water and Gas Separation* (pp. 1–28). Amsterdam: Elsevier. https://doi.org/10.1016/B978-0-12-816710-6.00001-8

27. Li, X., Mo, Y., Qing, W., Shao, S., Tang, C. Y., & Li, J. (2019). Membrane-based technologies for lithium recovery from water lithium resources: A review. *Journal of Membrane Science,591*, 117317.https://doi.org/10.1016/j.memsci.2019.117317

28. Munirasu, S., Haija, M. A., & Banat, F. (2016). Use of membrane technology for oil field and refinery produced water treatment—A review. *Process Safety and Environmental Protection,100*, 183–202. https://doi.org/10.1016/j.psep.2016.01.010

29. Alzahrani, S., & Mohammad, A. W. (2014). Challenges and trends in membrane technology implementation for produced water treatment: A review. *Journal of Water Process Engineering, 4*, 107–133. https://doi.org/10.1016/j.jwpe.2014.09.007

30. Evans, F. L. (1979). *Equipment Design Handbook for Refineries and Chemical Plants* (Vol. 1). Houston: Gulf Publishing Company. https://openlibrary.org/books/OL4426468M

31. Kose, B., Ozgun, H.,Ersahin, M. E., Dizge, N., Koseoglu-Imer, D. Y., Atay, B., …& Koyuncu, I. (2012). Performance evaluation of a submerged membrane bioreactor for the treatment of brackish oil and natural gas field produced water. *Desalination,285*, 295–300.https://doi.org/10.1016/j.desal.2011.10.016

32. Gondal, M. A., Sadullah, M. S., Dastageer, M. A., McKinley, G. H., Panchanathan, D., & Varanasi, K. K. (2014). Study of factors governing oil–water separation process using TiO2 films prepared by spray deposition of nanoparticle dispersions. *ACS Applied Materials &Interfaces, 6*(16), 13422–13429. https://doi.org/10.1021/am501867b

33. Abadi, S. R. H.,Sebzari, M. R., Hemati, M., Rekabdar, F., & Mohammadi, T. (2011). Ceramic membrane performance in microfiltration of oily wastewater. *Desalination,265*(1–3), 222–228. https://doi.org/10.1016/j.desal.2010.07.055

34. Zhu, X., Tu, W., Wee, K. H., & Bai, R. (2014). Effective and low fouling oil/water separation by a novel hollow fiber membrane with both hydrophilic and oleophobic surface properties. *Journal of Membrane Science,466*, 36–44. https://doi.org/10.1016/j.memsci.2014.04.038

35. Pancharoen, U., Poonkum, W., & Lothongkum, A. W. (2009). Treatment of arsenic ions from produced water through hollow fiber supported liquid membrane. *Journal of Alloys and Compounds,482*(1–2), 328–334. https://doi.org/10.1016/j.jallcom.2009.04.006

36. Ebrahimi, M., Schmitz, O., Kerker, S., Liebermann, F., & Czermak, P. (2013). Dynamic cross-flow filtration of oilfield produced water by rotating ceramic filter discs. *Desalination and Water Treatment,51*(7–9), 1762–1768. https://doi.org/10.1080/19443994.2012.694197

37. Kim, E. S., Liu, Y., & El-Din, M. G. (2011). The effects of pretreatment on nanofiltration and reverse osmosis membrane filtration for desalination of oil sands process-affected water. *Separation and Purification Technology,81*(3), 418–428. https://doi.org/10.1016/j.seppur.2011.08.016

38. Alpatova, A., Kim, E. S., Dong, S., Sun, N., Chelme-Ayala, P., & El-Din, M. G. (2014). Treatment of oil sands process-affected water with ceramic ultrafiltration membrane: Effects of operating conditions on membrane performance. *Separation and Purification Technology,122*, 170–182. https://doi.org/10.1016/j.seppur.2013.11.005

39. Miller, D. J., Huang, X., Li, H., Kasemset, S., Lee, A., Agnihotri, D., ...& Freeman, B. D. (2013). Fouling-resistant membranes for the treatment of flowback water from hydraulic shale fracturing: A pilot study. *Journal of Membrane Science, 437,* 265–275.https://doi.org/10.1016/j.memsci.2013.03.019

40. Goldblatt, M. E., Gucciardi, J. M., Huban, C. M., Vasconcellos, S. R., & Liao, W. P. (2014). New polyelectrolyte emulsion breaker improves oily wastewater cleanup at lower usage rates. https://docplayer.net/1846395

41. Wang, Y. S., Hsieh, S. H., Lee, C. H., & Horng, J. J. (2013). Adsorption of complex pollutants from aqueous solutions by nanocomposite materials. *CLEAN–Soil, Air, Water, 41*(6), 574–580. https://doi.org/10.1002/clen.201200093

42. Omer, N. H. (2020). *Water Quality Parameters, Water Quality-Science, Assessments and Policy.* http://dx.doi.org/10.5772/intechopen.89657

43. Gherghel, A., Teodosiu, C., & De Gisi, S. (2019). A review on wastewater sludge valorisation and its challenges in the context of circular economy. *Journal of Cleaner Production, 228,* 244–263. https://doi.org/10.1016/j.jclepro.2019.04.240

44. Kirchherr, J., Reike, D., & Hekkert, M. (2017). Conceptualizing the circular economy: An analysis of 114 definitions. *Resources, Conservation and Recycling, 127,* 221–232.https://doi.org/10.1016/j.resconrec.2017.09.005

45. Bassandeh, M., Antony, A., Le-Clech, P., Richardson, D., & Leslie, G. (2013). Evaluation of ion exchange resins for the removal of dissolved organic matter from biologically treated paper mill effluent. *Chemosphere,90*(4), 1461–1469. https://doi.org/10.1016/j.chemosphere.2012.09.007

46. Puig, S., Baeza, J. A., Colprim, J., Cotterill, S., Guisasola, A., He, Z., ...& Pous, N. (2017). *Niches for Bioelectrochemical Systems in Sewage Treatment Plants*(pp. 96–107). London: IWA Publishing. www.researchgate.net/publication/316472558

47. Zhidong, L., Na, L., Honglin, Z., & Dan, L. (2009). Study of an A/O submerged membrane bioreactor for oil refinery wastewater treatment. *Petroleum Science and Technology,27*(12), 1274–1285. https://doi.org/10.1080/10916460802455228

48. Fan, L., Thompson, J. W., & Robinson, J. R. (2010,October). Understanding gas production mechanism and effectiveness of well stimulation in the Haynesville Shale through reservoir simulation. In *Canadian Unconventional Resources and International Petroleum Conference.* OnePetro.https://doi.org/10.2118/136696-MS

49. Zhang, C., Zhang, W., Huang, Y., &Gao, X. (2017). Analysing the correlations of long-term seasonal water quality parameters, suspended solids and total dissolved solids in a shallow reservoir with meteorological factors. *Environmental Science and Pollution Research, 24,* 6746–6756. https://doi.org/10.1007/s11356-017-8402-1

50. Cañedo-Argüelles, M., Kefford, B. J., Piscart, C., Prat, N., Schäfer, R. B., &Schulz, C. J. (2013). Salinisation of rivers: An urgent ecological issue. *Environmental Pollution, 173,* 157–167. https://doi.org/10.1016/j.envpol.2012.10.011

51. Iowa, D. N. R. (2009). Water quality standards review: Chloride, sulfate and total dissolved solids. *Iowa Department of Natural Resources Consultation Package*.www. iowadnr.gov/portals/idnr/uploads/water/standards/ws_review.pdf

52. Scanlon, B. R., Reedy, R. C., Xu, P., Engle, M., Nicot, J. P., Yoxtheimer, D., ...& Ikonnikova, S. (2020). Can we beneficially reuse produced water from oil and gas extraction in the US?*Science of The Total Environment, 717,* 137085.https://doi.org/ 10.1016/j.scitotenv.2020.137085

53. Vikström, H., Davidsson, S., & Höök, M. (2013). Lithium availability and future production outlooks. *Applied Energy, 110,* 252–266. https://doi.org/10.1016/j.apene rgy.2013.04.005

54. Kumar, A., Fukuda, H., Hatton, T. A., & Lienhard, J. H. (2019). Lithium recovery from oil and gas produced water: A need for a growing energy industry. *ACS Energy Letters, 4*(6), 1471–1474. https://doi.org/10.1021/acsenergylett.9b00779

55. Sarode, S., Upadhyay, P., Khosa, M. A., Mak, T., Shakir, A., Song, S.,& Ullah, A. (2019). Overview of wastewater treatment methods 6 wastewater treatment and reuse best practices in Morocco: Targeting circular economy 123 with special focus on biopolymer chitin–chitosan. *International Journal Biological Macromolecules, 121,* 1086–1100. https://doi.org/10.1016/j.ijbiomac.2018.10.089

56. Baudino, L., Santos, C., Pirri, C. F., La Mantia, F., & Lamberti, A. (2022). Recent advances in the lithium recovery from water resources: From passive to electrochemical methods. *Advanced Science, 9*(27), 2201380.https://doi.org/10.1002/advs.202201380

57. Seip, A., Safari, S., Pickup, D. M., Chadwick, A. V., Ramos, S., Velasco, C. A., ...& Alessi, D. S. (2021). Lithium recovery from hydraulic fracturing flowback and produced water using a selective ion exchange sorbent. *Chemical Engineering Journal,426,* 130713. https://doi.org/10.1016/j.cej.2021.130713

58. Stringfellow, W. T., & Dobson, P. F. (2021). Technology for the recovery of lithium from geothermal brines. *Energies, 14*(20), 6805. https://doi.org/10.3390/en1 4206805

59. Wen, X., Ma, P., Zhu, C., He, Q., & Deng, X. (2006). Preliminary study on recovering lithium chloride from lithium-containing waters by nanofiltration. *Separation and Purification Technology,49*(3), 230–236. https://doi.org/10.1016/ j.seppur.2005.10.004

60. Li, Y., Zhao, Y., & Wang, M. (2017). Effects of pH and salinity on the separation of magnesium and lithium from brine by nanofiltration. *Desalination and Water Treatment, 97,* 141–150. https://doi.org/10.5004/dwt.2017.21606

61. Zhang, Y., Paepen, S., Pinoy, L., Meesschaert, B., & Van der Bruggen, B. (2012). Selectrodialysis: Fractionation of divalent ions from monovalent ions in a novel electrodialysis stack. *Separation and Purification Technology,* 88, 191–201. https://doi. org/10.1016/j.seppur.2011.12.017

62. Bi, Q., & Xu, S. (2018). Separation of magnesium and lithium from brine with high Mg^{2+}/Li^+ ratio by a two-stage nanofiltration process. *Desalination and Water Treatment, 129,* 94–100. https://doi: 10.5004/dwt.2018.23062

63. Guo, Y., Ying, Y., Mao, Y., Peng, X., & Chen, B. (2016). Polystyrene sulfonate threaded through a metal–organic framework membrane for fast and selective lithium-ion separation. *Angewandte Chemie, 128*(48), 15344–15348. https://doi.org/10.1002/ ange.201607329

64. Sharma, A. D., Patil, N. D., Patwardhan, A. W., Moorthy, R. K., & Ghosh, P. K. (2016). Synergistic interplay between D2EHPA and TBP towards the extraction of lithium using hollow fiber supported liquid membrane. *Separation Science and Technology, 51*(13), 2242–2254. https://doi.org/10.1080/01496395.2016.1202280

65. Nie, X. Y., Sun, S. Y., Sun, Z., Song, X., & Yu, J. G. (2017). Ion-fractionation of lithium ions from magnesium ions by electrodialysis using monovalent selective ion-exchange membranes. *Desalination*, *403*, 128–135. https://doi.org/10.1016/j.desal.2016.05.010

66. Sridhar, S., Smitha, B., & Aminabhavi, T. M. (2007). Separation of carbon dioxide from natural gas mixtures through polymeric membranes: A review. *Separation & Purification Reviews*, *36*(2), 113–174. https://doi.org/10.1080/15422110601165967

67. Sridhar, S., & Khan, A. A. (1999). Simulation studies for the separation of propylene and propane by ethylcellulose membrane. *Journal of Membrane Science*, *159*(1–2), 209–219. https://doi.org/10.1016/S0376-7388(99)00061-7

14 Computational Work on Designing of Membrane Materials Useful for Petroleum Industry

Anwesh Pandey and Swapnali Hazarika

14.1 INTRODUCTION

The application of membranes in the petroleum industry has been emphasised over the past decade instead of the low energy consumption and ease of implications. The petroleum industry often needs technological advancements for the treatment of natural gases, carbon dioxide capture, purification of hydrogen and separation of hydrocarbons [1–4]. For such dynamic processes, membranes have proven their potential implications and have eliminated the need for conventional approaches, such as scrubbing, swing adsorption, distillations, etc. [2]. Membranes in the petrochemical industry are vastly implemented for the separation of hydrogen sulphides, capturing carbon dioxide, recovery of hydrogen, separation of air, dehydration of various gases, removal of different volatile organic compounds (VOCs) and in the recovery of LPG [5–7]. Various types of polymeric membranes are used for the separation phenomena, such as H_2S separation is carried out through cellulose acetate membrane, for CO_2 capturing a polyamide membrane is implemented. Cellulose membranes may also be vastly used for a variety of separation phenomena, such as removal of acid gas, recovery of hydrogen and air separation [8,9]. However, the presence of water (moisture) adversely affects the membrane performance and may lead to spills and damages; also, membranes need to be tested for their stability for long-term operations [10–13]. Nanoscience and nanotechnology have played a pivotal role in the development of membranes for applications in petroleum and petrochemical industries [14–17]. For example, nanocomposite-based membranes are developed by incorporating nanoparticles within the membrane matrix, which improves the membrane performance by enhancing its transport properties [18–20]. In addition, membranes that perform nanofiltration are developed with small pore size, which leads to effective separation of the impurities from the petrochemical products [21–23]. Modifying the chemical properties of the surface is crucial for creating effective

 DOI: 10.1201/9781003441359-14

membranes. Using this approach makes it possible to control the interactions at the interface accurately, which is essential for developing high-quality coatings and membranes that can resist fouling and improve separation [24]. Among the available nanomaterials for membrane fabrication, carbon-based nanomaterials have gained significant importance [25–30], as they may be easily tuned for desired applications, such as controlled permeability, enhanced thermochemical stability and increased prevention of fouling properties. They may further be manipulated by using advanced inorganic nanomaterials having tuneable pore sizes, high surface areas, unique surface chemistries, etc. [31–33]. Carbon-based nanomaterials are mainly separated into three categories, viz., graphene and graphene oxide (GO) [34,35], carbon nanotubes (CNT) [36] and carbon nanofibers (CNF) [37,38]. Several successful experimental strategies have been devised and implemented for the development of such membrane materials owing to low costs, increased durability and stability with minimal damages [20,23,34]. Computational strategies have also contributed adequately to the development and implementation of membranes and processes for petroleum and petrochemical applications [39–41]. Computational simulations in the petrochemical industry span the entire molecular, meso and continuum scales [42]. Artificial intelligence and machine learning techniques have also paved the way for themselves in the application domains of petroleum and petrochemical industries [43]. Theoretical and computational simulation protocols have addressed several crucial aspects regarding the structure–property–processing relationships for graphene or graphene-based nanocomposite materials [44,45]. Various simulation techniques and methods have been utilised across multiple levels, spanning from the quantum scale to the atomic, mesoscopic and finally to the macroscopic level [46]. Findings obtained from these theoretical and computational tools and techniques have contributed significantly to understanding the underlying interactions and mechanisms governing them. They have also enhanced our understanding of the macroscopically exhibited properties, viz., pressure, temperature, density, volume, viscosity, surface tension, etc., for applications in the petroleum industry [47]. Over the last few years, molecular modelling methods have been increasingly used to examine the properties of reservoir fluids and rocks, as well as the interactions between them and other phenomena, all at the atomic level [48]. These simulations involve studying the interfacial properties of oil and water systems containing surfactants, polymers, foams and nanoparticles. This field includes developing theoretical approaches for modelling simulation systems, analysing molecular arrangements at oil–water interfaces and adjusting research conditions, such as electronic and thermodynamic parameters [49,50]. Many tools and techniques have been implemented for molecular modelling studies for the membrane modelling. Molecular dynamics, in particular, has been advantageous for finding implications in science and engineering [51]. This is because it spans the entire length and time scales, i.e., ranging from relatively large simulations on the molecular scales (all-atom simulations) beginning from simple gases and liquids to complex materials (coarse-grained simulations). Several widely used programmes, such as LAMMPS [52], AMBER [53], DL-POLY [54], CHARMM [55], GROMACS [56] and NAMD [57], have been frequently used to model systems involving membranes. Monte Carlo (MC) simulations are a frequently utilised method that employs chance atomic movements that are accepted based on probability calculation from the ratio of a random number and the Boltzmann factor (which expresses the likelihood of a new configuration). This approach is utilised to solve problems and discover novel configurations [58]. Herein, this chapter discusses the implications of various computational modelling approaches in the design and development of membrane materials, which find applications in industries.

14.2 TYPE OF MEMBRANES USEFUL FOR PETROLEUM INDUSTRY

Membranes in the petroleum industry are useful for the treatment of various gases, and separation of hydrocarbons, waste gases, removal of VOCs, water treatment, etc., uses of which can replace old and conventional approaches, viz., scrubbing, swing adsorption, distillations, etc. [59,60]. In the petroleum industry, membranes find vast implications for the separation of H_2S, capturing CO_2, recovery of H_2, separation of air, dehydration of various gases, removal of VOCs and the recovery of liquified petroleum gases [61,62]. Various types of membranes are used in the petroleum industry, such as polymeric membranes, ceramic membranes, mixed matrix membranes (MMMs), etc. [63].

14.2.1 POLYMERIC MEMBRANES

These membranes are vastly used in the petroleum and petrochemical industry for gas separations, oil–water separations and desalination applications. The selection of the appropriate polymeric membrane material depends on the nature of the feed stream, operating conditions and separation efficiency. Some common polymeric membrane materials used are polyamides, polyethylenes, polypropylenes, polyvinylidene fluorides, cellulose acetates, polyimides, etc. [64,65].

14.2.2 CERAMIC MEMBRANES

Such membranes are widely used in the petroleum and petrochemical industry for high-temperature and harsh chemical applications. The selection of the appropriate ceramic membrane material also depends on the nature of the feed stream, operating conditions and separation efficiency. Ceramic membranes are more expensive than polymeric membranes but offer superior performance and durability in harsh environments. However, they are not flexible in nature. Some common ceramic membrane materials used in the industry include alumina (Al_2O_3), zirconia (ZrO_2), silicon carbide (SiC), titania (TiO_2), zeolites, etc. [66,67].

14.2.3 MIXED MATRIX MEMBRANES

MMMs are composite membranes that combine two or more types of materials: a polymeric matrix and inorganic fillers. These membranes are used in the petroleum and petrochemical industry for various gas separation applications, such as natural gas purification, carbon capture and hydrogen purification. MMMs offer advantages over traditional polymeric, ceramic and metal membranes, such as improved selectivity and permeability, enhanced stability and durability and reduced cost. The selection of the appropriate mixed matrix membrane material depends on the nature of the gas mixture, the desired separation efficiency and the operating conditions [68]. Some common mixed matrix membrane materials used in the industry include polymer–ceramic, polymer–metal, polymer–zeolite, etc. [69–71].

14.3 NOVEL MATERIALS FOR DESIGNING OF MEMBRANES USEFUL IN PETROLEUM INDUSTRY

Two-dimensional materials have gained popularity as membrane materials due to their specific characteristics. MXenes, graphenes, transition metal dichalcogenides, layered double hydroxides (LDHs), metal–organic framework (MOF)-based nanosheets, covalent organic framework (COF) based nanosheets and carbon nitrides are among the two-dimensional materials investigated for their potential across various fields [72–74]. Due to their high aspect ratios, these materials are assembled into ultrathin membranes with high permeation flux. Graphene-based membranes have gained significant attention due to their high thermal and mechanical stability, along with functional groups that allow for further membrane structure tuning. Consequently, graphene has been used as a filler material in mixed-matrix membrane fabrications or incorporated onto porous membrane supports to create thin-film composite membranes. Carbon-based materials such as graphene, CNTs and fullerene possess unique physical and chemical characteristics such as high mechanical strength, exceptional electrical and thermal conductivity, excellent gas barrier properties and high optical transparency. Due to these properties, carbon materials are used as nanofillers in composite materials across several fields like biology, energy storage, transport and aviation, optoelectronics, pharmaceutics, medicine and others [75,76]. For instance, fullerenes are utilised as electron acceptors in the creation of organic solar cells. At the same time, graphene's electrical and optical properties are used to develop biosensors that can efficiently detect specific target molecules [77]. Additionally, CNTs have enabled the production of high-performance supercapacitors with increased energy storage and power delivery capabilities and a relatively long life cycle as compared to traditional batteries [78]. We now discuss some of the important two-dimensional nanomaterials which are extensively used as membrane materials.

14.3.1 CARBON NANOSTRUCTURES

CNSs include graphenes, graphene oxides, reduced graphene oxides, fullerenes, CNTs, etc. (Figure 14.1(a)) and have a number of properties that make them attractive for membrane applications. These properties are high aspect ratio, mechanical strength, chemical and thermal stability, etc. In addition to these properties, CNSs can also be functionalised with various molecules to enhance their properties or add new functionality. For example, functionalising CNSs with biological molecules can make them suitable for biomedical applications such as drug delivery. Additionally, the cost of producing CNSs is relatively high, which limits their widespread use. CNSs are used as either free-standing membranes or as membranes integrated into a support structure. However, there are still some challenges associated with the use of CNSs as membrane materials. For example, CNSs tend to agglomerate, which can affect their performance in some applications. The unique properties of carbon nanostructures make them a promising material for a variety of membrane applications. However, more research is needed to fully understand their behaviour and optimise their performance for desired applications [79–81].

14.3.2 MXENES

MXenes are a relatively advanced class of two-dimensional nanomaterials. They are also extensively employed as carbon nanomaterials for a diverse range of membrane applications. MXenes have a number of properties that make them attractive for membrane applications. These properties are high surface area, tuneable pore size, high hydrophilicity, high selectivity, chemical stability, electrical conductivity, etc. MXenes tend to form a layered structure that can stack on top of each other, which can affect the performance of the membrane. However, researchers are exploring various methods to prevent or mitigate agglomeration, such as functionalising MXenes with polymers or introducing intercalants between the MXene layers. However, challenges such as the cost of production and scalability of MXene membranes still need to be addressed to enable their widespread use in industrial applications [82–84].

14.3.3 GO/rGO

Graphene oxide (GO) and reduced graphene oxide (rGO) have shown great potential for designing novel membranes with improved properties. GO is a two-dimensional nanomaterial that is produced by the oxidation of graphene and it contains functional groups such as hydroxyl, carboxyl and epoxy groups (Figure 14.1(a)). These functional groups make GO hydrophilic and enable it to form stable dispersions in water and other polar solvents. GO membranes are prepared by depositing GO sheets onto a porous support material, followed by chemical reduction to form rGO. rGO has a reduced oxygen content and a higher electrical conductivity compared to GO, making it more suitable for applications where electrical conductivity is required. The unique properties of GO and rGO make them ideal candidates for designing membranes with improved permeability, selectivity and durability. GO membranes have been used for water desalination and purification, where they showed high water flux and salt rejection. rGO membranes have also been used for gas separation, where they exhibited high selectivity for carbon dioxide and methane. Additionally, the ability to tune the properties of GO and rGO by controlling their synthesis and modification provides a means to tailor the performance of these membranes for specific applications. However, challenges such as the scalability of production, stability of the membranes under harsh conditions and cost-effectiveness of the synthesis methods still need to be addressed to enable their commercial use. Here are some examples of the usage of GO and rGO as membrane materials, such as water filtration, gas separation, energy storage, separation of organic compounds, desalination, gas sensing, electrocatalysis, antifouling, anticorrosion coatings, etc. [85–88].

14.3.4 IONIC LIQUIDS

Polymeric ionic liquids (PIL) based membranes are a class of materials that combine the properties of traditional ionic liquids with those of polymers (Figure 14.1(b)). PILs have gained attention in the petroleum and petrochemical industry due to their unique properties, such as high thermal stability, chemical resistance and tuneable properties. PILs are used as membrane materials for various gas and liquid separation applications, such as natural gas purification, CO_2 capture and desalination. Some advantages of PIL-based membranes include high selectivity, low energy consumption and low fouling propensity. The selection of the appropriate PIL-based membrane material depends on the nature of the feed stream, the desired separation efficiency and the operating conditions. PIL-based membranes are a relatively new technology and require further development to optimise their performance and cost-effectiveness [89–93]. PIL-based membranes have been used in various petroleum and petrochemical industry applications, including: CO_2 capture: PIL-based membranes can selectively capture CO_2 from flue gas or natural gas streams, making them useful for carbon capture and storage. In hydrogen purification, PIL-based membranes can selectively separate hydrogen from gas mixtures, such as natural gas, refinery gas or synthesis gas. In desalination, PIL-based membranes have the ability to perform desalination on seawater or brackish water, thus enabling the creation of clean, fresh water that is utilised in both domestic and industrial settings.

14.3.5 BIOPOLYMERS

Biopolymers have gained significant interest as a potential membrane material due to their biodegradability, biocompatibility, hydrophilicity, suitable for surface modification and sustainability [94–102] (Figure 14.1(c)). Several types of biomaterials are used as biopolymer-based membrane materials. Cellulose, chitosan and alginate are examples of biopolymers found in various resources. Due to their high tensile strength, chemical stability and biocompatibility, they have been used to produce membranes for water filtration and separation applications. They have been used as membrane material for various applications, including water filtration, wound healing and drug delivery. Polysaccharides are also used in diverse applications such as drug delivery, wound healing and tissue engineering.

14.3.6 LAYERED DOUBLE HYDROXIDES

LDHs have gained interest as potential membrane materials due to their unique properties. LDH membranes are typically prepared by intercalating organic or inorganic species into the interlayer space of the LDH structure. One common method is to use anion exchange to replace the interlayer anion with a desired species. For example, LDH membranes with interlayer carbonate ions are prepared by exchanging the original interlayer anion with carbonate ions. After the intercalation step, the LDH material is typically processed into a membrane form by various techniques such as filtration, spin coating, or dip coating. The resulting LDH membrane can be supported or unsupported, depending on the application. Supported LDH membranes are typically deposited on a porous support material, such as alumina or polymeric membranes, to enhance their mechanical stability. The performance of LDH membranes are evaluated by measuring their selectivity and permeability for specific applications. For example, the gas permeability and selectivity of LDH membranes are measured using a permeation cell system with a pressure difference across the membrane. The water permeability and rejection of LDH membranes are evaluated using a dead-end filtration setup with a feed solution containing a specific concentration of contaminants [103–108]. LDHs are used as membrane material for gas separation, water treatment, catalysis and drug delivery applications. The interlayer space of the LDH is used to incorporate drugs, and the LDH membrane is used to control the release of the drugs.

14.3.7 METAL ORGANIC FRAMEWORKS (MOFs)

MOFs are porous materials used as potential membrane materials due to their high porosity, tuneable pore size and chemical versatility (Figure 14.1(d)). MOFs are fabricated of metal ions or clusters linked via organic ligands, forming a three-dimensional porous structure. MOFs are used as membrane material for different applications like gas separation, water filtration, catalysis, drug delivery, sensing, etc. [109–113]. The MOFs tuneable pore size and high surface area allow for efficient gas separation. The MOFs high porosity and chemical versatility make them ideal for designing custom-tailored membranes for specific filtration applications. However, challenges such as the stability and scalability of MOFs still need to be addressed to enable their widespread use in industrial applications.

14.3.8 COVALENT ORGANIC FRAMEWORKS (COFs)

COFs are porous materials that have potential applications in membrane separation technology. COFs possess a crystalline structure that is tailored at the molecular level to achieve specific properties such as pore size, surface area and chemical functionality (Figure 14.1(d)). In the case of membrane separation technology, COFs offer several advantages over conventional materials. For instance, their high porosity and tuneable pore size distribution allow for the precise separation of molecules based on size and shape. Moreover, their chemical stability and robustness make them suitable for harsh environments and prolonged use. COFs are used in membrane separation processes in two ways: as membrane materials themselves or as fillers in composite membranes. In the former approach, COFs are directly synthesised on a support material, such as a porous polymer or ceramic, to form a thin film that acts as the membrane. In the latter approach, COFs are incorporated into a polymer matrix to enhance its separation properties. In both cases, COFs are designed to have specific chemical functionalities that enable the selective separation of molecules. COFs are functionalised with amine or carboxylic acid groups to allow for the separation of acidic or basic molecules. Additionally, COFs are functionalised with hydrophobic or hydrophilic groups to enhance their selectivity for polar or non-polar molecules. Some important properties of COFs are pore size and distribution, selectivity and stability. COFs have been investigated for a wide range of membrane separation

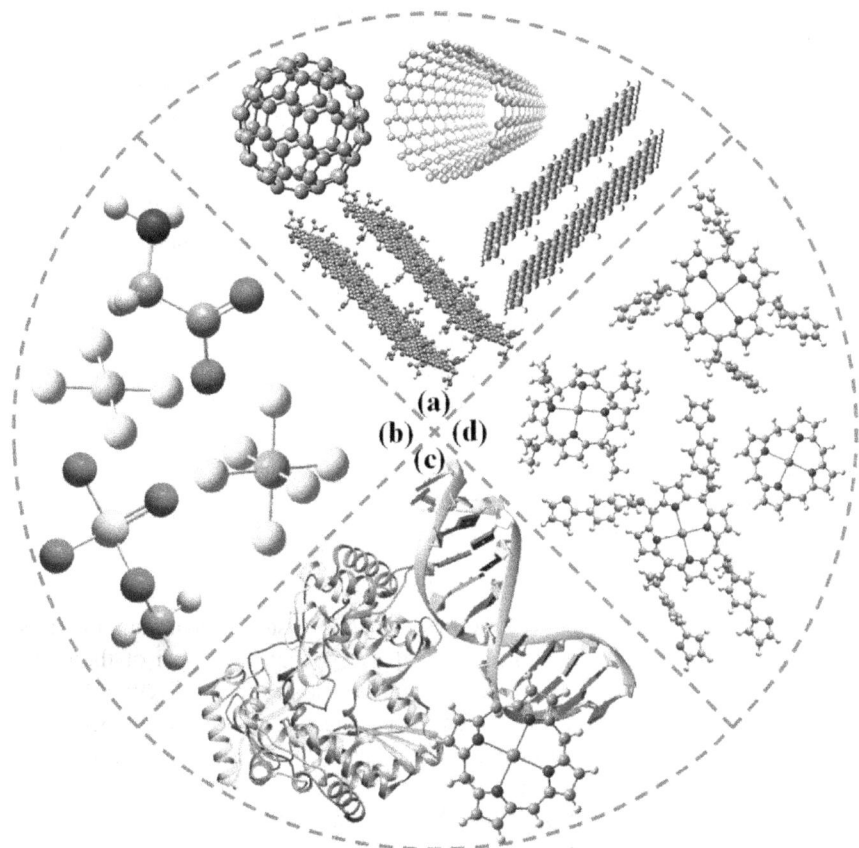

FIGURE 14.1   Novel Materials for designing of membranes useful in petroleum industry: (a) carbon nanostructures; (b) ionic liquids; (c) biopolymers and (d) MOFs and COFs.

applications, including gas separation, water desalination and organic solvent filtration [114–117].

14.4   DESIGNING A MEMBRANE: ROLE OF COMPUTATIONAL MODELLING

Computational tools and techniques play an important role in modelling membranes for the petroleum industry. These tools and techniques allow for the prediction of the behaviour of polymeric membranes, including their performance, selectivity and durability, without the need for expensive and time-consuming experimental trials. MD simulations are widely used in the petroleum and petrochemical industry to study the structure and behaviour of polymeric membranes at the molecular level. MD simulations can provide insights into the interactions between the membrane and the feed, which can help in designing membranes with better selectivity and permeability [118,119]. Computational fluid dynamics (CFD) simulations predict the flow of fluids through polymeric membranes. CFD simulations can help optimise the design of membrane modules, which can improve performance and reduce the energy consumption of membrane processes [120,121]. Monte Carlo Simulations (MCS) are used to study the thermodynamics of polymeric membranes. MCS can provide insights into the phase behaviour of polymer solutions, which can help in designing membranes with better stability and durability [122]. Density functional theory (DFT) is implemented to study the electronic structure of materials. DFT simulations can provide insights into the chemical properties of polymeric membranes, which can help in designing membranes with better selectivity and durability [123]. Certain optimisation algorithms are used to search for the optimal membrane design given certain constraints and objectives. This can save significant time and resources and lead towards the inclusion of AI/ML techniques in the conceptualisation and development of new membranes and their applications in petroleum industries. For example, neural networks are used to predict membrane selectivity and permeability based on the properties of the polymer and the operating conditions [124–126]. We now discuss these techniques in detail.

14.4.1 Molecular Dynamics Simulations

A typical MD simulation involves the following steps: system preparation, minimisation, equilibration and production [127–129]. Let us consider each of these steps more in-depth as follows: system preparation: the first step in a molecular dynamics simulation involves preparing the system to be analysed. The starting structure is obtained through experimental or computational means or a combination of both, with the goal being a structure that closely resembles the equilibrium configuration. Once the initial configuration has been obtained, additional steps are taken to refine the molecular system. System preparation involves setting up a simulation box, solvating the system and neutralising the system. To begin a simulation, we create a virtual box that contains our molecule and implement periodic boundary conditions (PBC) to simulate its bulk properties. PBC involves generating replicas of our box in all directions, which enables us to simulate a larger system starting from a smaller box. In the solvation step we need to solvate our system using some solvent of our choice or depending upon the choice of the problem. In the last stage, it is necessary to neutralise the charge of the system. This is achieved by introducing suitable counterions (such as Na^+ or Cl^-), which will replace solvent molecules and balance out the charges; minimisation: the purpose of this step is to decrease the system's energy by adjusting the system's coordinates to prevent any atom clashes. Each minimisation step will result in a new set of coordinates representing a lower potential energy state of the system. The steepest descent method is a common algorithm used for energy minimisation in MD simulations. In this method, the gradient of the potential energy with respect to the atomic coordinates is calculated, and the coordinates are updated by moving in the direction opposite to the gradient. This process is repeated until the potential energy of the system reaches a minimum. During the minimisation process, various stopping criteria are used to determine when the system has reached a minimum energy state. For example, the minimisation is stopped when the energy has converged to a certain level or when the maximum force acting on any atom in the system falls below a certain threshold. The use of energy minimisation depends on the specific needs and goals of the simulation, as well as the characteristics of the system being studied. The initial velocities (v_i) are selected from a Maxwell–Boltzmann distribution at the temperature (T).

$$P(v_i) = \sqrt{\frac{m_i}{2\pi k_b T}} \exp\left(-\frac{m_i v_i^2}{2k_b T}\right)$$
(14.1)

where $P(v_i)$ is the probability that an atom (i) with mass m_i has a velocity v_i at temperature (T). To avoid the simulation box from drifting, it is necessary to modify the velocities to ensure that the momentum of the system is zero; equilibration: after establishing the initial conditions for an MD experiment, the equilibration phase can begin. During this stage, the system is brought to a state of equilibrium by monitoring various parameters until they reach stable values, which depend on the chosen thermodynamic ensemble. Typically, the first equilibration run is performed at constant volume and temperature, followed by a simulation at constant pressure and temperature to ensure the correct density for the production run. Other parameters to consider include kinetic and potential energy, total energy and pressure. The RMSD is the main quantity used to determine if the system has reached equilibrium. By plotting RMSD measures as a function of time, the deviation of the system's structure from its original configuration is observed. Once the RMSD curve stabilises and fluctuates around constant values, the system is considered at equilibrium and the production phase can begin. Production: the production phase is the final step of an MD simulation and involves running the simulation in a specific ensemble, such as NPT or NVT. The NPT ensemble is generally preferred as it is more representative of reactions in a real laboratory setting where pressure is kept constant. Unlike the equilibration step, the goal of the production phase is not to equilibrate the system but rather to collect data on the molecule of interest. The outcome of the production run is the trajectory of the molecule, which can be visualised using molecular visualisation software to observe structural changes and analyse the behaviour and properties of the molecule. The length of the simulation is determined by the timescale of the molecular motions or the length of time necessary to observe the phenomenon of interest. In some cases, the length of the production phase can be several orders of magnitude longer than the equilibration phase. Additionally, during the production phase, various output files are generated that contain information on the system's energy, temperature, pressure and other relevant properties, which can be analysed to gain further insights into the behaviour of the system. Overall, the production phase is the most critical step in an MD simulation, as it produces the data that are used to validate theoretical models and make predictions about the system's behaviour.

14.4.2 COMPUTATIONAL FLUID DYNAMICS

Computational fluid dynamics (CFD) uses mathematical formulas and computing power to forecast the movement of fluids in the physical world. Despite its usefulness, evaluating aerodynamic efficiency during the initial design stage can be challenging. Nevertheless, as computer technology and computational power have advanced, CFD has become a commonly used tool for predicting real-world physics. CFD software examines the flow of fluids and their physical characteristics, such as speed, pressure, viscosity, density and temperature. To achieve precise results, it is essential to calculate these variables simultaneously while considering operational circumstances, numerical methods and physical principles. The most commonly used CFD tools are based on the Navier–Stokes (N–S) equations, but additional terms are added or removed based on the physics being considered. By conducting an accurate CFD analysis, engineers can quickly obtain performance insights, resulting in a more efficient and high-performing final product. There have been significant advancements in informatics, and CFD is widely used in virtually every sector worldwide [130–132].

14.4.3 MONTE CARLO SIMULATIONS

MCS are a powerful tool in the design and optimisation of membrane materials. The performance of these materials is often governed by complex physical and chemical interactions at the molecular level, which can be difficult to predict using traditional analytical and experimental techniques. MCS offer a way to model these interactions using statistical methods. These simulations generate many random samples to simulate the behaviour of molecules in the membrane material. The samples are then analysed to determine the properties of the material, such as permeability, selectivity and mechanical strength. One of the key advantages of MCS is their ability to incorporate a wide range of parameters, including molecular size, shape and interactions, as well as environmental factors such as temperature and pressure. This allows designers to explore various material properties and optimise their designs to meet specific performance requirements. MCS can also be used to predict the performance of membrane materials under different conditions, such as changes in temperature or pressure. This can help designers to identify potential weaknesses in the material and optimise their designs to improve performance and durability. MCS involve generating many random samples to simulate the behaviour of molecules in the membrane material. These samples are typically generated using a random number generator and are based on a probability distribution function that describes the behaviour of the molecules in the material. The simulations are performed using various software packages and algorithms, including Metropolis–Hastings and Gibbs sampling. These algorithms use a Markov chain to sample from the probability distribution function and generate the random samples. Once the samples are generated, they are analysed to determine the properties of the material. This analysis typically involves calculating statistical properties such as the mean and standard deviation of the samples. These properties are used to estimate the permeability, selectivity and mechanical strength of the material. MCS can also be used to perform sensitivity analysis on the material properties. Sensitivity analysis involves varying one or more parameters in the simulation to determine how they affect the material properties. This analysis can help designers to identify the most important parameters and optimise their designs accordingly. One of the challenges of MCS is that they are computationally expensive, particularly when simulating large molecules or complex systems. To address this challenge, researchers have developed a variety of techniques to improve the efficiency of the simulations, such as parallel computing, importance sampling and variance reduction [133–135].

14.4.4 DENSITY FUNCTIONAL THEORY

DFT is a computational tool used to understand and predict the properties of materials at the atomic and molecular levels (Figure 14.2). Recently, it has been increasingly used in the design of membrane materials. DFT can provide valuable insights into the underlying physical and chemical mechanisms that govern the properties

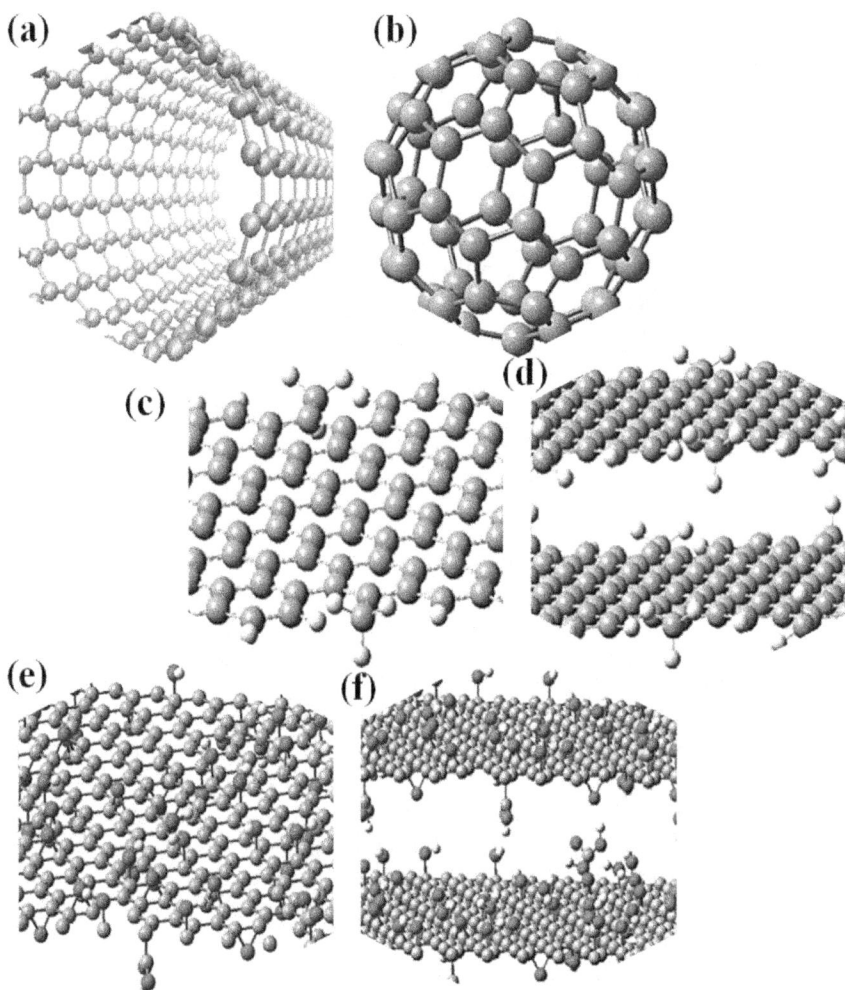

FIGURE 14.2 Several membrane materials that may be studied using DFT.

of membranes. DFT calculations can predict the adsorption energies, diffusion coefficients and transport properties of molecules through membranes. By combining DFT with molecular dynamics simulations, it is possible to simulate the behaviour of molecules in realistic membrane environments. The use of DFT in the design of membrane materials involves predicting the adsorption and diffusion of different gas and liquid molecules on the membrane surface. This is achieved by calculating the electronic structure of the membrane and its interaction with the target molecules. DFT can predict their selectivity, permeability and other transport properties by analysing the electronic properties of the membrane materials, such as their electron density and energy levels. This approach can help in identifying the promising membrane materials with the desired properties. DFT calculations can also be used to screen many potential membrane materials and predict their performance before experimental synthesis and testing. This can significantly reduce the time and cost of developing new membrane materials with desired properties. This information is crucial for the design of membrane materials that can withstand harsh operating conditions and have improved stability [136,137].

14.4.5 ARTIFICIAL INTELLIGENCE AND MACHINE LEARNING ALGORITHMS

Artificial intelligence (AI) and machine learning (ML) algorithms are increasingly being utilised in the design and development of membrane materials for the petroleum industry. AI and ML algorithms are used to predict the properties of membrane materials, such as permeability, selectivity, stability and optimise their performance for specific applications. These algorithms use large datasets of experimental and theoretical data to develop predictive models, which can be used to design new membrane materials. For example, ML algorithms are used to analyse the relationships between the chemical structure and properties of membrane materials and their performance in gas and liquid separations. By identifying the most critical structural features of the membrane materials that contribute to their performance, ML algorithms are used to design new materials with improved properties. Similarly, AI algorithms are used to simulate the molecular dynamics of membrane materials and predict their behaviour under different conditions. By integrating these simulations with experimental data, AI algorithms can provide insights into the mechanisms that govern the transport of molecules across the membrane and guide the design of new materials with improved performance [138–141]. Several models are developed using AI and ML algorithms like regression models, classification models, clustering models, neural network models, deep learning models, etc., to design membrane materials. Regression models are used to predict the relationship between the properties of the membrane materials and their performance in gas and liquid separation processes. It is developed using large datasets of experimental and theoretical data to identify the key structural features of the membrane materials that influence their performance. Classification models are used to classify membrane materials into different categories based on their performance in specific applications. It is developed to identify membrane materials that are best suited for gas separation versus liquid separation. Clustering models are used to group similar membrane materials based on their chemical and structural properties. It can help to identify materials with similar properties that may be suitable for specific applications. Neural network models are used to simulate the behaviour of the membrane materials under different conditions and predict their performance in specific applications. It is used in large datasets of experimental and theoretical data to identify the most critical structural features of the membrane materials. Deep learning models are used to simulate the molecular dynamics of the membrane materials and predict their behaviour under different conditions. It can also be used in large datasets of molecular simulations and experimental data to provide insights into the mechanisms that govern the transport of molecules across the membrane and guide the design of new materials with improved performance.

14.5   CASE STUDIES

A few case studies on the deployment of computational work on membrane chemistry and engineering have been found. A study conducted by Yang et al. [24] developed a method to control the surface of a graphene oxide membrane at a molecular level for oil–water separation. They used a hydrophobic chain engineering approach to assemble hydrophilic and hydrophobic acids in sequence on the membrane, resulting in an amphiphilic surface. By introducing perfluoroalkyl chains, they decreased the surface energy and adjusted the surface hydration by altering the hydrophobic chain length, which improved both fouling resistance and fouling-release properties. Their research demonstrated that the perfluoroalkyl chain length has a non-linear effect on the surface hydration capacity, with the highest hydration capacity at C6 due to the more uniform water orientation, as shown by molecular dynamics simulations. They also used a simple model system to examine the interfacial water on graphene oxide membranes functionalised with perfluoroalkyl chains of various lengths.

Pétuya et al. conducted a recent study using MD simulation to investigate how the aggregation of asphaltene is affected by molecular polydispersity [43]. The study found that simulations using a single asphaltene model revealed various aggregation behaviours influenced by different structural features, such as the size of the aromatic core, length of aliphatic chains and presence of heteroatoms. Using mixtures containing different asphaltene model molecules showed more complex effects of molecular polydispersity on the aggregation process. The simulations identified opposing and cooperative effects triggered by specific asphaltene model molecules. These results highlight the importance of considering molecular polydispersity when studying asphaltene aggregation and have allowed the development of a reliable method to evaluate the effectiveness of asphaltene inhibitors, which was demonstrated in the study's nonylphenol resin case.

14.6   FUTURE ASPECTS

Computational research into the design of membrane materials for the petroleum industry is a rapidly growing field and will be leading research in the near future. The research includes predictive modelling of membrane properties, high-throughput screening of materials, multiscale modelling, integration of machine learning and data analytics and development of hybrid materials. This emphasises the importance of such work for the petroleum industry, and how computational methods and data analytics are used to speed up the discovery of new membrane materials and to improve the efficiency and sustainability of petroleum industry processes. Despite this potential, numerous challenges and limitations exist, such as limited accuracy of models, limited understanding of structure–function relationships, limited availability of experimental data, complexity of the industry environment and the need for interdisciplinary collaboration. Therefore, while computational techniques have great promise for designing membrane materials, there are many obstacles to overcome. For this, ongoing innovations in computational methods and collaboration between disciplines are essential.

14.7 CONCLUSIONS

Computer simulation and modelling studies are extremely important when it comes to the design and development of membranes for the petroleum industries. These simulations can predict chemical interactions, new substances and their properties, pressure and temperature ranges. Although molecular modelling is already used in reservoir engineering and hydrocarbon processing, there is still much to learn about its potential. It is used to anticipate phase equilibria under process conditions, where tests would be impossible due to the rapid degradation of hydrocarbons. Furthermore, it may be suggested that computer simulations will soon be a major factor in applied thermodynamics, as they are more reliable than traditional models, faster and cheaper than experiments. Therefore, simulations can be used quickly, safely and affordably to explore the atomic and molecular details of complex events.

REFERENCES

1. Baker, R. (2012). *Membrane Technology and Applications*. John Wiley & Sons. https://doi.org/10.1002/0470020393
2. Ismail, A. F., & Matsuura, T. (2012). *Sustainable Membrane Technology for Energy, Water, and Environment*. John Wiley & Sons. https://doi.org/10.1002/9781118190180.ch2
3. Stewart, M., & Arnold, K. (2011). *Gas Sweetening and Processing Field Manual*. Gulf Professional Publishing. https://doi.org/10.1016/C2009-0-62122-X
4. Ismail, A., Khulbe, K., & Matsuura, T. (2015). *Gas Separation Membranes: Polymeric and Inorganic*. Springer. https://doi.org/10.1007/978-3-319-01095-3
5. Baker, R. W. (2002). Future directions of membrane gas separation technology. *Industrial and Engineering Chemistry Research*, *41*, 1393. https://doi.org/10.1016/S0958-2118(01)80332-3
6. Henis, J. M. S., & Tripodi, M. K. (1980). A novel approach to gas separations using composite hollow fiber membranes. *Separation Science and Technology*, *15*(4), 1059–1068. https://doi.org/10.1080/01496398008076287
7. George, G., Bhoria, N., AlHallaq, S., Abdala, A., & Mittal V. (2016). Polymer membranes for acid gas removal from natural gas. *Separation and Purification Technology*, *158*, 333–356. https://doi.org/10.1016/j.seppur.2015.12.033
8. Nandi, B. K., Uppaluri, R., & Purkait, M. K. (2008). Preparation and characterization of low-cost ceramic membranes for microfiltration applications. *Applied Clay Science*, *42*(1–2), 102–110. https://doi.org/10.1016/j.clay.2007.12.001

9. Araújo, T., Parnell, A. J., Bernardo, G., & Mendes, A. (2023). Cellulose-based carbon membranes for gas separations – Unraveling structural parameters and surface chemistry for superior separation performance. *Carbon, 204*, 398–410. https://doi.org/10.1016/j.carbon.2022.12.062

10. Rasouli, S., Rezaei, N., Hamedi, H., Zendehboudi, S., & Duan, X. (2021). Superhydrophobic and superoleophilic membranes for oil–water separation application: A comprehensive review. *Materials & Design, 204*, 109599. https://doi.org/10.1016/j.matdes.2021.109599

11. Loeb, S., & Sourirajan, S. (1962). Sea water demineralization by means of a semipermeable membrane. *Advances in Chemistry, 38*, 117–132. https://doi.org10.1021/BA-1963-0038.CH009

12. Suga, Y., Takagi, R., & Matsuyama, H. (2021). Effect of the characteristic properties of membrane on long-term stability in the vacuum membrane distillation process. *Membranes, 11*(4), 252. https://doi.org/10.3390/membranes11040252

13. Tanudjaja, H. J., Hejase, C. A., Tarabara, V. V., Fane, A. G., & Chew, J. W. (2019). Membrane-based separation for oily wastewater: A practical perspective. *Water Research, 156*, 347–365. https://doi.org/10.1016/j.watres.2019.03.021

14. Shingala, J., Shah, V., Dudhat, K., & Shah, M. (2020). Evolution of nanomaterials in petroleum industries: Application and the challenges. *Journal of Petroleum Exploration and Production Technology, 10*, 3993–4006. https://doi.org/10.1007/s13202-020-00914-4

15. Peng, B., Tang, J., Luo, J., Wang, P., Ding, B., & Tam, K. (2018). Applications of nanotechnology in oil and gas industry: Progress and perspective. *Canadian Journal of Chemical Engineering, 96*, 91–100. https://doi.org/10.1002/cjce.23042

16. Khraisheh, M., Elhenawy, S., AlMomani, F., Al-Ghouti, M., Hassan, M. K., & Hameed, B. H. (2021). Recent progress on nanomaterial-based membranes for water treatment. *Membranes, 11*(12), 995–1030. https://doi.org/10.3390/membranes11120995

17. Etim, U. J., Bai, P., & Yan, Z. (2018). Nanotechnology applications in petroleum refining. In: Saleh, T. (eds.), *Nanotechnology in Oil and Gas Industries. Topics in Mining, Metallurgy and Materials Engineering*. Springer. https://doi.org/10.1007/978-3-319-60630-9_2

18. Kononova, S. V., Gubanova, G. N., Korytkova, E. N., Sapegin, D. A., Setnickova, K., Petrychkovych, R., & Uchytil, P. (2021). Polymer nanocomposite membranes. *Applied Sciences, 8*, 1181–1222. https://doi.org/10.3390/app8071181

19. Goh, P. S., & Ismail, A. F. (2021). Advances in nanocomposite membranes. *Membranes, 11*, 158–160. https://doi.org/10.3390/membranes11030158

20. Yadav D., Borpatra, G. M., Karki, S., & Ingole, P. G. 2022, A novel approach for the development of low-cost polymeric thin film nanocomposite membranes for the biomacromolecule separation. *ACS Omega, 7*(51), 47967–47985. https://doi.org/10.1021/acsomega.2c05861

21. Sewerin, T., Elshof, M. G., Matencio, S., Boerrigter, M., Yu, J., & de Grooth, J. (2021). Advances and applications of hollow fiber nanofiltration membranes: A review. *Membranes, 11*(11), 890–923. https://doi.org/10.3390/membranes11110890

22. Chuah, C. Y., Lee, J., & Bae, T. H. (2020). Graphene-based membranes for H2 separation: Recent progress and future perspective. *Membranes, 10*(11), 336–371. https://doi.org/10.3390/membranes10110336

23. Darvishmanesh, S., Robberecht, T., Luis, P., Degrève, J., & Bruggen, B. V. D. (2011). Performance of nanofiltration membranes for solvent purification in the oil industry.

Journal of the American Oil Chemists' Society, 88(8), 1255–1261. https://doi.org/10.1007/s11746-011-1779-y

24. Yang, C., Long, M., Ding, C., Zhang, R., Zhang, S., Yuan, J., Zhi, K., Yin, Z., Zheng, Y., Liu, Y., Wu, H., & Jiang, Z. (2022). Antifouling graphene oxide membranes for oil–water separation via hydrophobic chain engineering. *Nature Communications, 13,* 7334–7342. https://doi.org/10.1038/s41467-022-35105-8

25. Shi, Z., Zhang, W., Zhang, F., Liu, X., Wang, D., Jin, J., & Jiang, L. (2013). Ultrafast separation of emulsified oil/water mixtures by ultrathin free-standing single-walled carbon nanotube network films. *Advanced Materials, 25,* 2422–2427. https://doi.org/10.1002/adma.201204873

26. Al-anzi, B. S., & Siang, O. C. (2017). Recent developments of carbon-based nanomaterials and membranes for oily wastewater treatment. *RSC Advances, 7,* 20981–20994. https://doi.org/10.1039/C7RA02501G

27. Gupta, S., & Tai, N. H. (2016). Carbon materials as oil sorbents: A review on the synthesis and performance. *Journal Materials Chemistry A, 4,* 1550–1565. https://doi.org/10.1039/C5TA08321D

28. Bai, L., Bossa, N., Qu, F., Winglee, J., Li, G., Sun, K., Liang, H., Wiesner, M. R. (2017). Comparison of hydrophilicity and mechanical properties of nanocomposite membranes with cellulose nanocrystals and carbon nanotubes. *Environmental Science & Technology, 51*(1), 253–262. https://doi.org/10.1021/acs.est.6b04280

29. Das, R., Ali, M. E., Hamid, S. B. A., Ramakrishna, S., & Chowdhury, Z. Z. (2014). Carbon nanotube membranes for water purification: A bright future in water desalination. *Desalination, 336,* 97–109. https://doi.org/10.1016/j.desal.2013.12.026

30. Liu, J., Li, X., Jia, W., Ding, M., Zhang, Y., & Ren, S. (2016). Separation of emulsified oil from oily wastewater by functionalized multiwalled carbon nanotubes. *Journal of Dispersion Science and Technology, 37*(9), 1294–1302. https://doi.org/10.1080/01932691.2015.1090320

31. Zhang, J., Zhang, F., Song, J., Liu, L., Si, Y., Yu, J., & Ding, B. (2019). Electrospun flexible nanofibrous membranes for oil/water separation. *Journal of Materials Chemistry A, 7*(35), 20075–20102. https://doi.org/10.1039/C9TA07296A

32. Ng, L. Y., Mohammad, A. W., Leo, C. P., & Hilal, N. (2013). Polymeric membranes incorporated with metal/metal oxide nanoparticles: A comprehensive review. *Desalination, 308,* 15–33. https://doi.org/10.1016/j.desal.2010.11.033

33. Liu, Y., Su, Y., Cao, J., Guan, J., Xu, L., Zhang, R., He, M., Zhang, Q., Fanab, L., & Jiang, Z. (2017). Synergy of the mechanical, antifouling and permeation properties of a carbon nanotube nanohybrid membrane for efficient oil/water separation. *Nanoscale, 9,* 7508–7518. https://doi.org/10.1039/C7NR00818J

34. Gai, J. G., Gong, X. L., Wang, W. W., Zhang, X., & Kang, W. L. (2014). An ultrafast water transport forward osmosis membrane: Porous graphene. *Journal of Materials Chemistry A, 2,* 4023–4028. https://doi.org/10.1039/C3TA14256F

35. Geim, A. K., & Novoselov, K. S. (2007). The rise of graphene. *Nature Material, 6,* 183–191. https://doi.org/10.1038/nmat1849

36. Duan, W., Dudchenko, A., Mende, E., Flyer, C., Zhua, X., & Jassby, D. (2014). Electrochemical mineral scale prevention and removal on electrically conducting carbon nanotube-polyamide reverse osmosis membranes. Environmental *Science: Processes Impacts, 16,* 1300–1308. https://doi.org/10.1039/C3EM00635B

37. Aung, A., Bhullar, I. S., Theprungsirikul, J., Davey, S. K., Lim, H. L., Chiu, Y. J., Ma, X., Dewan, S., Lo, Y. H., McCulloch, A., & Varghese, S. (2016). 3D cardiac μ tissues within a microfluidic device with real-time contractile stress readout. *Lab Chip, 16,* 153–162. https://doi.org/10.1039/C5LC00820D

38. Deng. D., Prendergast, D. P., MacFarlane. J., Bagatin. R., Stellacci. F., & Gschwend, P. M. (2013). Hydrophobic meshes for oil spill recovery devices. *ACS Applied Materials & Interfaces*, *5*, 774–781. https://doi.org/10.1021/am302338x

39. Dimian, A., Bildea, C., & Kiss, A. (2014). *Integrated Design and Simulation of Chemical Processes*. Elsevier Science. www.elsevier.com/books/integrated-design-and-simulation-of-chemical-processes/dimian/978-0-444-62700-1

40. Young, D. C. (2001). *Computational Chemistry: A Practical Guide for Applying Techniques to Real-World Problems*. Wiley-Interscience. ISBN: 978-0-471-33368-5.

41. Foo, D. C. Y. (2017). *Introduction to Process Simulation, Chapter1, Chemical Engineering Process Simulation*. ICHEM. ISBN: 9780323901680.

42. Zhang, H., Moh, D. Y., Wang, X., & Qiao, R. (2022). Review on pore-scale physics of shale gas recovery dynamics: Insights from molecular dynamics simulations. *Energy & Fuels*, *36, 24*, 14657–14672. https://doi.org/10.1021/acs.energyfuels.2c03388

43. Pétuya, R., Punase, A., Bosoni, E., de Oliveira Filho, A. P., Sarria, J., Purkayastha, N., Wylde, J. J., & Mohr, S. (2023). Molecular dynamics simulations of asphaltene aggregation: Machine-learning identification of representative molecules, molecular polydispersity, and inhibitor performance. *ACS Omega*, *8*(5), 4862–4877. https://doi.org/10.1021/acsomega.2c07120

44. Tariq, W., Ali, F., Arslan, C., Nasir, A., Gillani, S. H., & Rehman, A. (2022). Synthesis and applications of graphene and graphene-based nanocomposites: Conventional to artificial intelligence approaches. *Frontiers in Environmental Chemistry*, *3*, 890408. https://doi.org/10.3389/fenvc.2022.890408

45. Papageorgiou, D. G., Kinloch, I. A., & Young, R. J. (2017). Mechanical properties of graphene and graphene-based nanocomposites. *Progress in Materials Science*, *90*, 75–127. https://doi.org/10.1016/j.pmatsci.2017.07.004

46. Ibrahim, A., Klopocinska, A., Horvat, K., & Hamid, A. Z. (2021). Graphene-based nanocomposites: Synthesis, mechanical properties, and characterizations. *Polymers*, *13*(17), 2869. https://doi.org/10.3390/polym13172869

47. Miller, J. A., Sivaramakrishnan, R., Tao, Y., Goldsmith, C. F., Burke, M. P., Jasper, A. W., Hansen, N., Labbe, N. J., Glarborg, P., & Zádor, J. (2021). Combustion chemistry in the twenty-first century: Developing theory-informed chemical kinetics models. *Progress in Energy and Combustion Science*, *83*, 100886. https://doi.org/10.1016/j.pecs.2020.100886

48. Mehana, M., Kang, Q., Nasrabadi, H., & Viswanathan, H. (2021). Molecular modeling of subsurface phenomena related to petroleum engineering. *Energy & Fuels*, *35*(4), 2851–2869. https://doi.org/10.1021/acs.energyfuels.0c02961

49. Groß, A., & Sakong, S. (2022). Ab initio simulations of water/metal interfaces. *Chemical Reviews*, *122*(12), 10746–10776. https://doi.org/10.1021/acs.chemrev.1c00679

50. Gartner III, T. E., & Jayaraman, A. (2019). Modeling and simulations of polymers: A roadmap. *Macromolecules*, *52*(3), 755–786. https://doi.org/10.1021/acs.macromol.8b01836

51. Obotey Ezugbe, E., & Rathilal, S. (2020). Membrane technologies in wastewater treatment: A review. *Membranes*, *10*(5), 89. https://doi.org/10.3390/membranes10050089

52. Thompson, A. P., Aktulga, H. M., Berger, R., Bolintineanu, D. S., Brown, W. M., Crozier, P. S., in't Veld, P. J., Kohlmeyer, A., Moore, S. G., Nguyen, T. D. and Shan, R., & Plimpton, S. J. (2022). LAMMPS: A flexible simulation tool for particle-based materials modeling at the atomic, meso, and continuum scales. *Computer Physics Communications*, *271*, 108171. https://doi.org/10.1016/j.cpc.2021.108171

53. Case, D. A., Aktulga, H. M., Belfon, K., Ben-Shalom, I. Y., Berryman, J. T., Brozell, S. R., Cerutti, D. S., Cheatham, T. E., Cisneros, G. A., Cruzeiro, V. W. D. and Darden, T., & Kollman P. (2022). *Amber.* University of California, San Francisco, CA, USA.

54. Smith, W., Yong, C. W., & Rodger, P. M. (2002). DL_POLY: Application to molecular simulation. *Molecular Simulation, 28*(5), 385–471. https://doi.org/10.1080/089270 20290018769

55. Jo, S., Kim, T., Iyer, V. G., & Im, W. (2008). CHARMM-GUI: A web-based graphical user interface for CHARMM. *Journal of Computational Chemistry, 29*(11), 1859–1865. https://doi.org/10.1002/jcc.20945

56. Abraham, M. J., Murtola, T., Schulz, R., Páll, S., Smith, J. C., Hess, B., & Lindahl, E. (2015). GROMACS: High performance molecular simulations through multi-level parallelism from laptops to supercomputers. *SoftwareX, 1,* 19–25. https://doi.org/ 10.1016/j.softx.2015.06.001

57. Phillips, J. C., Hardy, D. J., Maia, J. D., Stone, J. E., Ribeiro, J. V., Bernardi, R. C., Buch, R., Fiorin, G., Hénin, J., Jiang, W. and McGreevy, R., & Tajkhorshid, E. (2020). Scalable molecular dynamics on CPU and GPU architectures with NAMD. *The Journal of Chemical Physics, 153*(4), 044130. https://doi.org/10.1063/5.0014475

58. H. L. Anderson. (1986). Metropolis, Monte Carlo and the MANIAC. *Los Alamos Science, 14,* 96–108. http://library.lanl.gov/cgi-bin/getfile?00326886.pdf

59. Chuah, C. Y. (2022). Membranes for gas separation and purification processes. *Membranes, 12*(6), 622. https://doi.org/10.3390/membranes12060622

60. Ji, G., & Zhao, M. (2017). Membrane separation technology in carbon capture. *Recent Advances in Carbon Capture and Storage,* 59–90. https://doi.org/10.5772/65723

61. Huang, W., Jiang, X., He, G., Ruan, X., Chen, B., Nizamani, A. K., Li, X., Wu, X., & Xiao, W. (2020). A novel process of H_2/CO_2 membrane separation of shifted syngas coupled with gasoil hydrogenation. *Processes, 8*(5), 590. https://doi.org/10.3390/ pr8050590

62. Lin, H., He, Z., Sun, Z., Kniep, J., Ng, A., Baker, R. W., & Merkel, T. C. (2015). CO_2-selective membranes for hydrogen production and CO_2 capture: Part II: Techno-economic analysis. *Journal of Membrane Science, 493,* 794–806. https://doi.org/ 10.1016/j.memsci.2015.02.042

63. Chen, G., Wang, T., Zhang, G., Liu, G., & Jin, W. (2022). Membrane materials targeting carbon capture and utilization. *Advanced Membranes, 2,* 100025. https://doi. org/10.1016/j.advmem.2022.100025

64. Hussain, A., & Al-Yaari, M. (2021). Development of polymeric membranes for oil/ water separation. *Membranes, 11*(1), 42. https://doi.org/10.3390/membranes11010042

65. Ravanchi, M. T., Kaghazchi, T., & Kargari, A. (2009). Application of membrane sep-aration processes in petrochemical industry: A review. *Desalination, 235*(1–3), 199–244. https://doi.org/10.1016/j.desal.2007.10.042

66. Issaoui, M., & Limousy, L. (2019). Low-cost ceramic membranes: Synthesis, classifications, and applications. *Comptes Rendus Chimie, 22*(2–3), 175–187. https:// doi.org/10.1016/j.crci.2018.09.014

67. Rani, S. L. S., & Kumar, R. V. (2021). Insights on applications of low-cost ceramic membranes in wastewater treatment: A mini-review. *Case Studies in Chemical and Environmental Engineering, 4,* 100149. https://doi.org/10.1016/j.cscee.2021.100149

68. Katare, A., Kumar, S., Kundu, S., Sharma, S., Kundu, L.M., Mandal, B. (2023, May 15). Mixed matrix membranes for carbon capture and sequestration: Challenges and scope. *ACS Omega, 8*(20), 17511–17522. doi: 10.1021/acsomega.3c01666

69. Clarizia, G., & Bernardo, P. (2021). A review of the recent progress in the development of nanocomposites based on poly(ether-block-amide) copolymers as membranes for CO_2 separation. *Polymers*, *14*(1), 10. https://doi.org/10.3390/polym14010010

70. Kardani, R., Asghari, M., Mohammadi, T., & Afsari, M. (2018). Effects of nanofillers on the characteristics and performance of PEBA-based mixed matrix membranes. *Reviews in Chemical Engineering*, *34*(6), 797–836. https://doi.org/10.1515/revce-2017-0001

71. Yave, W., Car, A., & Peinemann, K. V. (2010). Nanostructured membrane material designed for carbon dioxide separation. *Journal of Membrane Science*, *350*(1–2), 124–129. https://doi.org/10.1016/j.memsci.2009.12.019

72. Lu, Y., Liu, W., Liu, J., Li, X., & Zhang, S. (2021). A review on 2D porous organic polymers for membrane-based separations: Processing and engineering of transport channels. *Advanced Membranes*, *1*, 100014. https://doi.org/10.1016/j.advmem.2021.100014

73. Hyun, T., Jeong, J., Chae, A., Kim, Y. K., & Koh, D. Y. (2019). 2D-enabled membranes: Materials and beyond. *BMC Chemical Engineering*, *1*, 1–26. https://doi.org/10.1186/s42480-019-0012-x

74. Zheng, W., Tsang, C. S., Lee, L. Y. S., & Wong, K. Y. (2019). Two-dimensional metal–organic framework and covalent–organic framework: Synthesis and their energy-related applications. *Materials Today Chemistry*, *12*, 34–60. https://doi.org/10.1016/j.mtchem.2018.12.002

75. Slepičková Kasálková, N., Slepička, P., & Švorčík, V. (2021). Carbon nanostructures, nanolayers, and their composites. *Nanomaterials*, *11*(9), 2368. https://doi.org/10.3390/nano11092368

76. Patel, D. K., Kim, H. B., Dutta, S. D., Ganguly, K., & Lim, K. T. (2020). Carbon nanotubes-based nanomaterials and their agricultural and biotechnological applications. *Materials*, *13*(7), 1679. https://doi.org/10.3390/ma13071679

77. Paukov, M., Kramberger, C., Begichev, I., Kharlamova, M., & Burdanova, M. (2023). Functionalized fullerenes and their applications in electrochemistry, solar cells, and nanoelectronics. *Materials*, *16*(3), 1276. https://doi.org/10.3390/ma16031276

78. Chen, T., & Dai, L. (2013). Carbon nanomaterials for high-performance supercapacitors. *Materials Today*, *16*(7–8), 272–280. https://doi.org/10.1016/j.mattod.2013.07.002

79. Jariwala, D., Sangwan, V. K., Lauhon, L. J., Marks, T. J., & Hersam, M. C. (2013). Carbon nanomaterials for electronics, optoelectronics, photovoltaics, and sensing. *Chemical Society Reviews*, *42*(7), 2824–2860. https://doi.org/10.1039/C2CS35335K

80. Tiwari, S. K., Kumar, V., Huczko, A., Oraon, R., Adhikari, A. D., & Nayak, G. C. (2016). Magical allotropes of carbon: Prospects and applications. *Critical Reviews in Solid State and Materials Sciences*, *41*(4), 257–317. https://doi.org/10.1080/10408436.2015.1127206

81. Terrones, M. (2003). Science and technology of the twenty-first century: Synthesis, properties, and applications of carbon nanotubes. *Annual Review of Materials Research*, *33*(1), 419–501. https://doi.org/10.1146/annurev.matsci.33.012802.100255

82. Pogorielov, M., Smyrnova, K., Kyrylenko, S., Gogotsi, O., Zahorodna, V., & Pogrebnjak, A. (2021). MXenes: A new class of two-dimensional materials: Structure, properties and potential applications. *Nanomaterials*, *11*(12), 3412. https://doi.org/10.3390/nano11123412

83. Berkani, M., Smaali, A., Almomani, F., & Vasseghian, Y. (2022). Recent advances in MXene-based nanomaterials for desalination at water interfaces. *Environmental Research*, *203*, 111845. https://doi.org/10.1016/j.envres.2021.111845

84. Zhu, J., Ha, E., Zhao, G., Zhou, Y., Huang, D., Yue, G., Hu, L., Sun, N., Wang, Y., Lee, L. Y. S., Xu, C., Wong, K. Y., Astruc, D., & Zhao, P. (2017). Recent advance in MXenes: A promising 2D material for catalysis, sensor and chemical adsorption. *Coordination Chemistry Reviews*, *352*, 306–327. https://doi.org/10.1016/j.ccr.2017.09.012

85. Smith, A. T., LaChance, A. M., Zeng, S., Liu, B., & Sun, L. (2019). Synthesis, properties, and applications of graphene oxide/reduced graphene oxide and their nanocomposites. *Nano Materials Science*, *1*(1), 31–47. https://doi.org/10.1016/j.nanoms.2019.02.004

86. Ghulam, A. N., Dos Santos, O. A., Hazeem, L., Pizzorno Backx, B., Bououdina, M., & Bellucci, S. (2022). Graphene oxide (GO) materials: Applications and toxicity on living organisms and environment. *Journal of Functional Biomaterials*, *13*(2), 77. https://doi.org/10.3390/jfb13020077

87. Majumder, P., & Gangopadhyay, R. (2022). Evolution of graphene oxide (GO)-based nanohybrid materials with diverse compositions: An overview. *RSC Advances*, *12*(9), 5686–5719. https://doi.org/10.1039/D1RA06731A

88. Bellier, N., Baipaywad, P., Ryu, N., Lee, J. Y., & Park, H. (2022). Recent biomedical advancements in graphene oxide-and reduced graphene oxide-based nanocomposite nanocarriers. *Biomaterials Research*, *26*(1), 65. https://doi.org/10.1186/s40824-022-00313-2

89. Bhavsar, R. S., Kumbharkar, S. C., Rewar, A. S., & Kharul, U. K. (2014). Polybenzimidazole based film forming polymeric ionic liquids: Synthesis and effects of cation–anion variation on their physical properties. *Polymer Chemistry*, *5*(13), 4083–4096. https://doi.org/10.1039/C3PY01709E

90. Eftekhari, A., & Saito, T. (2017). Synthesis and properties of polymerized ionic liquids. *European Polymer Journal*, *90*, 245–272. https://doi.org/10.1016/j.eurpolymj.2017.03.033

91. Mecerreyes, D. (2011). Polymeric ionic liquids: Broadening the properties and applications of polyelectrolytes. *Progress in Polymer Science*, *36*(12), 1629–1648. https://doi.org/10.1016/j.progpolymsci.2011.05.007

92. Lei, Z., Chen, B., Koo, Y. M., & MacFarlane, D. R. (2017). Introduction: Ionic liquids. *Chemical Reviews*, *117*(10), 6633–6635. https://doi.org/10.1021/acs.chemrev.7b00246

93. Tomé, L. C., Gouveia, A. S., Freire, C. S., Mecerreyes, D., & Marrucho, I. M. (2015). Polymeric ionic liquid-based membranes: Influence of polycation variation on gas transport and CO_2 selectivity properties. *Journal of Membrane Science*, *486*, 40–48. https://doi.org/10.1016/j.memsci.2015.03.026

94. Baranwal, J., Barse, B., Fais, A., Delogu, G. L., & Kumar, A. (2022). Biopolymer: A sustainable material for food and medical applications. *Polymers*, *14*(5), 983. https://doi.org/10.3390/polym14050983

95. Gheorghita, R., Anchidin-Norocel, L., Filip, R., Dimian, M., & Covasa, M. (2021). Applications of biopolymers for drugs and probiotics delivery. *Polymers*, *13*, 2729. https://doi.org/10.3390/polym13162729

96. Mohan, S., Oluwafemi, O. S., Kalarikkal, N., Thomas, S., & Songca, S. P. (2016). Biopolymers – Application in nanoscience and nanotechnology. InTech. https://doi.org/10.5772/62225

97. Dassanayake, S. R., Acharya, S., & Abidi, N. (2019). Biopolymer-based materials from polysaccharides: Properties, processing, characterization and sorption applications. IntechOpen. https://doi.org/10.5772/intechopen.80898

98. Moradali, M. F., & Rehm, B. H. (2020). Bacterial biopolymers: From pathogenesis to advanced materials. *Nature Reviews Microbiology, 18*(4), 195–210. https://doi.org/10.1038/s41579-019-0313-3

99. Das, A., Ringu, T., Ghosh, S., & Pramanik, N. (2022). A comprehensive review on recent advances in preparation, physicochemical characterization, and bioengineering applications of biopolymers. *Polymer Bulletin*, 1–66. https://doi.org/10.1007/s00289-022-04443-4

100. Redondo-Gómez, C., Rodríguez Quesada, M., Vallejo Astúa, S., Murillo Zamora, J. P., Lopretti, M., & Vega-Baudrit, J. R. (2020). Biorefinery of biomass of agroindustrial banana waste to obtain high-value biopolymers. *Molecules, 25*(17), 3829. https://doi.org/10.3390/molecules25173829

101. Abral, H., Dalimunthe, M. H., Hartono, J., Efendi, R. P., Asrofi, M., Sugiarti, E., Sapuan, S. M., Park, J. W., & Kim, H. J. (2018). Characterization of tapioca starch biopolymer composites reinforced with micro scale water hyacinth fibers. *Starch-Stärke, 70*(7–8), 1700287. https://doi.org/10.1002/star.201700287

102. Lauer, M. K., Tennyson, A. G., & Smith, R. C. (2021). Inverse vulcanization of octenyl succinate-modified corn starch as a route to biopolymer–sulfur composites. *Materials Advances, 2*(7), 2391–2397. https://doi.org/10.1039/D0MA00948B

103. Tang, L., Xie, X., Li, C., Xu, Y., Zhu, W., & Wang, L. (2022). Regulation of structure and anion-exchange performance of layered double hydroxide: Function of the metal cation composition of a brucite-like layer. *Materials, 15*(22), 7983. https://doi.org/10.3390/ma15227983

104. Liu, Z., Ma, R., Osada, M., Iyi, N., Ebina, Y., Takada, K., & Sasaki, T. (2006). Synthesis, anion exchange, and delamination of Co–Al layered double hydroxide: Assembly of the exfoliated nanosheet/polyanion composite films and magneto-optical studies. *Journal of the American Chemical Society, 128*(14), 4872–4880. https://doi.org/10.1021/ja0584471

105. Pálinkó, I., Sipos, P., Berkesi, O., & Varga, G. (2022). Distinguishing anionic species that are intercalated in layered double hydroxides from those bound to their surface: A comparative IR study. *The Journal of Physical Chemistry C, 126*(36), 15254–15262. https://doi.org/10.1021/acs.jpcc.2c03547

106. Alnaqbi, M. A., Samson, J. A., & Greish, Y. E. (2020). Electrospun polystyrene/LDH fibrous membranes for the removal of Cd^{2+} ions. *Journal of Nanomaterials, 2020*, 1–12. https://doi.org/10.1155/2020/5045637

107. Awassa, J., Soulé, S., Cornu, D., Ruby, C., & El-Kirat-Chatel, S. (2022). Understanding the role of surface interactions in the antibacterial activity of layered double hydroxide nanoparticles by atomic force microscopy. *Nanoscale, 14*(29), 10335–10348. https://doi.org/10.1039/D2NR02395D

108. Islam, D. A., & Acharya, H. (2022). Pd-nanoparticles@ layered double hydroxide/reduced graphene oxide (Pd NPs@ LDH/rGO) nanocomposite catalysts for highly efficient green reduction of aromatic nitro compounds. *New Journal of Chemistry, 46*(11), 5346–5354. https://doi.org/10.1039/D1NJ05377A

109. Rowsell, J. L., & Yaghi, O. M. (2004). Metal–organic frameworks: A new class of porous materials. *Microporous and Mesoporous Materials, 73*(1–2), 3–14. https://doi.org/10.1016/j.micromeso.2004.03.034

110. Liu, X., Zhang, L., & Wang, J. (2021). Design strategies for MOF-derived porous functional materials: Preserving surfaces and nurturing pores. *Journal of Materiomics, 7*(3), 440–459. https://doi.org/10.1016/j.jmat.2020.10.008

111. Feng, L., Wang, K. Y., Lv, X. L., Yan, T. H., & Zhou, H. C. (2020). Hierarchically porous metal–organic frameworks: Synthetic strategies and applications. *National Science Review*, 7(11), 1743–1758. https://doi.org/10.1093/nsr/nwz170

112. Baumann, A. E., Burns, D. A., Liu, B., & Thoi, V. S. (2019). Metal–organic framework functionalization and design strategies for advanced electrochemical energy storage devices. *Communications Chemistry*, 2(1), 86. https://doi.org/10.1038/s42004-019-0184-6

113. Xiong, Q., Chen, Y., Yang, D., Wang, K., Wang, Y., Yang, J., Li, L., & Li, J. (2022). Constructing strategies for hierarchically porous MOFs with different pore sizes and applications in adsorption and catalysis. *Materials Chemistry Frontiers*, 6, 2944–2967. https://doi.org/10.1039/D2QM00557C

114. Wang, Z., Zhu, Q., Wang, J., Jin, F., Zhang, P., Yan, D., Cheng, P., Chen, Y., & Zhang, Z. (2022). Industry-compatible covalent organic frameworks for green chemical engineering. *Science China Chemistry*, 65, 2144–2162. https://doi.org/10.1007/s11426-022-1391-0

115. Yuan, S., Li, X., Zhu, J., Zhang, G., Van Puyvelde, P., & Van der Bruggen, B. (2019). Covalent organic frameworks for membrane separation. *Chemical Society Reviews*, 48(10), 2665–2681. https://doi.org/10.1039/C8CS00919H

116. Zhao, X., Pachfule, P., & Thomas, A. (2021). Covalent organic frameworks (COFs) for electrochemical applications. *Chemical Society Reviews*, 50(12), 6871–6913. https://doi.org/10.1039/D0CS01569E

117. Díaz, U., & Corma, A. (2016). Ordered covalent organic frameworks, COFs and PAFs. From preparation to application. *Coordination Chemistry Reviews*, 311, 85–124. https://doi.org/10.1016/j.ccr.2015.12.010

118. Xu, Q., & Jiang, J. (2020). Molecular simulations of liquid separations in polymer membranes. *Current Opinion in Chemical Engineering*, 28, 66–74. https://doi.org/10.1016/j.coche.2020.02.001

119. Ouinten, M. L., Szymczyk, A., & Ghoufi, A. (2023). Molecular dynamics simulation study of organic solvents confined in PIM-1 and P84 polyimide membranes. *The Journal of Physical Chemistry B*, 127(5), 1237–1243. https://doi.org/10.1021/acs.jpcb.2c05796

120. Chen, L., & Wu, B. (2021). Research progress in computational fluid dynamics simulations of membrane distillation processes: A review. *Membranes*, 11(7), 513. https://doi.org/10.3390/membranes11070513

121. Tabrizi, A. B., & Wu, B. (2019). Review of computational fluid dynamics simulation techniques for direct contact membrane distillation systems containing filament spacers. *Transfer*, 24, 30–34. https://doi.org/10.5004/dwt.2019.24322

122. Weyman, A., Mavrantzas, V. G., & Öttinger, H. C. (2022). Direct calculation of the functional inverse of realistic interatomic potentials in field-theoretic simulations. *The Journal of Chemical Physics*, 156(22), 224115. https://doi.org/10.1063/5.0090333

123. Van Mourik, T., Bühl, M., & Gaigeot, M. P. (2014). Density functional theory across chemistry, physics and biology. *Philosophical Transactions of the Royal Society A: Mathematical, Physical and Engineering Sciences*, 372(2011), 20120488. https://doi.org/10.1098/rsta.2012.0488

124. Rizki, Z., & Ottens, M. (2023) Model-based optimization approaches for pressure-driven membrane systems. *Separation and Purification Technology*, 315, 123682. https://doi.org/10.1016/j.seppur.2023.123682

125. Koopialipoor, M., & Noorbakhsh, A. (2020). Applications of artificial intelligence techniques in optimizing drilling. *Emerging Trends in Mechatronics*, 89. https://doi.org/10.5772/intechopen.85398

126. Wang, Y., Seo, B., Wang, B., Zamel, N., Jiao, K., & Adroher, X. C. (2020). Fundamentals, materials, and machine learning of polymer electrolyte membrane fuel cell technology. *Energy and AI*, *1*, 100014. https://doi.org/10.1016/j.egyai.2020.100014

127. Alder, B. J., & Wainwright, T. E. (1959). Studies in molecular dynamics. I. General method. *The Journal of Chemical Physics*, *31*(2), 459–466. https://doi.org/10.1063/1.1730376

128. Rahman, A. (1964). Correlations in the motion of atoms in liquid argon. *Physical Review*, *136*(2A), A405. https://doi.org/10.1103/PhysRev.136.A405

129. Wang, X., Ramírez-Hinestrosa, S., Dobnikar, J., & Frenkel, D. (2020). The Lennard–Jones potential: When (not) to use it. *Physical Chemistry Chemical Physics*, *22*(19), 10624–10633. https://doi.org/10.1039/C9CP05445F

130. Yeoh, G. H., Liu, C., Tu, J., & Timchenko, V. (2011). Advances in computational fluid dynamics and its applications. *Modelling and Simulation in Engineering*, *2011*. https://doi.org/10.1155/2011/304983

131. Calzolari, G., & Liu, W. (2021). Deep learning to replace, improve, or aid CFD analysis in built environment applications: A review. *Building and Environment*, *206*, 108315. https://doi.org/10.1016/j.buildenv.2021.108315

132. Ferziger, J. H., Perić, M., & Street, R. L. (2002). *Computational Methods for Fluid Dynamics* (Vol. 3, pp. 196–200). Berlin: *Springer*. ISBN: 978-3-319-99693-6.

133. Rubinstein, R. Y., & Kroese, D. P. (2016). *Simulation and the Monte Carlo Method*. Wiley. ISBN: 9781118632383.

134. Kroese, D. P., Brereton, T., Taimre, T., & Botev, Z. I. (2014). Why the Monte Carlo method is so important today. *Wiley Interdisciplinary Reviews: Computational Statistics*, *6*(6), 386–392. https://doi.org/10.1002/wics.1314

135. Hastings, W. K. (1970). Monte Carlo sampling methods using Markov chains and their applications. https://doi.org/10.1093/biomet/57.1.97

136. Politzer, P., Murray, J. (2002). The fundamental nature and role of the electrostatic potential in atoms and molecules. *Theoretical Chemistry Accounts*, *108*, 134–142. https://doi.org/10.1007/s00214-002-0363-9

137. Music, D., Geyer, R. W., & Schneider, J. M. (2016). Recent progress and new directions in density functional theory-based design of hard coatings. *Surface and Coatings Technology*, *286*, 178–190. https://doi.org/10.1016/j.surfcoat.2015.12.021

138. Abduljabbar, R., Dia, H., Liyanage, S., & Bagloee, S. A. (2019). Applications of artificial intelligence in transport: An overview. *Sustainability*, *11*(1), 189. https://doi.org/10.3390/su11010189

139. Siebes, A. (2000). Data mining and statistics. In: Della Riccia, G., Kruse, R., & Lenz, H. J. (eds.), *Computational Intelligence in Data Mining. International Centre for Mechanical Sciences, 408*. Vienna: Springer. https://doi.org/10.1007/978-3-7091-2588-5_1

140. Turing, A. (1950). Computing machinery and intelligence, *Mind*, *LIX*, 433. https://doi.org/10.1093/mind/lix.236.433

141. Tao, J., & Tan, T. (2005). Affective computing: A review. In *Affective Computing and Intelligent Interaction: First International Conference, ACII 2005*, Beijing, China, October 22–24, 2005. Proceedings 1, 981–995. Springer Berlin Heidelberg. https://doi.org/10.1007/11573548_125

15 Future Perspective of Membrane Technology in Petroleum Industry

Chinmoy Bhuyan and Swapnali Hazarika

15.1 INTRODUCTION

The petroleum industry being the world's fast-growing energy engine has been significantly contributing to the world's economic development and employment generation. The petroleum industry has been facing problems like loss of useful products during extraction and refining, treatment of oily wastewater, separation of gases, including olefin paraffin separation, etc. Conventional technology for addressing these challenges often suffers from low selectivity, large-scale separation, high energy consumption and generation of secondary pollutants. Among the existing technologies for mitigating these problems, membrane technology has been considered as one of the best techniques due to its remarkable selectivity, facile operational procedure, less energy consumption and ease of scaling up.

In the petroleum industry, a huge amount of wastewater is generated during various processes like extraction and refining of crude oil [1]. These wastewaters contain a large amount of petroleum hydrocarbons in a liquid state, as well as oil and grease with other organics and inorganics that harm the ecosystem [2]. Various technologies have been developed to remove or treat these harmful components from wastewater [3]. Membrane technology is one of the most advantageous techniques among them due to its remarkable selectivity as well as ease of large-scale separation in the small and simple separation process. Membrane separation has been crucial in separating numerous harmful chemicals at the molecular level [4]. As there is a provision for hybridising various processes like photodegradation, and selective evaporation with membrane technology for more separation in a single operation, this has always been considered superior to all the other separation techniques. Moreover, most of the operations in refining and separating products in the chemical process of the industries are highly important as they contribute to 30% of the capital cost. Membrane technology has the provision of hybridizing with other separation techniques like adsorption and others which is one of its advantages. The hybridized system of membrane and adsorption-based separation has been used as an alternative to the state-of-the-art cryogenic process that was applied earlier. It has emerged as a dominant

DOI: 10.1201/9781003441359-15

area of research in recent decades, with numerous applications. It is a scientific field that deals with modelling and management that allows identifying the most suitable operation pathway in an industrial sector. However, due to some issues like fouling of the membrane due to agglomeration of the organic molecules and pore broadening due to the action of organic solvents present in petroleum waste, membrane separation also sometimes fails in terms of its durability [5]. However, a novel membrane with a hybrid structure may work in such cases. This chapter emphasises the current approaches and prospects of membrane technology for making it more error-free and efficient for the separation of various components present in industrial wastewater in a much better way.

15.2 MEMBRANE TECHNOLOGY FOR PURIFICATION OF NATURAL GASES

Natural gas purification is a very demanding work that needs an energy-efficient and easy process. Since tonnes of natural gases are produced by the petroleum industries daily, they need to undergo different treatments before further utilisation. This type of treatment, such as syngas removal and gas sweating, can be done efficiently by membrane separation. Membrane technology can separate unwanted gases from natural gas mixtures based on their different properties. Till now, membranes have been extensively studied for carbon capture, desulfurization, hydrogen purification, etc. [6]. In most cases, the polymeric membranes face a trade-off between permeability and selectivity of gases. Thus, to face this problem, many modifications in membrane research have been developed, such as chemical and structural modifications with nanomaterial incorporation in the membrane during membrane fabrication, selective material doping, etc. Research on metal–organic frameworks (MOFs) and zeolite imidazolate frameworks (ZIFs) and their possible application in the purification of natural gases has been going on due to their high affinity for gas adsorption and the application of these materials in storage. Future research on the development of these materials for their use in applications as membrane materials for gas separation is still a challenge, and modification or functionalisation of these materials is required for a relatively more feasible process [7]. Hollow fibre membranes reinforced with braid are the newest technique, which has been done on a small scale. Though this type of membrane purifies liquid mixtures, many challenges have arisen while employing them in gaseous mixture separation. This field has numerous advantages, like the ease of scaling up and the requirement of small, easily accessible materials. Many newer materials are still under research and could show better gas separation and capture ability when incorporated into the membrane. Shortly, membrane technology can become the best technology for natural gas treatment with novel and highly efficient materials.

15.3 MEMBRANE TECHNOLOGY FOR OLEFIN AND PARAFFIN SEPARATION

Olefin and paraffin separation also come under gas separation when molecular weight is low (C1–C4). Olefins with a double bond have a chemically different structure than paraffin. Based on this difference, membrane technology has utilised a metal-based separation technique for olefin/paraffin separation, which has been proven to be a fruitful. Metals such as Ag, Cu, Zn, etc., can show preferential interaction with olefins but not with paraffin. Different porous materials such as MOFs, PAFs, COFs and ZIFs are being introduced into membranes nowadays for olefin/paraffin separation purposes [8]. Due to its tunable nature, membrane technology has a bright future in this type of azeotrope separation. One can change the membrane according to their needs with different compatible materials, which proved a great advantage. In recent times, the separation of olefin and paraffin has been a trending research topic as the petroleum industries have encountered a lot of problems regarding the separation of olefin and paraffin due to their similar physicochemical properties like molecular size, boiling point and solubility. Membrane-based separation is the only way that components in a mixture can be efficiently bound by various mechanisms like adsorption while permeating the others through facilitated transport. Thus, membranes are a great tool for gas separation purposes with a large scale-up scope. However, it is much more challenging to scale up the laboratory-based membrane design for large-scale separation. One must choose the method, materials and separation procedure that can be easily transformed for efficient separation.

15.4   MEMBRANE TECHNOLOGY FOR SEPARATION OF AROMATICS AND ALIPHATIC

Aromatics and aliphatic compounds are often found in the petroleum industry in mixture form. The separation of aromatic from aliphatic is also a matter of concern. Compared to the conventional methods of separation like liquid extraction, extractive distillation, etc., which are not preferred due to the high flow rate, non-conventional methods of separation, like membrane-based separation, are preferred for the separation of aromatics/aliphatics. The extra purification step [is] needed in this conventional technique [techniques] [that] consumes extra energy and costs more. Membrane technology can be a suitable alternative with high energy efficiency and low cost. Membranes prepared from different polymers and copolymers have been utilised to achieve high separation performance of aromatic and aliphatic mixtures [9,10]. With surface modification of the membrane by incorporating different nano-sized materials, coatings, etc., membrane technology has also been improving day by day in terms of performance. Aliphatic/aromatic separation via pervaporation, polymeric membranes and hybrid membranes are studied in this book, which will provide insight into the research already being done. Thus, membranes have tremendous scope for expanding their applications in this era of research.

15.5   MEMBRANE TECHNOLOGY FOR TREATMENT OF PETROLEUM INDUSTRY WASTEWATER

The wastewater generated from petroleum industries contains mainly hydrocarbons. Membrane technologies have been proven to be the most efficient for the separation of these priority pollutants in order to minimise environmental problems. Though it is much more difficult to complete removal of the hydrocarbons in a single method

[11], the combination of membranes with biological, chemical or photochemical methods can be considered a much greener approach. Moreover, membrane durability is another issue to be considered. Therefore, for improving the efficiency of the membranes, selection of polymer and development of highly solvent-resistant polymer derivatives are the major challenges. The wastewater of the petroleum industry contains high concentrations of organic solvents, and the problem of membrane pore broadening and swelling has a significant impact on membrane efficiency. Hence, the production of solvent-resistant membranes is one of the trending research topics in recent days. Polymers that are under examination as organic solvent resistant include commercial PEEK (poly(ether ether ketone)s), polybenzimidazole, laboratory-derived polymers like poly(arylene sulfide sulfone), etc. One of the biggest difficulties with solvent-resistant polymers is that it is difficult to solubilize them in common solvents. Polyethylene terephthalate (PET), is a trending copolymer that has been used in designing organic solvent-resistant polymeric membranes, is soluble only in trifluoroacetic acid. Hence, the production of PET-based membranes is still a challenging task, though laboratory-level development is going on in our research group. Thus, it is important to conduct research on the suitable solvents that can be used in ambient conditions with ease of handling.

Developing hybrid membrane systems by impregnating nanomaterials over the membrane surface or adding them to the polymer matrix is one of the advanced techniques for increasing the efficiency of the membrane system for the treatment of oil-based wastewater. Nanomaterials impregnation on membrane surfaces is another way to separate oil-based contaminants from other hydrocarbon-based harmful chemicals and is the newest approach at the laboratory scale. Hybrid membrane systems like high-performing photocatalytic membrane systems for separation and degradation of toxic hydrocarbons present in oil-based wastewater are also done at a lab scale in our group. However, scaling up these techniques is still a challenge [12]. Fouling minimisation and the development of self-cleaning membrane systems may be another approach for the durable and highly efficient separation of all types of materials from oily wastewater [13]. The development of newer technologies and scaling up of the existing lab-based studies are highly needed for the effective treatment of industry-based wastewater to nullify the effects of harmful industry-derived chemicals in the ecosystem. In the future, it is believed that membrane technology will be considered the most efficient technology for the separation of petroleum-based hydrocarbons from oil industry wastewater.

15.6 MEMBRANE TECHNOLOGY FOR HEAVY METAL SEPARATION AND PHENOL SEPARATION

Heavy metals have always been associated with petroleum industry sludge and waste. Metals such as Pb, Cd, Ni, Cr, Cu, Zn, V, As, etc., are often found in more or less amounts along with other organics such as phenol [14]. Phenol is also an unavoidable organic compound that can cause harm to living beings if it comes into contact with them. Wastewater containing these heavy metals and organics, if not properly treated, may cause different health issues in animals and humans. As a result, membranes have been widely used to treat wastewater efficiently. Out of the different conventional technologies used, such as flocculation, floatation, chemical precipitation, coagulation, etc., membrane-based separation is the most suitable process since it does not cause any environmental damage or secondary pollution. This application makes extensive use of membranes made of various materials. Complexation ultrafiltration membranes, micellar-enhanced ultrafiltration membranes, nanofiltration membranes and reverse osmosis membranes are examples of membranes used for heavy metal separation purposes. Based on size exclusion, solution diffusivity, absorption, etc., heavy metals can be separated with the help of a membrane. Materials with a high affinity for heavy metals or the ability to form complexes with heavy metals can be incorporated inside membranes to achieve high separation efficiency. Membranes can serve this purpose in a very eco-friendly manner. Inorganic layered materials, due to their unique properties like high accessible surface area, tuneable interlayer spacing and memory effect, have been found suitable for applications in the simultaneous adsorption of heavy metals and phenols in a single operation [15]. Moreover, the functionalisation of metal-based layered materials can also enhance the activity of the materials and make them an efficient candidate as photocatalysts in the photocatalytic degradation of harmful organics.

15.7 MEMBRANE TECHNOLOGY FOR ENHANCED OIL RECOVERY

In enhanced oil recovery, membranes can also play a crucial role with their advantageous properties [16]. Enhanced oil recovery is very demanding in the oil sector and is key to economic growth. This area of research has not been exposed widely. The membrane can be used as a potential candidate to separate the emulsified oil droplets from water based on size exclusion as well as the hydrophilicity of the membrane. Membranes have been used to tune the pore size according to which molecules need to be retained on the membrane and which ones need to be permeated. Pore size tuning has been discovered to play a critical role in balancing sulphate concentration and divalent ions in the injecting water membrane. Research and development of high-performing membrane systems for enhanced oil recovery are still challenging. The combination of membrane material with the water injection system may be one of the ways to achieve the maximum possible result for controlling the concentration of salt as well as parameters like pH, acidity, etc. Enhanced oil recovery is a trending research area as minimisation of oil wastage is highly recommended during the extraction process, considering the availability of crude oil and membranes are thought to be an excellent option for this purpose.

15.8 MEMBRANE TECHNOLOGY FOR OTHER IMPORTANT APPLICATIONS

Apart from the above-mentioned applications, membrane-based technology has applications in other related processes. For example, in refining natural gas, methane is the principal component and is separated from the other gaseous components present in small amounts. After the separation of these gaseous components, they again need to be separated from each other for purification and reuse. Membrane technology can be successfully applied to the separation of almost all types of gaseous molecules. Researchers are also constantly trying to extend the application of membrane-based technology in the petroleum industry. Recently, a group of scientists led by the Georgia Institute of Technology, USA, in collaboration with ExxonMobil, developed a membrane that is capable of fractionating crude oil. Using N-Aryl-linked spirocyclic polymers, they could develop membranes with varying pore sizes in the range of 2–12 Å that showed fractionation of molecules from a synthetic replica of a hydrocarbon mixture [17]. The development of membranes for fractionating crude oil can change the future scenario of oil refineries.

15.9   SUSTAINABLE SOLVENT, NON-SOLVENT SELECTION FOR DESIGNING POLYMERIC MEMBRANE

Most of the solvents that have been used in dissolving polymers for the fabrication of membranes are hazardous to human health. Using these solvents in the large-scale preparation of membranes is risky and is not recommended under various environmental protection acts. Therefore, it is of the utmost importance for the research and development of such solvents, which have no issue with environmental sustainability. Green solvents like DMSO (dimethylsulfoxide) are an organic solvent with no hazardous effects [18]. Moreover, vapour-induced phase separation is another technique by which membranes with tunable pore sizes can be obtained, and this procedure minimises the chance of mixing harmful solvents with water during membrane fabrication [19].

Non-solvents play a vital role in the design of membranes by the non-solvent-induced phase separation method. However, most of the membrane fabrication process uses various toxic non-solvents (like methanol, corrosive acids, etc.). Replacement of conventional and costly non-solvents with less costly non-solvents like water is highly recommended. Moreover, it is also possible to use water as the solvent and a non-solvent in NIPS. De Vos and his coworkers have designed a technique for the polymers that dissolve in water under different conditions like pH and temperature. After casting the polymer, the group tried to coagulate the polymer at the opposite pH, where the complexation of polyelectrolytes causes phase inversion [20]. A range of polymeric membranes from MF to NF was designed using the same mechanism.

15.10   MEMBRANE FOULING MITIGATION

Although membrane technology has become one of the most promising technologies for various industrial processes and for handling environmental issues, industrial applications in many areas still require much more development. The efficiency of a membrane is entirely dependent on its permeability and selectivity. The major issue with membrane-based technology is membrane fouling. Fouling of membranes reduces their efficiency with time. The reason for membrane fouling varies from process to process. For example, in oil–water separation, fouling occurs mainly because of oil deposition on the membrane surface, while in gas–water separation, fouling occurs because of swelling of the membrane material. Biofouling on membranes for wastewater treatment is another issue. Biofouling occurs due to the action of microbial pathogens present in the wastewater, which degrade the membrane surface and hence result in broadening the pores in the membrane, resulting in spoilage of the membrane. The development of membranes with anti-fouling properties and anti-biofouling membranes is highly recommended for the separation of wastewater where there is a possibility of high amounts of bacterial pathogens [21]. Successful implementation of a membrane-based separation process involves various factors, such as selection of proper membrane material, process, designing suitable module, setting up the process, etc. Computational modelling may play a significant role in designing an effective process by optimising all the associated parameters. Moreover, a membrane with self-cleaning ability is another approach for minimisation of the fouling issue. Various research has been going on for the design and development of membranes with self-cleaning and anti-fouling abilities by impregnating various chemically and biologically active components over the membrane surface. Combining the cross-flow filtration mechanism with the self-cleaning mechanism will be very fruitful for minimising the fouling issue in the membrane. Membrane fouling can also be alleviated by impregnating novel carbon-based nanomaterials, MOFs or inorganic layered nanomaterials over the membrane surface. Nanomaterials, due to their highly active surface area, significantly enhance the membrane selectivity and the anti-fouling properties of the membrane systems [22]. In the near future, the development of hybrid membrane systems for superior performance in membrane separation will be of the utmost importance.

15.11 MEMBRANE REUSABILITY

Membrane processes are the most sustainable and environment-friendly separation process, with low energy consumption and a minimum CO_2 footprint. However, reusability or more pronouncedly, the durability of the membrane processes is still problematic, as membrane systems start to lose their selectivity and permeability after lots of filtration through them. It is important to design the membrane module properly so that it can be reused after many cycles by simply backwashing the membrane system. A tubular membrane design is one such design that can purify large amounts of feed in a single operation with less space, time and energy requirements. Moreover, the tubular membrane systems can be used for many cycles and cleaning is easy. Advancements in the design of tubular membrane systems will definitely give rise to more durable membrane systems for efficient and long-term use of membranes for different types of treatments.

Various types of functionalised membranes have also been developed in recent years for their application in various fields like enantiomeric separation [23,24], biomolecule separation [25], natural dye extraction [26] and food packaging applications [27].

15.12 CONCLUSION

Membrane-based separation, an energy-efficient and environmental-friendly technology, has the potential to revolutionise the petroleum industry sector. Although there has been extensive research in this area, much of the advancement in membrane technology is still in the laboratory testing stage. Even after the development of suitable membranes for a particular process, industries are reluctant to take the risks associated with the new technology. Thus, only a few petroleum industries use this membrane-based separation, while others still rely on conventional technologies. Thus, to compete with these conventional technologies, membrane-based technologies must become cost effective, highly efficient and durable. New materials have constantly been searched to make the membranes durable and efficient. Since most of the polymeric membranes have poor stability in the presence of organic solvents and high temperatures, many different types of inorganic materials are being tested to make the polymeric membranes more stable. Two-dimensional nanomaterials like graphene and its derivatives are now being widely studied to make ultra-thin and stable membranes. New polymeric materials are also being developed for membrane fabrication. Modification of membrane surfaces and finding novel techniques for membrane fabrication are also aspects of this development. The design and development of a new membrane model to make the system more error-free has recently become a popular research topic.

REFERENCES

1. Xu, C., Lan, J., Ye, J., Yang, Y., Huang, K., & Zhu, H. (2023). Design of continuous-flow microwave reactor based on a leaky waveguide. *Chemical Engineering Journal*, *452*, 139690. https://doi.org/10.1016/j.cej.2022.139690
2. Denic-Roberts, H., Engel, L. S., Buchanich, J. M., Miller, R. G., Talbott, E. O., Thomas, D. L., ... & Rusiecki, J. A. (2023). Risk of longer-term neurological conditions in the deepwater horizon oil spill coast guard Cohort Study – Five years of follow-up. *Environmental Health*, *22*(1), 12. https://doi.org/10.1186/s12940-022-00941-0
3. Wang, G., Xu, Y., Zhang, R., Gai, S., Zhao, Y., Yang, F., & Cheng, K. (2023). Fire-resistant MXene composite aerogels for effective oil/water separation. *Journal of Environmental Chemical Engineering*, *11*(1), 109127. https://doi.org/10.1016/j.jece.2022.109127
4. Vandezande, P., Gevers, L. E., & Vankelecom, I. F. (2008). Solvent resistant nanofiltration: Separating on a molecular level. *Chemical Society Reviews*, *37*(2), 365–405. https://doi.org/10.1039/B610848M
5. Wang, B., Yan, J., Wang, H., Li, R., Fu, R., Jiang, C., ... & Xu, T. (2023). Ionic liquid-based pore-filling anion-exchange membranes enable fast large-sized metallic anion migration in electrodialysis. *Journal of Membrane Science*, 121348. https://doi.org/10.1016/j.memsci.2023.121348
6. Chen, D., Yang, F., Karousos, D. S., Lei, L., Favvas, E. P., & He, X. (2023). Process parametric testing and simulation of carbon membranes for H_2 recovery from natural gas pipeline networks. *Separation and Purification Technology*, *307*, 122842. https://doi.org/10.1016/j.seppur.2022.122842
7. Imtiaz, A., Othman, M. H. D., Jilani, A., Khan, I. U., Kamaludin, R., Iqbal, J., & Al-Sehemi, A. G. (2022). Challenges, opportunities and future directions of membrane technology for natural gas purification: A critical review. *Membranes*, *12*(7), 646. https://doi.org/10.3390/membranes12070646
8. Wei, R., Liu, X., & Lai, Z. (2022). MOF or COF membranes for olefin/paraffin separation: Current status and future research directions. *Advanced Membranes*, *2*, 100035. https://doi.org/10.1016/j.advmem.2022.100035

9. Tabbiche, A., & Aouinti, L. (2023). Preparation and characterization of nanocomposite membranes based on PVC/TiO$_2$ anatase for the separation of toluene/n-heptane mixtures via pervaporation. *Polymer Bulletin, 80*(1), 643–666. https://doi.org/10.1007/s00289-021-04062-5

10. Xiao, P., He, X., Ye, C., Zhang, S., Zheng, F., Lu, Q., & Ma, X. (2023). Tailoring the microporosity and gas separation property of soluble polybenzoxazole membranes derived from different regioisomer monomers. *Separation and Purification Technology, 311*, 123340. https://doi.org/10.1016/j.seppur.2023.123340

11. Xue, Y., Chen, L., Xiang, L., Zhou, Y., & Wang, T. (2023). Experimental investigation on electromagnetic induction thermal desorption for remediation of petroleum hydrocarbons contaminated soil. *Journal of Environmental Management, 328*, 117200. https://doi.org/10.1016/j.jenvman.2022.117200

12. Yang, J., He, T., Li, X., Wang, R., Wang, S., Zhao, Y., & Wang, H. (2023). Rapid dipping preparation of superhydrophobic TiO$_2$ cotton fabric for multifunctional highly efficient oil-water separation and photocatalytic degradation. *Colloids and Surfaces A: Physicochemical and Engineering Aspects, 657*, 130590. https://doi.org/10.1016/j.colsurfa.2022.130590

13. Li, Z., Yin, L., Jiang, S., Chen, L., Sang, S., & Zhang, H. (2023). A photocatalytic degradation self-cleaning composite membrane for oil–water separation inspired by light-trapping effect of moth-eye. *Journal of Membrane Science, 669*, 121337. https://doi.org/10.1016/j.memsci.2022.121337

14. Bharath, G., Rambabu, K., Alqassem, B., Morajkar, P. P., Haija, M. A., Nadda, A. K., … & Banat, F. (2023). Fabrication of gold nanodots decorated on 2D tungsten sulfide (Au-WS2) photoanode for simultaneous oxidation of phenol and arsenic (III) from industrial wastewater. *Chemical Engineering Journal, 456*, 141062. https://doi.org/10.1016/j.cej.2022.141062

15. He, T., Li, Q., Lin, T., Li, J., Bai, S., An, S., … & Song, Y. F. (2023). Recent progress on highly efficient removal of heavy metals by layered double hydroxides. *Chemical Engineering Journal, 142041*. https://doi.org/10.1016/j.cej.2023.142041

16. Rezaeian, M. S., Mousavi, S. M., Saljoughi, E., & Amiri, H. A. A. (2020). Evaluation of thin film composite membrane in production of ionically modified water applied for enhanced oil recovery. *Desalination, 474*, 114194. http://dx.doi.org/10.1016/j.desal.2019.114194

17. Thompson, K. A., Mathias, R., Kim, D., Kim, J., Rangnekar, N., Johnson, J. R., … & Finn, M. G. (2020). N-Aryl–linked spirocyclic polymers for membrane separations of complex hydrocarbon mixtures. *Science, 369*(6501), 310–315. www.science.org/doi/abs/10.1126/science.aba9806

18. Xie, W., Li, T., Chen, C., Wu, H., Liang, S., Chang, H., … & Crittenden, J. C. (2019). Using the green solvent dimethyl sulfoxide to replace traditional solvents partly and fabricating PVC/PVC-g-PEGMA blended ultrafiltration membranes with high permeability and rejection. *Industrial & Engineering Chemistry Research, 58*(16), 6413–6423. https://doi.org/10.1021/acs.iecr.9b00370

19. Nunes, S. P., Culfaz-Emecen, P. Z., Ramon, G. Z., Visser, T., Koops, G. H., Jin, W., & Ulbricht, M. (2020). Thinking the future of membranes: Perspectives for advanced and new membrane materials and manufacturing processes. *Journal of Membrane Science, 598*, 117761. https://doi.org/10.1016/j.memsci.2019.117761

20. Virga, E., de Grooth, J., Zvab, K., & de Vos, W. M. (2019). Stable polyelectrolyte multilayer-based hollow fiber nanofiltration membranes for produced water treatment. *ACS Applied Polymer Materials, 1*(8), 2230–2239. https://doi.org/10.1021/acsapm.9b00503

21. Bhuyan, C., Konwar, A., Bora, P., Rajguru, P., & Hazarika, S. (2023). Cellulose nanofiber-poly (ethylene terephthalate) nanocomposite membrane from waste materials for treatment of petroleum industry wastewater. *Journal of Hazardous Materials*, *442*, 129955. https://doi.org/10.1016/j.jhazmat.2022.129955

22. Sasikumar, B., Krishnan, S. G., Afnas, M., Arthanareeswaran, G., Goh, P. S., & Ismail, A. F. (2023). A comprehensive performance comparison on the impact of MOF-71, HNT, SiO_2, and activated carbon nanomaterials in polyetherimide membranes for treating oil-in-water contaminants. *Journal of Environmental Chemical Engineering*, *11*(1), 109010. https://doi.org/10.1016/j.jece.2022.109010

23. Gogoi, M., Goswami, R., Ingole, P. G., & Hazarika, S. (2020). Selective permeation of L-tyrosine through functionalized single-walled carbon nanotube thin film nanocomposite membrane. *Separation and Purification Technology*, *233*, 116061. https://doi.org/10.1016/j.seppur.2019.116061

24. Gogoi, M., Goswami, R., Borah, A., Sarmah, H., Rajguru, P., & Hazarika, S. (2021). Amide functionalized DWCNT nanocomposite membranes for chiral separation of the racemic DOPA. *Separation and Purification Technology*, *279*, 119704. https://doi.org/10.1016/j.seppur.2021.119704

25. Borah, A., Gogoi, M., Goswami, R., Sarmah, H., Hazarika, K. K., & Hazarika, S. (2022). Thin film nanocomposite membrane incorporated with clay-ionic liquid framework for enhancing rejection of epigallocatechin gallate in aqueous media. *Journal of Environmental Chemical Engineering*, *10*(3), 107423. https://doi.org/10.1016/j.jece.2022.107423

26. Goswami, R., Gogoi, M., Borah, A., Sarmah, H., Ingole, P. G., & Hazarika, S. (2021). Functionalized activated carbon and carbon nanotube hybrid membrane with enhanced antifouling activity for removal of cationic dyes from aqueous solution. *Environmental Nanotechnology, Monitoring & Management*, *16*, 100492. https://doi.org/10.1016/j.enmm.2021.100492

27. Hazarika, K. K., Konwar, A., Borah, A., Saikia, A., Barman, P., & Hazarika, S. (2023). Cellulose nanofiber mediated natural dye based biodegradable bag with freshness indicator for packaging of meat and fish. *Carbohydrate Polymers*, *300*, 120241. https://doi.org/10.1016/j.carbpol.2019.115030

Index

Note: Page locators in **bold** and *italics* represents tables and figures, respectively.

For Product Safety Concerns and Information please contact our EU
representative GPSR@taylorandfrancis.com
Taylor & Francis Verlag GmbH, Kaufingerstraße 24, 80331 München, Germany